Structural Concepts and Systems for Architects and Engineers

Structural Concepts and Systems for Architects and Engineers

T. Y. Lin
T. Y. Lin International
San Francisco, California

Sidney D. Stotesbury
College of Architecture and Design
Kansas State University
Manhattan, Kansas

JOHN WILEY & SONS
New York   Brisbane   Chichester   Toronto

Library of Congress Cataloging in Publication Data:

Lin, T'ung-yen, 1911-
    Structural concepts and systems for architects and
engineers.

    Bibliography: p.
    Includes index.
    1. Structural design.   I. Stotesbury, Sidney D.,
joint author.   II. Title.
TA658.L47   721   79-23458
ISBN 0-471-05186-1

Printed in the United States of America

10 9 8 7 6 5 4 3 2 1

PREFACE

As the technological and architectural complexity of building design has increased, there has been a tendency for the engineering and architectural professions to diverge. Unfortunately, the result has been an education and communication gap that limits the potential for creative interaction between architects and engineers. Such a gap is always undesirable, but it is especially so at the formative levels of a design project. It obscures the overall relationship between space-form and technological thinking and increases the chance for major conflict at subsequent levels.

This book is intended to bridge the gap at schematic and preliminary levels of design and is dedicated to students of architectural and structural engineering. It focuses on identifying and explaining the more basic, rather than detailed, issues of conceiving and manipulating structural design options. It does so through a unique *overall approach* that emphasizes a total-to-subsystem hierarchy for learning about and projecting structural thinking into architectural design, and vice versa. It is unique because it makes it possible to introduce 1) structural design principles as basic form determinants, and 2) simplified methods for approximate analysis in both conceptual and specific terms.

The primary educational goal is to foster understanding of, as well as the ability to control, the overall relationship between the structural and space-form properties of architectural schemes. Hence, the competency objective is to enable designers *first* to conceptualize schematic options for providing total-system integrity and *second* to quickly compare alternatives for designing major subsystems. This involves some quantitative analysis, but the use of formulas and calculations is limited to those needed for *approximate* design. Design of elemental components is treated only to a level sufficient to establish the feasibility of subsystem layouts and the relative efficiency of main components.

The stress throughout the book is on optimizing the interaction of key force and geometric properties of subsystems at a preliminary design level. Therefore, students using this text need only have a basic understanding of geometry, simple statics, and strength of materials, a brief review of which is presented in Appendices A-1 and A-2.

The book uses a *three-cycle* learning strategy. The cycles are designed to illustrate how a body of basic design principles can be learned and applied at any scale of design thinking from total schematic to preliminary design of subsystems, or and their key elemental components. Materials may change, but the same basic principles always apply.

The *first cycle* (Chapters 1 to 5), takes a total-system look at the relationship between the schematic properties of prototypical building forms and the basic types of total-system behavior that are expected of them. Chapter 1 introduces the idea of an overall approach by exploring the role of engineers and architects in building design and by establishing conceptual linkages between architectural and structural-system design thinking. Chapters 2 and 3 focus on a schematic exploration of the relationship between types of building form options and the fundamental types of subsystem interactions that are required to give the forms total-system integrity. Chapters 4 and 5 provide basic concepts and data for estimating overall loads on buildings and give practical examples to illustrate a schematic-level analysis of several building designs as total structural systems.

The *second cycle* (Chapters 6 to 9) elaborates on the discussion of the first cycle by introducing the requirements for design of specific subsystems. Here, a greater depth of analysis is presented to reveal the issues of component efficiency and to supply calculations that are sufficient to arrive at specific layouts and preliminary dimensioning of key members.

The *third cycle* (Chapters 10 to 14) treats some of the more special types of problems that must be considered in the design of buildings. Chapters 10 and 11 focus on high-rise and long-span designs. Then Chapter 12 concentrates on foundation subsystems, Chapter 13 on construction, and Chapter 14 on economics. These final chapters round out the text discussions by treating areas of structural form determinants that are perhaps less obvious but should be recognized at schematic and preliminary levels if one is to arrive at building designs that can be implemented.

The scope of the book is broad, and it should be understood that it is not intended to enable final design of structural components and their connections. Rather, it is intended to provide a working introduction to the basic concepts and issues of total-system building design. Nevertheless, readers who study this text should be able to understand and communicate clearly about basic types of structural subsystems and their interactions as architectural form determinants. In addition, they should know how to quickly analyze and compare various total or subsystem layouts in approximate terms that are adequate for schematic and preliminary design purposes.

In short, individuals who use this book will be able to apply an overall approach at the formative stages of a design project to assure that, at final stages, a fundamental compatibility exists between architectural and engineering design thinking. Those students who move on to more specialized studies (or who have already done so) will find that this text provides a broad and insightful background that promotes both consolidation and creative application of more in-depth learning.

The authors believe that this book will serve useful technological teaching purposes by providing a common foundation of knowledge for students of both architecture and engineering. It will also promote mutual respect for the difference in their design responsibilities and improve their potential for creative collaboration as future professionals. Thus, practicing architects and engineers (and their staff employees) will find this book useful as a refresher on overall thinking and as a means of exploring better ways for the two professions to interact during all stages of design projects.

T. Y. Lin
Sidney D. Stotesbury

CONTENTS

5 Structural Loads and Responses

6 Overall Design of Horizontal Subsystems

7 Overall Design of Vertical Subsystems

8 Horizontal Linear Components

9 Vertical Linear Components

10 High-Rise Buildings

11 Arch, Suspension, and Shell Systems

12 Foundation Subsystems

13 Construction

14 The Cost of Building Structures

Structural Concepts and Systems for Architects and Engineers

1
Introduction: Structure in Design of Architecture

SECTION 1: Overall Thinking

Because of modern technology, the capabilities of architectural and engineering designers are linked. As a result, architecture should be the product of a creative collaboration of architects and engineers. But such collaboration is often difficult. In contrast to most physical products, architecture is intended to perform in spatial terms and to be experienced as a total environment. This complicates matters by making the design responsibility at the same time comprehensive and specific, tangible and intangible.

To generate effective architecture, a designer must deal with the space-form implications of a broad range of interacting performance needs: for instance, activity-associated, physical, and symbolic (Figure 1-1). The challenge is to organize the many properties of a building in such a way that they fill these needs in a collectively optimum manner.

Activity-associated needs are operational and derive from the human desire to conduct activities in a controlled environment. For a given project and site, the interaction of a variety of activity spaces must be organized in terms of unique requirements for physical definition, enclosure, association, climatological control, and services. Of course, this implies that there are physical needs that can be viewed as primarily constructive in nature if they

FIGURE 1-1
ARCHITECTURAL DESIGN IS A COMPREHENSIVE SPATIAL ORGANIZATION PROBLEM.

THE DESIGNER MUST
ORGANIZE THE PERFORMANCE
PROPERTIES OF BUILDINGS TO
FILL A BROAD RANGE OF USER NEEDS :

1. ACTIVITY-ASSOCIATED (OPERATIONAL)
2. PHYSICAL (CONSTRUCTIVE)
3. SYMBOLIC (EXPERIENTIAL)

IN SPATIAL
TERMS

3

are considered alone. That is, a designer must examine the issues of providing for energy, mechanical equipment, structure, and construction. But, to become architecturally relevant, these physical needs must be considered in the context of a total scheme for organizing the interaction of activity spaces.

In addition, if the designer is to generate the operational and constructive properties of a space-form scheme as a total environmental system, he or she will have to consider the symbolic needs of its future users. This is of basic importance because users will experience, as well as operate in, a built environment. For users, a building is viewed as a symbol of their life context, society's attitude toward them, and the owner's respect for their social and aesthetic values. The designer of architecture must respond by insuring that an expression of these experiential values is made an integral part of his design proposals. In fact, one can say that the fundamental challenge of all architectural design projects is to provide a humanistic setting for the pursuit of activities, one that is inherently inspiring instead of indifferent or, even, degrading to the user.

For architects, the above-mentioned needs represent an interactive set of design problems that must be dealt with in a comprehensive manner. As a result, architects usually stress an overall, rather than an elemental, approach to design thinking. This is especially true early in the design process, since they must conceptualize a space-form scheme as a total system that insures overall compatibility between activity, physical, and symbolic performance expectations. They then use this holistic picture to guide their subsequent efforts, and those of collaborating designers, to elaborate and refine the scheme in terms of relevant parts and details.

The interactive nature of architectural projects makes a comprehensive approach necessary. Thus, if creative collaboration of architects and engineers is to be possible at schematic stages, both participants must be capable of overall thinking about technological issues. Unfortunately, the specialized mode of engineers' education often leads them to think in the reverse, starting with details, and without sufficient regard for the overall picture. This causes a conceptual gap to exist that acts, in general, to limit creative interaction between architects and engineers at all levels of design. Moreover, a gap of this kind is especially limiting as it applies to the schematic and preliminary stages of a design process. And, since architects' education in structures is often a condensed version of engineering courses, they are often in no position to help bridge the gap. As a result, a tendency exists for engineers to wait for an architect to initiate a (nonstructural) space-form scheme, and then to try to find a way of implementing it. Not only is this an inefficient use of knowledge, energy, and time; it also produces conflicts.

The existence of the gap and its limiting effect on designers has been recognized by many creative engineers and architects for some time. They acknowledge the importance of learning how to conceptualize the constructive implications of a space-form scheme along with, instead of after, the generation of its basic operational and experiential characteristics. This is also true for educators in both fields, despite the fact that the content of an engineer's education will, of necessity, be more specialized and in depth than the architect's. As a result, many educators in architecture and in engineering now agree that any field of technological knowledge, such as structure, must be understood in overall terms before it can be applied with creativity at the formative stages of environmental design thinking. They do not deny that specialized learning can be useful, but they recognize a central challenge for educators: A means must be found for teaching students of both architecture and engineering how to conceptualize technological knowledge in a total-system context. It has become clear that the potential for creative collaboration of future environmental designers will be improved to the degree that this capability can be developed.

One may find it easy to agree with the general argument that an overemphasis on educational specialization can produce creativity problems

FIGURE 1.2
OVERALL THINKING ENABLES INTEGRATION OF ENGINEERING WITH ARCHITECTURAL SKILLS.

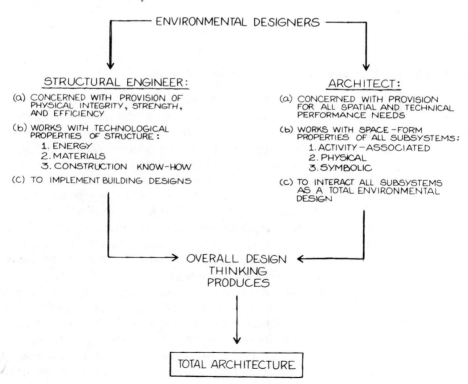

for environmental designers. But the essential reason that a capacity for overall thinking will support creativity is that it can provide a natural bridge for linking the more comprehensive, space-form aspects of architectural design thinking, with the more specialized and physical concerns of the engineer (Figure 1-2). It can enable the two professionals to work at the same level to recognize, and to resolve, structural-spatial design conflicts in broad schematic terms *before* they attempt to tackle the problems associated with the more detailed levels of final design. As a result designers can be confident that they will always be able to communicate easily about the more basic structural implications of architectural concepts and vice versa. This makes creative collaboration not only possible, but welcomed, at early design stages, because it can support rather than interfere with the generation of total architecture.

Accordingly, this book advocates that overall thinking should form the basis of one's introduction to structural knowledge so that one can always view future specialized studies in a total-system context. Hence, the first four chapters will emphasize an overall approach to structural learning by introducing the structural problems of building design from a total-system, rathern than elemental, point of view. As the total-system picture becomes clear, subsequent chapters will turn to the problems of preliminary design of basic subsystems and their key components.

SECTION 2: **The Architectural Design Process**

We have said that the architect must deal with the spatial aspects of activity, physical, and symbolic needs in such a way that overall performance integrity is assured. Hence, he or she will want to think of evolving a building environment as a total system of interacting and space-forming subsystems (Figure 1-3). This represents a complex challenge, and to meet it the architect will need a hierarchic design process that provides at least three levels of feedback thinking: schematic, preliminary, and final (Figure 1-4). Such a hierarchy is necessary if he or she is to avoid being confused, at conceptual stages of design thinking, by the myriad detail issues that can distract attention from more basic considerations. In fact, we can say that an architect's ability to distinguish the more basic from the more detailed issues is essential to his success as a designer.

The object of the schematic feedback level is to generate and evaluate overall site-plan, activity-interaction, and building-configuration options. To do so the architect must be able to focus on the interaction of the basic attributes of the site context, the spatial organization, and the symbolism as determinants of physical form. This means that, in schematic terms, the architect may first conceive and model a building design as an

FIGURE 1.3
AN INTERACTION MODEL OF THE ARCHITECTURAL DESIGN CHALLENGE

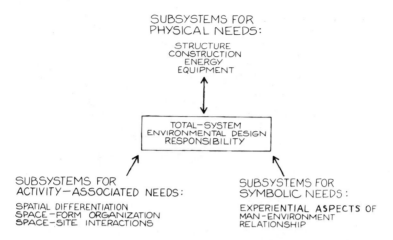

FIGURE 1.4
FEEDBACK AND THE HIERARCHY OF ARCHITECTURAL DESIGN THINKING

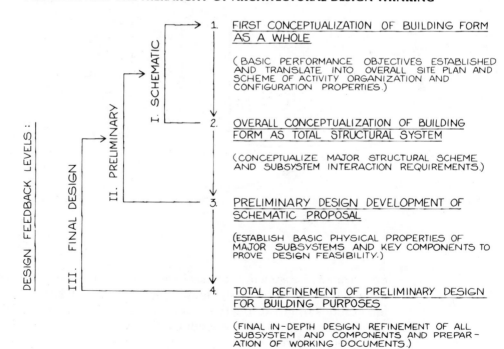

FIGURE 1-5
**AT SCHEMATIC LEVELS THE THINKING IS OVERALL AND FROM THE
ABSTRACT (a) TO THE PHYSICAL MODEL (b).**

(a) AN ORGANIZATIONAL ABSTRACTION OF BASIC PERFORMANCE OBJECTIVES

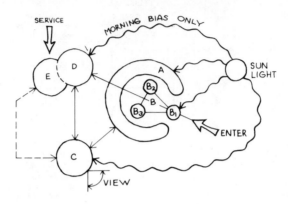

(b) BUILDING CONFIGURATION SCHEME ON SITE
(OVERALL PHYSICAL FORM PROPERTIES IMPLIED BY (a)).

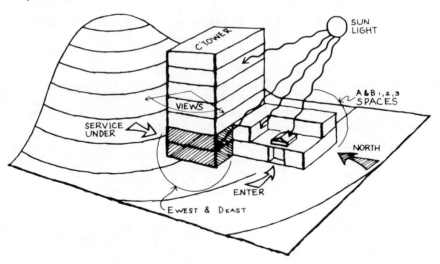

organizational abstraction of essential performance-space interactions
(Figure 1-5a). Then he or she may explore the overall space-form
implications of the abstraction. As an actual building configuration option
begins to emerge, it will be modified to include consideration for basic site
conditions (Figure 1-5*b*).

At the schematic stage, it would also be helpful if the designer could visualize his or her options for achieving overall structural integrity and consider the constructive feasibility and economics of his or her scheme. But this will require that the architect and/or a consultant be able to conceptualize total-system structural options in terms of basic types of subsystem interaction schemes and not in terms of elemental details. Such overall thinking can be easily fed back to improve the space-form scheme.

At the preliminary level, the architect's emphasis will shift to the elaboration of his or her more promising schematic design options. Here the architect's structural needs will shift to approximate design of specific subsystem options. At this stage the total structural scheme is developed to a middle level of specificity by focusing on identification and design of major subsystems to the extent that their key geometric, component, and interactive properties are established. Basic subsystem interaction and design conflicts can thus be identified and resolved in the context of total-system objectives. Consultants can play a significant part in this effort, but consideration of details should still be highly selective. Of course, these preliminary-level decisions may also result in feedback that calls for refinement or even major change in schematic concepts.

When the designer and the client are satisfied with the feasibility of a design proposal at the preliminary level, it means that the basic problems of overall design are solved and details are not likely to produce major changes. The focus shifts again, and the design process moves into the final level. At this stage the emphasis will be on the detailed development of all subsystem specifics. Here the role of specialists from various fields, including structural engineering, is much larger, since all details of the preliminary design must be worked out. Decisions made at this level may produce feedback into Level II that will result in changes. However, if Levels I and II are handled with insight, the relationship between the overall decisions, made at the schematic and preliminary levels, and the specifics of the final level should be such that gross redesign is not in question. Rather, the entire process should be one of moving in an evolutionary fashion from creation and refinement (or modification) of the more general properties of a total-system design concept, to the fleshing out of requisite elements and details.

To summarize: At Level I, the architect must first establish, in conceptual terms, the overall space-form feasibility of basic schematic options. At this stage, collaboration with specialists can be helpful, but only if in the form of overall thinking. At Level II, the architect must be able to identify the major subsystem requirements implied by the scheme and substantiate their interactive feasibility by approximating key component properties. That is, the properties of major subsystems need be worked out only in sufficient depth to verify the inherent compatibility of their basic form-related and behavioral interactions. This will mean a somewhat more specific form of collaboration with specialists than that in Level I. At Level III, the architect

FIGURE 1-6
**A DESIGNER'S ATTITUDE TOWARD ANALYSIS AND FEEDBACK INDICATES HIS
OR HER ATTITUDE TOWARD THE IMPORTANCE OF TOTAL SYSTEM DESIGN.**

	MIN. ← EMPHASIS ON TOTAL DESIGN → MAX.			
DESIGNER ATTITUDE	ACCUMULATION	COMPOSITION	SYNTHESIS	INTEGRATION
PROCESS ATTRIBUTE	PARTS ADDED AS NEED IS RECOGNIZED	SUBORDINATION OF PARTS TO DOMINANT FORM CONCEPT	COORDINATION OF PARTS FOR OPTIMUM INTERACTION	INTEGRATION OF PARTS INTO ORGANIC TOTAL SYSTEM
ORGANIZATIONAL INTENTION	REACTIVE	FORM FOR OWN SAKE	FORM FOLLOWS FUNCTION	FORM & FUNCTION ARE ONE
	MIN. ← EMPHASIS ON ANALYSIS & FEEDBACK → MAX.			

and the specialists must follow through by providing for all of the elemental design specifics required to produce biddable construction documents. The strategy is to insure that levels after the first represent a progressive elaboration of the previous level. The need is to recognize that there *is* an issue hierarchy, and to rely on overlap and feedback between levels to insure that the property of wholeness is built into the final treatment of details. Thus, the potential for interaction of architectural and engineering thinking can be high at *all three levels.*

Such a strategy is logical because it recognizes a natural need to establish a comprehensive context for dealing with the specifics of architectural design. It is simply an extension of the total-system attitude that the whole of a design problem should control the effectiveness of solution details and not vice versa. For example, one can imagine a range of attitudes in dealing with an environmental design problem that emphasizes a minimal to maximal degree of emphasis on wholeness from the start: accumulation, dominance, synthesis, and integration or simultaneity (Figure 1-6). At one pole we have a reactionary attitude that emphasizes simple addition of components as the need arises (the whole will be the sum of the parts). At the other pole we have a predictive attitude that emphasizes integration of overall form, performance, and the means of achieving it (the whole is more than the sum of its parts).

Eliel Saarinen, father of architect Eero Saarinen, put the basic concept well when he admonished his son to "always think of the next bigger thing."[1] In this spirit we recommend the total-system attitude as being productive for both the architect and the engineer. However, we do not mean that details

[1]*Eero Saarinen* by Alan Temko, George Braziller Publishers, New York, 1962.

are unimportant. In fact, structural failures and other troubles in buildings can arise because of poor detailing. But in general, both the architect and the engineer must be able to see the overall space-form design problem from a holistic point of view, first to understand the overall performance context *and then* to creatively explore the relationship between total, subsystem, and elemental design implications. In this way, the power of specialized knowledge can be harnessed.

SECTION 3: An Overall Approach to Structural Education

We have said that the objective of architectural design is to create an effective environmental whole, a total system of interacting environmental subsystems. Hence the designer will naturally wish to begin to deal with the structural aspect of building design by identifying and manipulating overall schemes for interacting major subsystems rather than elements or details. But, in contrast to the natural effectiveness of overall thinking, students of architecture and engineering are often taught to deal with engineering knowledge by means of a study of elemental components and their specific design and construction factors. This approach is based on the assumption that the students will somehow be able to work backwards (on their own) to discover how the parts can be integrated to act holistically (Figure 1-7). Unfortunately, this assumption is seldom substantiated in practice, because the mode of learning is the reverse of the natural flow of architectural design thinking. That is, discontinuity between the mode of one's actual design experience and the mode of learning about technological information will make it difficult for the student to apply technological information at the formative stages of design thinking.

Today, many design educators are recognizing that this has always been a problem, but it has been grossly exacerbated by an explosion of technological information and its corollary, specialized learning. As a result, many of today's students of architecture and engineering have not learned how to apply their specialized education in a variety of total-problem contexts. Specialization has made it possible for students to be exceptionally capable of solving predetermined problems that are clearly presented to them. But they are often incapable of analyzing a complex system of problems to identify basic versus detail issues and formulate a hierarchic plan for dealing with them. Thus we can see why a capability for overall design thinking about specialized knowledge is essential to the creative interaction of architects with engineers (or sociologists with economists, etc.).

Since the architectural challenge is to deal, in a coherent way, with organizational, symbolic, and constructive complexity, fragmentation of technical knowledge does not contribute to a creative response by

FIGURE 1-7
THE ELEMENTAL VERSUS THE OVERALL APPROACH TO DESIGN EDUCATION.

I ELEMENTAL APPROACH:

(a) EMPHASIS ON DESIGN OF SPECIFIC
STRUCTURAL ELEMENT

(b) EDUCATIONAL PROBLEM:
KNOWLEDGE IS DEVOID OF BASIS
FOR RELATING TO TOTAL SYSTEM
OBJECTIVES. EMPHASIS ON PARTS,
NOT WHOLE SYSTEMS

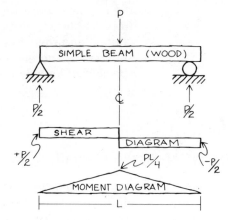

II OVERALL APPROACH:

(a) EMPHASIS ON DESIGN OF TOTAL
SPACE–FORM STRUCTURAL
SYSTEM (e.g. CONCEPT FOR
BLYTH ARENA, SQUAW VALLEY,
CALIFORNIA)

(b) EDUCATIONAL ADVANTAGE:
AN OVERALL CONTEXT IS PROVIDED
FOR LEARNING ABOUT AND APPLYING
STRUCTURAL CONCEPTS AND SYSTEMS
TO ARCHITECTURAL DESIGN. THE
WHOLE DETERMINES THE ELEMENTS

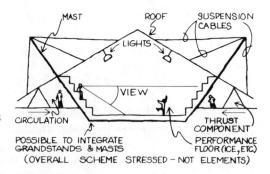

designers. This leads to an educational conclusion that the learner must
never be allowed to forget that his ability to conceptualize overall
space-form interactions will allow him to control the need for details,
and not vice versa. It also suggests that a common educational strategy for
students of both engineering and architecture would be to move
deductively: *from* an introduction to structures that considers the schematic
implications of buildings viewed as space-form wholes, *to* a logical
elaboration of this basic understanding. The basic understanding focuses on
consideration of *major* structural subsystems and discrimination of *key*
elements, whereas the act of elaboration involves attention to the details
required to realize the whole.

The good sense of such an overall approach to education can be vividly
characterized by considering what are often termed the nonstructural space
enclosure and subdivision aspects of architectural design. The spatial

organization and articulation of the various properties of activity spaces calls for control of their external and internal adjacency and interface potentials. Horizontal and vertical surfaces in the form of floors, walls, roofs, and penetrations through these surfaces must be provided to establish varying degrees of spatial differentiation, enclosure, access, and geometric definition.

Imagine that the physical components of a spatial organization scheme were designed with no thought for their structural implications. The probability for major revision of early concepts due to structural requirements will be high. Now, in contrast, imagine that these components of spatial organization were organized from the beginning with overall structural implications of the schematic space-form system in mind. The probability for major revision would be minimized, and the symbolic and physical integration of the structure with the overall architectural scheme would be insured.

It becomes apparent that an ability for overall thinking can make it possible to apply structural knowledge to the total architectural design effort from the very beginning and with a minimum of distraction by lower-level details. It alone can enable the architect to think of the physical issues of a space-structure in a context that is inherently compatible with his mode of dealing with the many organizational and symbolic issues of space-forming. Thus, it can assure that the emphasis is on components conceived as acting together as total systems rather than separately, as independent parts. It is also apparent that much can be gained from applying this overall-to-specifics model of educational management to a reconsideration of teaching and writing strategies in many specialized fields of design-related knowledge.

SECTION 4: Structure and Other Subsystems

There are other important reasons for suggesting that structural thinking should be introduced at the very earliest stages of the design process. These derive from the need to provide buildings with mechanical and other environmental service subsystems that support horizontal and vertical movement of men and materials as well as provide for heating, ventilation, air-conditioning, power, water, and waste disposal. In addition, provision for acoustical and lighting needs is often influenced by structural design.

Vertical movement of objects through a building requires rather large shafts, and overall thinking can result in the use of these service components as major structural subsystems (Figure 1-8a). The requirements for provision of heating, ventilation, air-conditioning, power, water, and waste services can be visualized in the form of a tree diagram (Figure 1-8b). These services usually originate at a centralized location and must trace their

FIGURE 1-8
MAJOR MOVEMENT AND SERVICE SUBSYSTEMS CAN HAVE OVERALL STRUCTURAL IMPLICATIONS.

(a) VERTICAL MOVEMENT SUB-SYSTEMS CAN PLAY BASIC STRUCTURAL ROLES

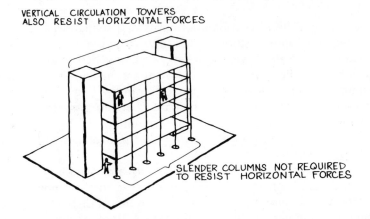

VERTICAL CIRCULATION TOWERS ALSO RESIST HORIZONTAL FORCES

SLENDER COLUMNS NOT REQUIRED TO RESIST HORIZONTAL FORCES

(b) SERVICES ARE DISTRIBUTED FROM A CENTRAL SOURCE THROUGHOUT BUILDINGS. THE NEED FOR HORIZONTAL AND VERTICAL CHASE SPACES CAN BE AN ASSET OR PROBLEM

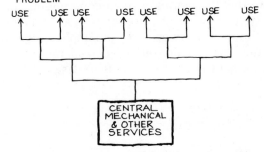

USE   USE USE   USE USE   USE USE   USE

CENTRAL MECHANICAL & OTHER SERVICES

way horizontally and vertically throughout the structure in order to serve the activity spaces. Large trunk-chase spaces may be required, and their structural implications should be considered early in the design process.

In terms of acoustics, it is clear that the structural shape of a spatial organization can directly influence acoustical properties (Figure 1-9). In addition, if a spatial organization calls for heavy equipment to be located such that it impinges on a flexible structure, vibration and acoustical disturbances can be transmitted throughout the space because of an incompatible interface between machines and structure.

The requirement for artificial and natural light brings up other considerations. Artificial lighting often calls for integrating consideration of structural subsystems with considerations of the spatial qualities of light and of the spatial requirements for housing the lighting fixtures (Figure 1-10 and 1-11). The structural implications of natural lighting are even more obvious.

FIGURE 1-9
SOUND AND STRUCTURE INTERACT.

DOME ROOF
CONCENTRATES

SOUND DISTRIBUTION IS IN-
FLUENCED BY THE OVERALL
SHAPE OF SPACES

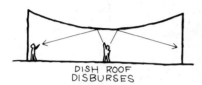

DISH ROOF
DISBURSES

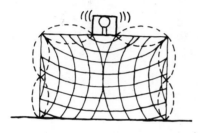

MECHANICAL EQUIPMENT SOUND
IS TRANSMITTED THROUGH
STRUCTURE. WHEN THE STRUCTURE
IS FLEXIBLE, VIBRATIONS ARE ALSO
TRANSMITTED

FIGURE 1-10
ARTIFICIAL LIGHT AND STRUCTURE INTERACT AT SUBSYSTEM LEVEL.

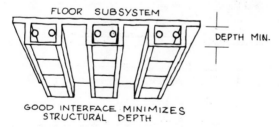

FLOOR SUBSYSTEM

DEPTH MIN.

GOOD INTERFACE MINIMIZES
STRUCTURAL DEPTH

LIGHTING SYSTEMS SHOULD BE
MADE TO INTERFACE WELL WITH
STRUCTURAL SUB-SYSTEMS

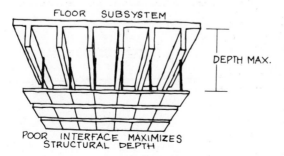

FLOOR SUBSYSTEM

DEPTH MAX.

POOR INTERFACE MAXIMIZES
STRUCTURAL DEPTH

FIGURE 1-11

SCHEMATIC OF VERTICAL SUBSYSTEM FOR INTEGRATING STRUCTURAL AND LIGHTING REQUIREMENTS IN THE JOHNSON WAX BUILDING (FRANK LLOYD WRIGHT).

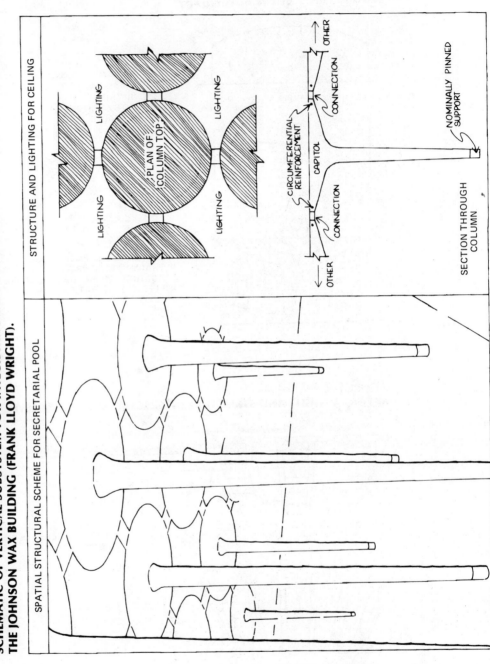

FIGURE 1-12
NATURAL LIGHT AND STRUCTURE INTERACT AT OVERALL LEVEL.

(a) FULLY ENCLOSED BOX REPRESENTS SIMPLE STRUCTURAL PROBLEMS BUT PROVIDES NO NATURAL LIGHT

(b) FULLY TRANSPARENT ROOF PROVIDES NATURAL LIGHT BUT POSES MORE COMPLEX STRUCTURAL DESIGN PROBLEMS

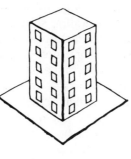

(c) BEARING AND SHEAR WALL DESIGN WITH FEW WINDOWS IS SIMPLE BUT ADMITS LITTLE LIGHT

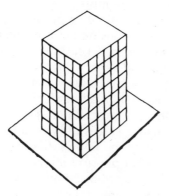

(d) FRAME DESIGN IS MORE COMPLEX BUT ALLOWS UP TO 80% OF THE WALL TO BE TRANSPARENT FOR LIGHT AND VIEW

For example, consider a fully enclosed space-form with all lighting provided artificially (Figure 1-12a). Then consider an open-top spatial organization with a heavy reliance on natural lighting throughout the space (Figure 1-12b). Figures 1-12c and 1-12d illustrate the same principle in terms of a **high-rise** building form.

With an overall approach, structural design may be considered along with the space-form and service requirements. It will be easier to recognize opportunities for optimum interfacing of service components with the total structural scheme and with major subsystems.

SECTION 5: Summary

We have shown that, in the earliest stages of architectural design thinking, overall space-form options will tend to emerge as more-or-less abstract entities in that few specific structural decisions are committed or even desired. This is because, at the formative stages, the basic design problem is to organize the overall interactions of a broad range of space-form determinants, and maximum freedom is required to explore and synthesize schematic options. Therefore, in generating schematic design proposals, it is the more fundamental space-form interaction and implementation potentials, rather than the detailed properties of a building's physical design, that will be of immediate importance to the designer.

For example, in initial exploration, the designer may be totally involved in making space-form decisions about which activity settings will be created as separate spatial entities, what their external and internal environmental conditions are, and how they must be organized to insure effective interaction between activities and with respect to the site-planning context. The strategy is to emphasize only the more *primary levels* of functional and aesthetic and economic requirements until a promising total system design scheme has emerged. In this way, consideration of the more detailed specifics of each entity (the *secondary-level* issues) is postponed so that they may be influenced by the total-system context. Such an approach enables the designer to identify overall issues with clarity and to deal with the design of each spatial subsystem in a hierarchic fashion.

But this is not to say that the designer should ignore the importance of structural specifics at the formative stages of his spatial thinking. Indeed, if the designer is experienced, he or she knows that the ultimate provision of both overall and specific means for physical integrity is essential. But he or she will need to have a suitable way of discriminating and then dealing with *only* the overall issues of structural design at the formative stages of design thinking. When this is not the case, it is not surprising that structural (and other technical) concerns will be more or less overlooked at schematic levels.

Therefore, the remainder of this book will be devoted to illustrating how the overall approach, which is successful in dealing with the nonstructural aspects of schematic design thinking, can also be applied to consideration of the structural aspects of schematic and preliminary design of buildings. The goal is to enable both architects and structural engineers to engage the full power of their knowledge in a creative manner.

2
Schematic Building Forms as Total Structural Systems

SECTION 1: The Assumption of Integrity

To maximize the integration of structural with architectural knowledge, a designer must be able to analyze schematic building forms in terms of their overall implications for structural behavior, strength, and efficiency. To do so, he or she can assume total physical integrity in the building forms that emerge from his or her schematic analysis of basic organizational and symbolic performance needs. When this is done, a given form option is assumed to behave as a structural whole, and can be analyzed as such to determine its overall load and resistance design implications.

For example, an architect's first schematic conceptualization of a building may be more for the purpose of manipulating its space-organizing properties, as a total space-form, than for specifying its physical structure. It could even be conceived as pure form, floating in space (Figure 2-1), until site-planning, climatic, service, and symbolic objectives are optimized. Even so, an assumption of total form integrity allows one to take into account certain simple but important physical observations related to the load-resistance implications of a building form (Figure 2-2):

1. The form will have to be fixed to the ground.
2. The form will have mass that must be supported by the ground.
3. The form will have to resist horizontal wind and earthquake forces.

FIGURE 2-1
IN ABSTRACTED SPACE ORGANIZING TERMS, A BUILDING FORM MAY BE CONCEIVED AS FLOATING IN SPACE.

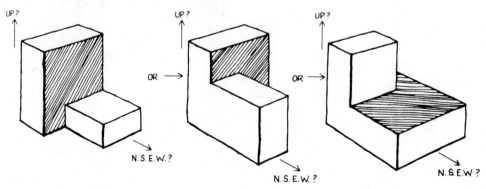

22

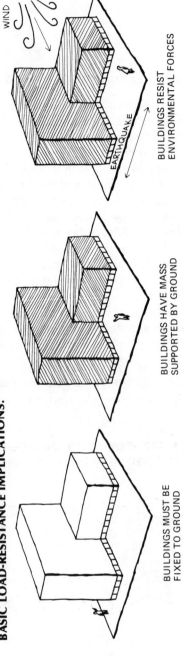

FIGURE 2-2
TAKEN AS A SPACE-FORM WHOLE, ANY SCHEMATIC DESIGN OPTION WILL HAVE
BASIC LOAD-RESISTANCE IMPLICATIONS.

BUILDINGS MUST BE
FIXED TO GROUND

BUILDINGS HAVE MASS
SUPPORTED BY GROUND

BUILDINGS RESIST
ENVIRONMENTAL FORCES

WIND

EARTHQUAKE

The grounding of a building is important to both the vertical and the horizontal stability of a building form. Since a building is to be constructed of physical components that are in themselves massive, the structure must transfer its own weight to the ground. This overall mass-weight (W) acts vertically and is usually described as a building's total dead weight or dead load ($W = \Sigma DL$). In addition, the structure must support occupancy loads, normally termed live loads (LL), which are more-or-less temporary. But dead loads are generally the dominant vertical loads and, at schematic stages, it is possible to draw important conclusions from observations and assumptions about the distribution of dead loads alone. In fact, as will be shown in Chapter 5, live loads are normally translated into equivalent dead loads, and the ratio of live load to dead load determines the relative importance of considering live loads at schematic stages.

Since the dead weight of a structure is always tending to push down into the earth, a fundamental schematic design requirement is for exploration of optional systems for interfacing these forces with the bearing capacity of the earth (Figure 2-3). Conceptually, dead load forces will be concentrated at the foundation along lines (for walls) or at points (for columns). These lines

FIGURE 2-3
THE FOUNDATION DISTRIBUTES THE WEIGHT OF A BUILDING ACCORDING TO EARTH RESISTANCE CAPABILITIES.

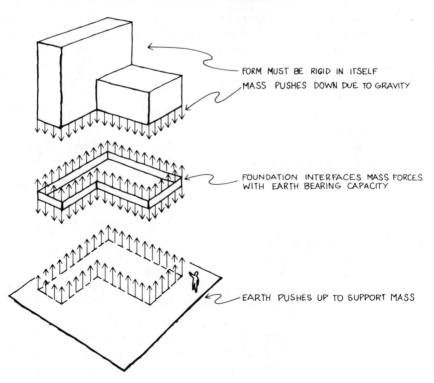

FORM MUST BE RIGID IN ITSELF
MASS PUSHES DOWN DUE TO GRAVITY

FOUNDATION INTERFACES MASS FORCES WITH EARTH BEARING CAPACITY

EARTH PUSHES UP TO SUPPORT MASS

FIGURE 2-4
OVERALL APPROXIMATION OF BUILDING FOOTPRINT LOADS.

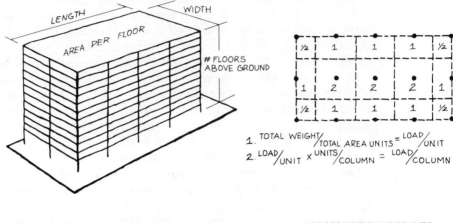

APPROXIMATE TOTAL DEAD WEIGHT GIVEN BY:
ESTIMATED LOAD/FLOOR × # OF FLOORS

FOOTPRINT DIVIDED INTO
TRIBUTARY AREAS AND
DESIGNATED AS RELATIVE
LOAD UNITS

or points of support may be at the plan edges or distributed throughout the entire base.

Thus, even at the earliest stages of design, overall assumptions can be made about the number and distribution of main column and wall supports. A plan diagramming these lines or points of support can be called a support *footprint*. These footprints make it possible to get a rough idea of how the total dead load will be distributed to the walls and columns (Figure 2-4).

For example, a 10-story building, 50 ft × 200 ft in plan, with an average dead load weight assumed to be 100 lb per sq ft, will weigh 10,000,000 lb. (@ 1000 kips per floor). If there are 15 columns layed out as in Figure 2-4, the load (and thus the size) per column can be quickly estimated by relating tributary footprint area units to column load units as shown:

load per unit = 1000 kips per floor × 10 floors ÷ 16 area units = 625 kips per unit.

Therefore, a 2-unit column would have to carry 1250 kips; a 1-unit column, 625 kips; and a $\frac{1}{2}$-unit column, 312 kips.

Note that the load per column in Figure 2-4 can be easily determined by observing the overall footprint and designating the relative size of tributary areas in units of load-carrying responsibility. The total number of units must carry the total dead load. Here, an assumption of one-half the distance between column points was used to indicate the tributary grid, the ratio $(2:1:\frac{1}{2})$ of footprint areas determining the load units per column. The ratios $(4:2:1)$ would work just as well. The actual tributary grid can be somewhat different as determined by the specifics of floor design, but this simple assumption is sufficient as a start.

It is important to understand that a dead load assumption of 100 lb per sq ft is arbitrary. A base 100 figure can be used to simplify the initial computation and get a base value per unit of tributary area. This makes it easy to consider the effect on column loads when the *DL* is other than the base assumption of 100 lb per sq ft. For example, what would be the column load at 80 lb per sq ft dead load, at 150 lb per sq ft, and so on? Similarly, consider the effect of varying the pattern of columns. Couldn't the ground-level layout shown in Figure 2-4 be but one of several options for that schematic building form? If it were the typical layout of a 20-story building, what would happen if there were nontypical floors?

Mass becomes a horizontal load design problem when the foundation is suddenly moved beneath a building as in an earthquake. This is because the mass of a building, owing to inertia, tends to remain in place when the foundation is translated horizontally by an earthquake. This phenomenon results in a distribution of horizontal forces along the height of the form. However, as shown in Figure 2-5, the resultant of these forces (H) will not act directly on the support system. It must be transferred some distance (a) to the foundation by horizontal shear resistance ($-H$), and this results in an overturning moment ($M = Ha$). Thus, to achieve equilibrium, the resultant of *DL* resistance ($-W$) must become eccentric with the *DL* resultant (W) such that $Ha = (-W)e$.

In addition to earthquake forces, horizontal loads can arise from the atmospheric environment in the form of normal and stormy winds that push against exterior surfaces. Winds also produce shearing forces and overturning moments that must be resisted by total-system action.

In either case of horizontal load the overturning moment could be offset by $-W$ as shown in Figure 2-5. Thus, it can be useful to investigate the overall implications of horizontal loads for footprint design. This can also be accomplished by assuming overall integrity of the form as a total structural system and leaving the specific questions of actually providing internal resistance until later. For example, it is clear that if overturn stability is to be provided by the weight of a form, the eccentricity of a symmetric form cannot exceed ($d/2$) without serious design implications, and frequently should be kept under ($d/4$) or ($d/6$). Thus, at the start one can explore the relationship between horizontal load and form characteristics to anticipate the basic requirements for overturn resistance at the support footprint.

Normally, overturn results from horizontal forces such as wind or earthquake, and the severity of the design problem is greater for tall or slender forms than for short or stubby forms. But, it should be recognized that some building forms will also produce overturning tendencies because of asymmetry between the centroid of the total dead load acting downward (W) and that of the upward resisting support system ($-W$) as illustrated in Figure 2-6. In this case, horizontal shear resistance is not involved.

FIGURE 2-5

OVERALL EQUILIBRIUM REQUIRES THE RESULTANT OF *DL* RESISTANCE (− W) TO BECOME ECCENTRIC WITH THE *DL* RESULTANT (*W* = Σ*DL*).

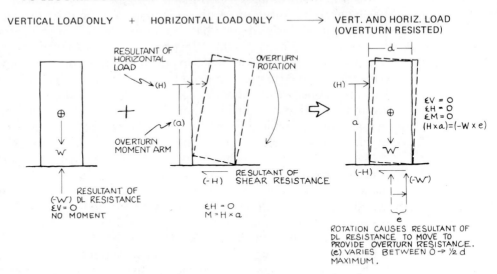

VERTICAL LOAD ONLY + HORIZONTAL LOAD ONLY ⟶ VERT. AND HORIZ. LOAD (OVERTURN RESISTED)

RESULTANT OF HORIZONTAL LOAD

OVERTURN ROTATION

(H)

OVERTURN MOMENT ARM

(a)

RESULTANT OF (−W) DL RESISTANCE
ΣV = O
NO MOMENT

RESULTANT OF (−H) SHEAR RESISTANCE

ΣH = O
M = H × a

d

(H)

ΣV = O
ΣH = O
ΣM = O
(H×a)=(−W × e)

a

W

(−H)   (−W)

e

ROTATION CAUSES RESULTANT OF DL RESISTANCE TO MOVE TO PROVIDE OVERTURN RESISTANCE. (e) VARIES BETWEEN O → ½ d MAXIMUM.

FIGURE 2-6

ASYMMETRY BETWEEN CENTROID OF BUILDING MASS AND CENTROID OF FOOTPRINT RESULTS IN *DL* OVERTURN MOMENT.

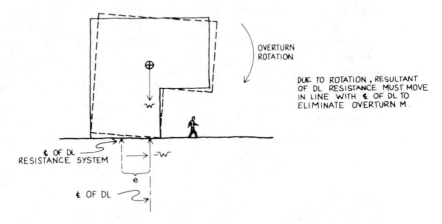

OVERTURN ROTATION

DUE TO ROTATION, RESULTANT OF DL RESISTANCE MUST MOVE IN LINE WITH ₡ OF DL TO ELIMINATE OVERTURN M.

₡ OF DL RESISTANCE SYSTEM

−W

e

₡ OF DL

However, the problem of designing for overturn resistance will be the same; only the degree of significance will vary.

Horizontal loads can also produce torsional (twisting) moments on a building when there is eccentricity in plan between the resultant of load and the resultant of shear resistance (Figure 2-7). Conversely, symmetry between resultant of horizontal load and the footprint layout eliminate overall

FIGURE 2-7
**ASYMMETRY BETWEEN THE RESULTANT OF HORIZONTAL LOAD AND
THE RESULTANT OF SHEAR RESISTANCE (S) REQUIRES TORSIONAL RESISTANCE.**

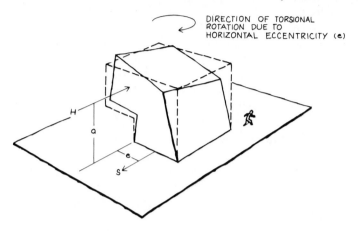

DIRECTION OF TORSIONAL
ROTATION DUE TO
HORIZONTAL ECCENTRICITY (e)

torsion. Since, in general, the distribution of shear resistance can be
assumed to vary with the distribution of columns or walls, a footprint layout
can be designed to minimize this problem. This will be discussed further in
Chapter 7, and torsion resistance in the building itself will be treated
separately in Chapter 10.

SECTION 2:　Estimating Overall Forces on Building Forms

To actually analyze building forms as total structural systems, it is useful to
be able to approximate and to compare the overall vertical and horizontal
forces acting on a form. Such approximations will always be very rough, but
they are a necessary first step if the designer wishes to get a realistic idea of
the significance of overall horizontal and vertical resistance requirements.

Because of gravity, the mass of a building produces vertical loads that are
additive. As a result, it is generally a simple matter to estimate the overall
weight (W) of any building form. This is so because, when all DLs are
averaged, the magnitude of W will depend more on the amount of floor
area and the type of construction than on a building's form properties.
Thus, for a given type of construction, an appropriate average unit DL value
(in pounds per square foot of floor area) can be estimated and multiplied by
the total floor area to roughly approximate the total vertical mass force (W)
of a building.[1] Form becomes a determining factor only when eccentricity

---

[1]Chapter 5 treats the specifics of estimating unit DL, WL, and EQ values. This section is
intended to explain how those values can be used.

between the resultant of vertical load and the centroid of the support system produces a significant overturning moment.

In contrast, earthquake produces inertial mass loads which act horizontally to create both shear and overturn design problems. Further, both the magnitude and overturn arm of the overall *EQ* force (H_{EQ}) are determined by the distribution of mass from top to bottom of a building form. Generally, the effect is greatest for mass at the top and diminishes to zero for mass at or below ground level. However, in most building designs, the distribution of floor area, and thus the distribution of mass, will be indicated by the overall form properties of a building.[2] As a result, it is possible to estimate the overall force and arm values by relating the (*W*) values to the overall properties of a building's form. Figure 2-8 illustrates the relationship by comparing two forms of equal mass to reveal how differences in form properties determine differences in magnitude and distribution of *EQ* forces

Figure 2-8 also illustrates that horizontal wind load (*WL*) effects are determined by overall form properties. That is, the overall wind force (H_{wind}) is determined by the amount of exposed surface area and can be estimated by simply multiplying an appropriate unit *WL* value (in pounds per square foot of exposed surface) by the amount of exposed area. However, the distribution of H_{wind} will correspond with the distribution of surface area, and the overturn arm will be determined accordingly. The total amount of wind force can be equal for two different forms, but the distribution (and thus the *M*-arm) may vary.

As it turns out, basic form types represent unique surface and mass distribution properties. Therefore, it is possible to compare the overall properties of prototypical forms and learn much about the amount and distribution of earthquake and wind load forces.

Figure 2-9 compares five equal mass prototypical forms to reveal their unique wind and earthquake force and overturn-arm characteristics. The reader should study the prototypes carefully, since real building forms that approximate or lie between two pure types will have their basic mass and surface distribution properties bracketed. Thus, for schematic purposes, the patterns of load distribution and the centroids of application shown for idealized types of building forms can be applied to real forms by analogy or interpolation to enable a rough approximation of overall *W*, H_{wind} and H_{EQ} design conditions.

It is recognized that real building forms will often be more complex than

---

[2]This assumption is sufficient for most buildings of more than one floor level above ground. For single-floor buildings it is less so, but will yield a rough idea of what the magnitude of *EQ* forces will be, providing the ratio of roof mass to wall mass is less than about 1:1. At greater ratios, the roof mass effect will not be adequately expressed by the overall assumption. In case of doubt, the mass effects of floor, roof, and wall subsystems can be considered separately, as shown in Chapter 5.

FIGURE 2-8

THE MAGNITUDE AND DISTRIBUTION OF EQ AND WIND FORCES ARE DETERMINED BY DISTRIBUTION OF OVERALL FORM PROPERTIES.

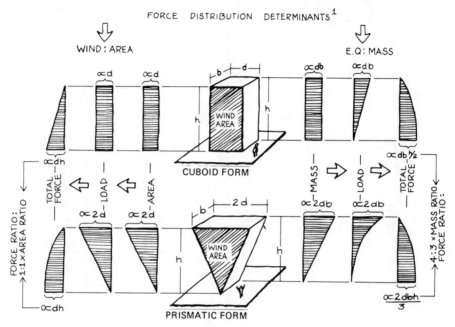

1. NOTE THAT THE EXPOSED AREA & MASS IS ASSUMED TO BE EQUIVALENT FOR BOTH FORMS.

the forms illustrated in Figure 2-9. But it is also possible to factor more complex forms into more-or-less pure components. For example, Figure 2-10 illustrates how truncated versions of the prototypical forms may be factored to reveal idealized components. The components may then be analyzed as pure types and their structural implications summed. The factoring concept may also be applied to more complex forms as will be shown in Section 5.

The following procedure (suggested for Figure 2-10) assumes that, for a given building form, the overall mass and surface area values have been estimated:

1. Determine component volume or surface area ratios.
2. Apportion total mass or exposed area accordingly.
3. Determine the *EQ* or wind effects of each component.
4. Sum component effects to reveal overall effect.

The basic idea is to apportion the total mass or surface area, evaluate the horizontal load effect of each component, and sum their effects to understand the overall effect. To do so, it can be helpful to first treat each component as though it were a cuboid form. The effects of change to other

FIGURE 2-9
A COMPARISON OF WIND AND EQ FORCES AND MOMENT ARMS FOR PROTOTYPE FORMS

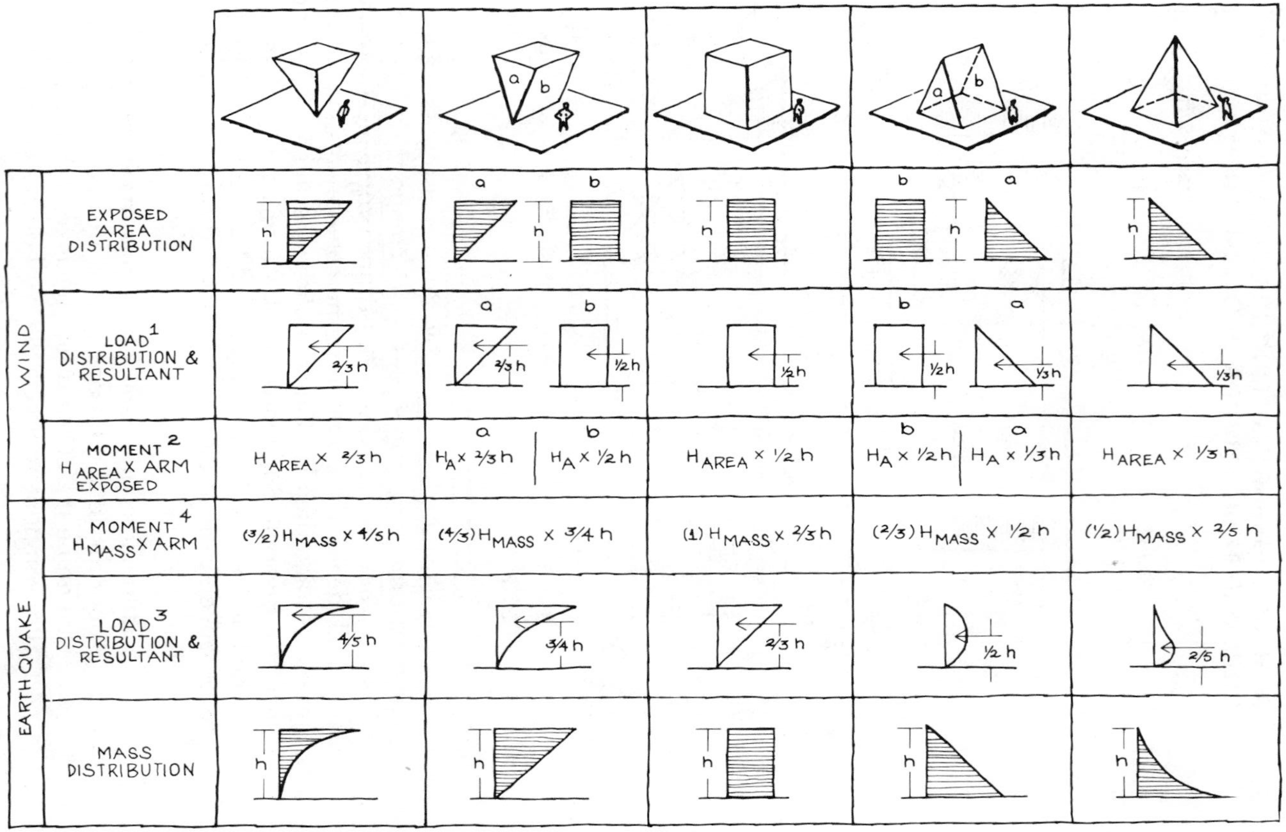

	Form 1	Form 2	Form 3	Form 4	Form 5
WIND — EXPOSED AREA DISTRIBUTION		a / b		a / b	
WIND — LOAD[1] DISTRIBUTION & RESULTANT	2/3 h	a — 2/3 h / b — 1/2 h	1/2 h	a — 1/3 h / b — 1/2 h	1/3 h
MOMENT[2] H_{AREA} × ARM EXPOSED	H_{AREA} × 2/3 h	H_A × 2/3 h | H_A × 1/2 h	H_{AREA} × 1/2 h	H_A × 1/2 h | H_A × 1/3 h	H_{AREA} × 1/5 h
MOMENT[4] H_{MASS} × ARM	$(3/2) H_{MASS}$ × 4/5 h	$(4/3) H_{MASS}$ × 3/4 h	$(1) H_{MASS}$ × 2/3 h	$(2/3) H_{MASS}$ × 1/2 h	$(1/2) H_{MASS}$ × 2/5 h
EARTHQUAKE — LOAD[3] DISTRIBUTION & RESULTANT	4/5 h	3/4 h	2/3 h	1/2 h	2/5 h
EARTHQUAKE — MASS DISTRIBUTION					

1. WIND LOAD DISTRIBUTION EQUIVALENT TO DISTRIBUTION OF EXPOSED AREA.
   WIND FORCE (H_{AREA}) = UNIT WL × EXPOSED AREA. SEE CH.5 FOR DETERMINATION OF UNIT WL VALUES.
2. E.Q. LOAD DISTRIBUTION (H_{MASS}) IS RELATED, BUT NOT EQUIVALENT, TO MASS DISTRIBUTION. (SEE CH.5)
3. FRACTIONAL COEFFICIENTS RELATE THE TOTAL E.Q. FORCE OF PRISM & PYRAMID FORMS TO THE TOTAL E.Q. FORCE OF CUBOID FORMS.
4. FRACTIONAL COEFFICIENTS RELATE THE TOTAL E.Q. FORCE OF CUBOID FORMS. (ASSUMING EQUAL MASS FOR ALL FORMS). SEE CH.5 FOR DETERMINATION OF F MASS VALUES FOR CUBOID FORMS.

FIGURE 2-10
COMPLEX FORMS MAY BE FACTORED TO DETERMINE OVERALL WIND AND EQ LOAD EFFECTS.

(a) TRUNCATED PRISM COMPONENTS

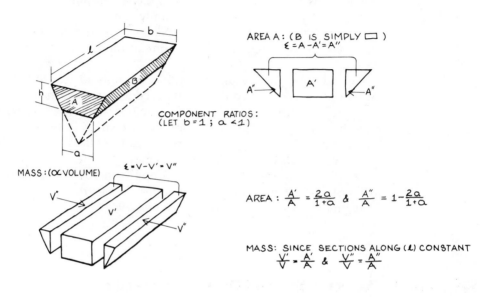

AREA A: (B IS SIMPLY ☐)
$$\xi = A - A' = A''$$

COMPONENT RATIOS:
(LET $b = 1$; $a < 1$)

MASS: ($\propto$ VOLUME)

$$\xi = V - V' = V''$$

AREA: $\dfrac{A'}{A} = \dfrac{2a}{1+a}$ & $\dfrac{A''}{A} = 1 - \dfrac{2a}{1+a}$

MASS: SINCE SECTIONS ALONG (ℓ) CONSTANT
$$\frac{V'}{V} = \frac{A'}{A} \quad \& \quad \frac{V''}{V} = \frac{A''}{A}$$

(b) TRUNCATED PYRAMID COMPONENTS

AREA A: (B TREATED SIMILARLY)
$$\xi = A - A' = A''$$

COMPONENT RATIOS:
(LET $b = 1$; $a < 1$)

MASS: ($\propto$ VOLUME)

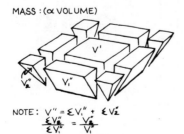

NOTE: $V'' = \xi V_1'' + \xi V_2''$
$$\frac{\xi V_2''}{\xi V_1''} = \frac{V_2''}{V_1''}$$

AREA: $\dfrac{A'}{A} = \dfrac{B'}{B}$ & $\dfrac{A''}{A} = \dfrac{B''}{B}$; $\left(\begin{array}{c}\text{REFER TO PRISM}\\ \text{RATIO ABOVE}\end{array}\right)$

MASS: $\dfrac{V'}{V} = \dfrac{3(a^2 - a^3)}{(1 - a^3)}$ & $\dfrac{V''}{V} = 1 - \dfrac{V'}{V}$
$$\frac{V_2''}{V_1''} = \frac{1-a}{3a}$$

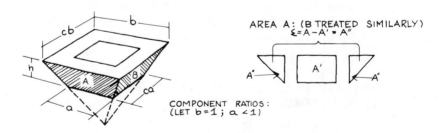

form types may then be approximated by referring to the comparative form factors shown in Figure 2-9.

In fact, if the distribution of mass and volume can be assumed to be in rough correspondence, it is possible to compare the overall structural implications of equal volume form options without even calculating the actual W, H_{EQ}, or H_{wind} values by simply comparing the form properties shown in Figure 2-9. Such an approach will not reveal specific force values, but it will reveal force, arm, and moment ratios for the form options under review.

For example, given equal mass, it is possible to realize that when one option produces a force that is 50 percent larger and has an overturn arm 33 percent larger than another option, the overall moment comparison will then be $1.5 \times 1.33 = 2:1$. If the options are varied in terms of either the amount of mass or exposed surface area, the effect on the ratio shown above will be direct (i.e., if the mass ratio is $\frac{2}{3}$, the moment ratio becomes $2 \times \frac{2}{3} = \frac{4}{3}$). Put another way, if form property factors are known, a variation in mass or surface area will result in a direct exacerbation or mitigation of the equal mass tendencies shown in Figure 2-9.

SECTION 3: Aspect Ratios and Overturn Resistance

Section 1 introduced the concept that to resist overturn by overall cantilever action, the resultant of a building's DL support system must become eccentric with the DL resultant. We shall now see that this results from the fact that the overturn moment (M) must be resisted by a vertical force couple ($\pm V$) that acts through the DL support system (Figure 2-11). As a

FIGURE 2-11
AN OVERTURN MOMENT MUST BE RESISTED BY A VERTICAL RESISTING FORCE COUPLE.

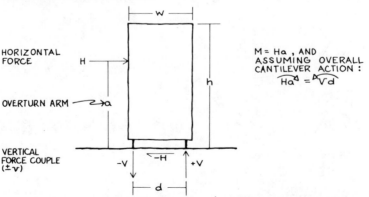

result, overall design of vertical support subsystems requires that the combined effect of *DL* and overturn resistance forces be considered. In order to do so, the concept of the structural aspect ratio of a building design is now introduced.

The aspect ratio is defined as h/d, where (h) is the gross height of a building and (d) is the overall depth of the *support subsystem* as measured in the direction of overturn action. Since the overturn moment arm (a) will vary with (h), and the resistance arm will vary with (d), the aspect ratio is an overall measure of the relationship between the overturn force (H) and the required resistance force couple $(\pm V)$.

Figure 2-11 illustrates the relationship in its simplest form, since, in a two-column elevation, the overturning resistance forces must act over the full depth of the support systems. By assuming that the building acts as a vertical cantilever supported at the base, it is clear that $Ha = Vd$. This means that $\pm V$ will vary directly with the overturn moment, and inversely with the depth of the support subsystem $(V = Ha/d)$. Now, if we let $a = ch$, where (c) represents an overturn moment arm factor $(Hch = Vd)$, and the required resistance force is given by $V = Hc(h/d)$.

This is useful since we know from Section 2 that the total horizontal force (H) and the overturn-arm factor (c) are determined by the overall form properties of a building design (Figure 2-9). Thus, for a given design, it is possible to compare the resistance requirements of various two-column footprint options by comparing the resulting aspect ratios only (Figure 2-12).

Similar comparisons can also be used to control the combined effect of overturn and *DL* resistance forces on a building's vertical support subsystem. For example, it will often be necessary to balance the overturn and *DL* forces by designing the footprint so that no uplift forces will result under horizontal load (Figure 2-13). Since the two-column elevation is symmetric, the available *DL* forces are $W/2$, and design for no uplift requires that $\pm V \leq \pm W/2$. This means that for balanced design $[Hch \leq (W/2)d]$ while, for a given (d), the actual amount of resisting force varies with the overturn moment.

Note that we can also write $[Hch \leq W(d/2)]$ to illustrate that overall resistance to overturn is accomplished by the resultant of *DL* resistance (W) moving to become eccentric with the resultant of *DL* (Figure 2-13d). That is, for a given (W), the required eccentricity (e) varies with the amount of overturn moment $(e = M/W)$. But the maximum balanced design eccentricity for a symmetric two-column elevation is $(e_b = d/2)$ where $[\pm V = \pm(W/2)]$.

To express the required amount of eccentricity for any overturn moment *in terms of* the maximum balanced design eccentricity, one can use a design eccentricity ratio $(e_r = e/e_b)$ and write: $[Hch = We_r(d/2)]$. Thus, (e_r) is given by the ratio of overturn moment to maximum balanced design moment $[e_r = Hch/W(d/2)]$, where $e_r \leq 1$ for balanced design. And, since the vertical force produced in a 2-column elevation is known to be $\pm W/2$ when the overturn moment equals the maximum balanced design moment, the

FIGURE 2-12
OVERTURN RESISTANCE FORCES VARY WITH CHANGES IN THE ASPECT RATIO.

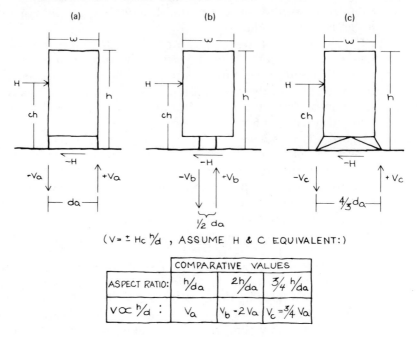

$(V = \pm Hc \frac{h}{d} , \text{ ASSUME } H \& C \text{ EQUIVALENT:})$

	COMPARATIVE VALUES		
ASPECT RATIO:	$\frac{h}{da}$	$\frac{2h}{da}$	$\frac{3}{4}\frac{h}{da}$
$V \propto \frac{h}{d}$:	V_a	$V_b = 2 V_a$	$V_c = \frac{3}{4} V_a$

FIGURE 2-13
OVERTURN RESISTANCE FORCES CAN BE BALANCED AGAINST DEAD LOAD FORCES.

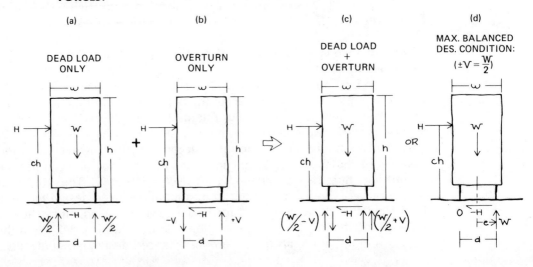

relative forces for larger or smaller overturn moments are also indicated by (e_r), that is,

$$V = \frac{Hch}{d} = e_r(W/2)$$

For example, if in Figure 2-13c, $e_r = 3/2$,

$$\left[\pm V = \pm \left(\frac{3}{2}\right)\frac{W}{2} = \pm \left(\frac{3}{4}\right)W \right]$$

and the columns would have to be designed to take combined forces of $-(W/4)$ and $+(\frac{5}{4})W$. If this $-(W/4)$ uplift is not desirable, the depth of the footprint would have to be made at least $\frac{3}{2}$ larger. Conversely, if $(e_r = 1)$, and it is also possible to design for an uplift force of $-W/4$, the designer will have the option of reducing the depth to $\frac{2}{3}$ the trial value.

Similarly, if $(e_r = \frac{1}{2})$, the required eccentricity will be only $\frac{1}{2}$ of the maximum value for balanced design, $\pm V = \pm(\frac{1}{2})W/2 = \pm W/4$, and the combined vertical force in the columns will be $+W/4$ and $+(\frac{3}{4})W$. Alternatively, the designer could reduce the column spacing by $\frac{1}{2}$ without resulting in any uplift.

The above example illustrates that for a given building design, the ratio (e_r) should be considered when designing the footprint. But the reader should also note from the above definition of (e_r) that if we let $(\beta = H/W)$, $e_r = 2\beta c(h/d)$, where the integer (2) is characteristic of a symmetric two-column elevation.[3] Hence, it becomes clear that a designer can control the eccentricity ratio by varying the load ratio (β), the overturn-arm factor (c), and the aspect ratio (h/d) of a building design. The (β) and (c) factors can be controlled by manipulating overall form properties, while the aspect ratio is controlled by manipulating (h) or (d).

For example, if (e_r) is to be unity, the maximum aspect ratio for balanced design is given by $(h/d = 1/2\beta c)$. As a result, when the overall form properties of a schematic building design are established, (β), (c) and (h) can be approximated as discussed previously, and a minimum value for (d) is given. Figure 2-14 illustrates this for a wind-loaded building that is 100′ tall and must be provided with columns spaced so that the dead load and overturn forces will be in balance (i.e., one column row under zero force). We know from Section 2 that $(c = \frac{1}{2})$ for a cuboid form, and if the exposed area produces a total wind force of, say, $(H = 0.25W)$, the maximum balanced design aspect ratio is

$$\left(\frac{h}{d} = \frac{1}{(2\beta c)} = \frac{4}{1} \right)$$

and (d) must be at least 25 ft. If (d) turned out to be less, columns would

---

[3]This characteristic will be explained for other footprints in Section 4.

FIGURE 2-14
**FOR A GIVEN BUILDING DESIGN, THE BALANCED DESIGN ASPECT RATIO
DETERMINES MINIMUM COLUMN SPACING.**

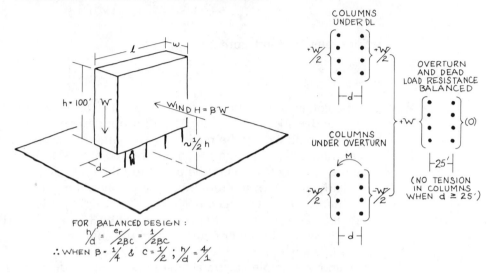

FOR BALANCED DESIGN :
$$\frac{h}{d} = \frac{e_r}{2\beta c} = \frac{1}{2\beta c}$$
∴ WHEN β = ¼ & c = ½ ; $\frac{h}{d} = \frac{4}{1}$

have to take tension (uplift at the foundation); if more, all columns would be in compression, the amount being determined by $[\pm V = e_r(W/2)]$.

In cases of schemes where the value of (d) is more-or-less fixed, we can write

$$\left(\pm V = e_r \frac{W}{2} = W\beta c \cdot \frac{h}{d} \right)$$

to illustrate that a designer has the option of manipulating the overall form properties to insure balanced design. For example, if we assume Figure 2-15a to be balanced, Figure 2-15b will not be, since its aspect ratio will be twice as great, while (β) and (c) remain constant. As a result, if the designer wishes to achieve balanced design while adopting the more closely spaced footprint option, he or she could consider changing the plan while holding the form type, volume, and surface exposure properties constant, to reduce the height of the design to ½h. Thus, in Figure 2-15c, the values for (β) and (c) will remain unchanged while the aspect ratio becomes equivalent to that of Figure 2-15a. On the other hand, a compromise can be achieved if it is possible to design for some tensile forces in the columns.

One should recognize that Figure 2-15c assumes that volume constancy implies that the total floor area and (W) values would tend to remain constant. Hence, for a given form type, it is possible to manipulate only the plan and height to control the combined effect of the load factor and aspect ratio. Further, other combinations of plan and height are possible that could

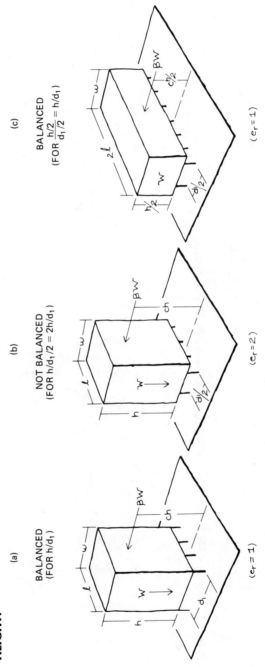

FIGURE 2-15
FOR A GIVEN BUILDING DESIGN, A CHANGE IN ASPECT RATIO MAY REQUIRE A CHANGE IN PLAN AND HEIGHT.

(a)

BALANCED
(FOR h/d_1)

$(e_r = 1)$

(b)

NOT BALANCED
(FOR $h/d_1/2 = 2h/d_1$)

$(e_r = 2)$

(c)

BALANCED
(FOR $\dfrac{h/2}{d_1/2} = h/d_1$)

$(e_r = 1)$

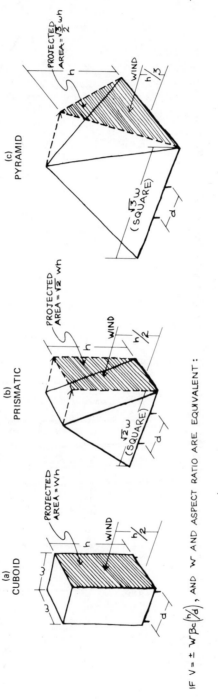

FIGURE 2-16
OVERTURN FORCES CAN ALSO BE CONTROLLED BY VARYING FORM TYPE.

result in a reduction or increase in load ratio (β) as well as in the aspect ratio.

The overall load ratio (β) and the overturn arm factor (c) can be manipulated by changing from one type of form to another. For earthquake design, a change in form type will always change (β) and (c). For wind design, (c) will change for at least one horizontal direction, while (β) may change, depending on the plan and (h). For example, Figure 2-16 illustrates how the requirement for wind load overturn resistance varies with the type of form. If the total weight (W) and the aspect ratio are assumed constant, a change in form type produces a change in (β) and (c), and the overturn resistance force requirements vary accordingly. If the aspect ratio is also varied, the effect would be equally direct. Refer to Figure 2-9, (page 30), to see how earthquake load (β) and overturn arm (c) factors will vary with form type.

Up to now, our intention has been only to introduce the general concepts of how overall form properties can influence structural design. Therefore, we have illustrated these concepts in the simplest terms by referring only to idealized two-column elevations. Nevertheless, it should now be clear that, at schematic levels, a designer *can* use an overall approach to quickly understand how the basic properties of form and footprint layout will tend to affect efficient structural performance in architecture.

The concluding sections of this chapter will now show how these concepts apply to other types of elevations. Then Chapters 3–5 will follow up by dealing with the specific questions of estimating the actual overturn load ratios and of designing for effective horizontal shear as well as overturn resistance.

SECTION 4: Strength and Stiffness of Buildings

When we analyze a building form by means of the assumption of overall integrity, we also make assumptions about sufficient strength and stiffness of the total structural system. Strength is the property of a structural system that enables it to resist loads without collapsing altogether. On the other hand, stiffness is the property of a structural system that enables it to limit deflection under load. For example, a fishing rod can be visualized as a structure that must be strong enough to avoid breakage but flexible enough to deflect a great deal. But a building should neither collapse nor deflect significantly under load. It is therefore necessary to discuss the requirement of strength and stiffness in building structures.

To illustrate the basic design issues, we can consider a building that is supported by springs instead of stiff columns (Figure 2-17). The springs may be strong enough to hold the building off the ground, but they may not

FIGURE 2-17
SPRINGLIKE COLUMNS WOULD ALLOW TOO MUCH DEFLECTION UNDER LOAD.

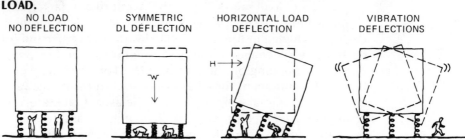

| NO LOAD
NO DEFLECTION | SYMMETRIC
DL DEFLECTION | HORIZONTAL LOAD
DEFLECTION | VIBRATION
DEFLECTIONS |

provide sufficiently stiff column action to resist noticeable deflections under varying types of loads.

Now, if each floor of a building were to be supported by springs, the whole building would be analogous to a springlike column. Under symmetric *DL* it would shorten significantly in an axial manner (Figure 2-18*a*). Under horizontal load, the building would act as a springlike cantilever, and significant horizontal deflections would occur at the top of the building due to bending (Figure 2-18*b*).

Note that axial deflections are the result of vertical loads that cause uniform shortening, or compression, of a building's form. And, since the shortening is uniform, the compressive stresses ($f_c = W$/area) will also be

FIGURE 2-18
OVERALL DEFLECTIONS ARE CONTROLLED BY OVERALL STIFFNESS OF A STRUCTURAL SYSTEM.

(a)   (b)

AXIAL DEFLECTIONS   BENDING DEFLECTION DUE
DUE TO DEAD LOAD   TO HORIZONTAL LOAD

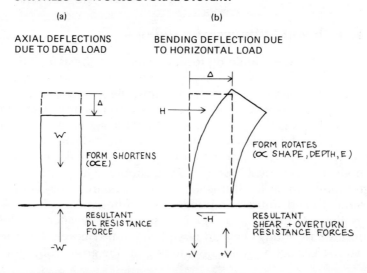

uniform across any horizontal section of the structural system. This means that axial stiffness is directly influenced by a fairly constant stress-to-strain property of the material called the modulus of elasticity (i.e., E = stress/unit strain). The lower the modulus, the more deflection will occur under a given axial load. Both the amount of material (sectional area) and modulus represent axial stiffness variables.

On the other hand, bending stiffness involves resistance to rotational, rather than axial, deflection of a building's form. That is, for each unit of rotational deflection the material in a horizontal section will be vertically elongated (in tension) on one side of a neutral axis of no deformation (N-A) and be shortened (in compression) on the other side, the amount varying with the distance of the material from the neutral axis (Figure 2-19). As a result, the tensile and compressive resistance stresses (f_t and f_c) will also vary with the distribution of the sectional area about the neutral axis. Evidently shape and depth act together with material properties to determine the overall resistance of a structural system to bending deflection.

Obviously the type of material is important. The higher the elastic modulus, the less the rotational deflection for a given load-moment, amount of material and footprint layout. But the reader should note that shape and depth are geometric variables that determine how effectively *any* type or amount of material will be utilized. Further, the overall effectiveness of various types of sectional shapes can be compared in terms of the product of effectiveness factors for area and depth. For example, Figure 2-19a represents an idealized case where both area (for $\pm V$) and depth (for moment-resisting arm) are fully effective, their product being unity for a maximum overall shape factor of (1). But the shape in Figure 2-19b (in direction of M) is less efficient with a shape factor of 1/3. For equivalent type, amount of material, and sectional depth, a given load-moment on case (b) will result in three times as much rotation (and three times higher stresses) as on 2-19a. Similarly, Figure 2-19b would require three times more material (or a type of material that has an elastic modulus three times higher) than used in 2-19a to be equally stiff. Note that efficient shapes are those that place material away from the N-A. And, it is possible to improve the bending stiffness, *but not the efficiency*, of a given sectional shape by manipulating the amount and type of building material used.

The efficiency of a given shape may only be increased by increasing the overall depth of the system rather than either the amount or type of material. However, this is an important option because an increase in depth will not only change the length of the moment-resisting arm but also the amount of overall resistance force per unit of rotation. That is, f_{max} will be higher per unit since part (or all) of the area will be more distant from the neutral axis. Therefore, the effect of a depth change from d_1 to d_2 will be given by $(d_2/d_1)^2$, and Figure 2-19b would be as stiff as 2-19a if the depth were increased by $\sqrt{3}$, the material properties being held constant. The

FIGURE 2-19
SECTIONAL SHAPE IS A FACTOR IN BENDING STIFFNESS.

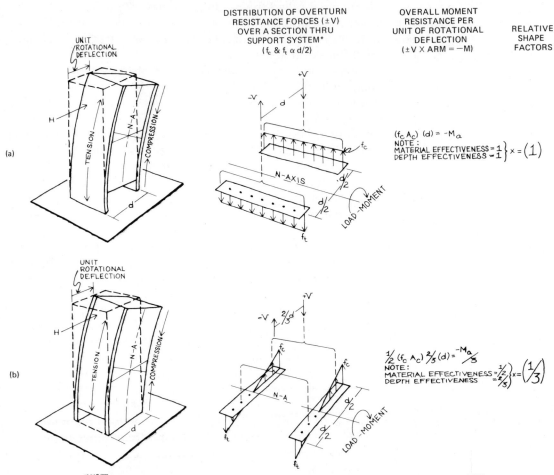

*NOTE:
1. THE OVERALL RESISTANCE FORCE (±V) VARIES WITH THE DISTRIBUTION OF THE SECTIONAL AREA RELATIVE TO THE N-A.
2. THE AMOUNT & TYPE OF MATERIAL IS ASSUMED EQUIVALENT.

combined effect of manipulating shape, depth, and area (amount of material), is illustrated in Figure 2-20 by assuming that the type of material is a constant. You should be able to anticipate the effect of a change in material type.

It should now be clear that stiffness and strength can be dealt with in overall terms by understanding how the shape effectiveness of footprint designs can be maximized and what effect manipulation of material and depth properties will have on overall bending resistance.

For example, the stiffness of any support-system layout can be compared with that of an idealized layout using the same amount and type of material

FIGURE 2-20
RELATIVE EFFECTIVENESS IN RESISTING BENDING DEFLECTION (i.e., RELATIVE M PER UNIT OF DEFLECTION) IS INDICATED BY THE PRODUCT OF RELATIVE SHAPE, MATERIAL QUANTITY, AND DEPTH FACTORS.

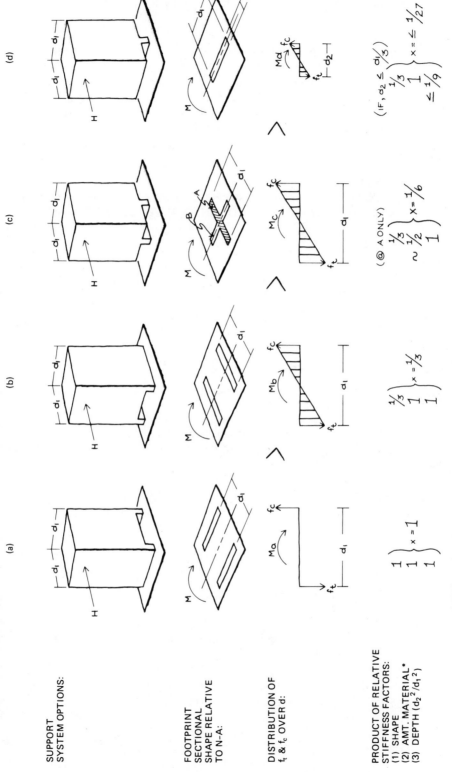

(Figure 2-21). Once a designer assigns column (or wall) locations, he or she can approximate their relative section sizes by means of the tributary area distribution of the overall *DL* to the column locations. This makes it obvious that basic decisions about the layout of *DL* support components should be influenced by consideration of overall bending stiffness properties of the building. When the footprint is symmetric as shown, the relative stiffness is easily indicated by treating the exterior and interior columns as pairs that

FIGURE 2-21
THE OVERALL STIFFNESS OF *DL* SUPPORT SYSTEMS WITH INTERIOR COLUMNS MAY BE COMPARED WITH THAT OF AN IDEALIZED TWO-COLUMN ELEVATION.

RELATIVE LOADS ON COLUMNS ARE INDICATED BY TRIBUTARY AREA ANALYSIS

BUILDING DESIGN

COLUMN LAYOUT AND RELATIVE LOAD AREAS

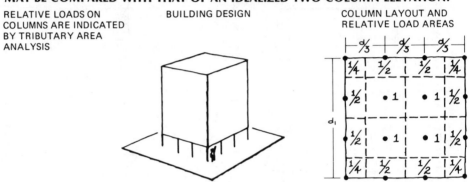

COMPARISON OF ACTUAL DESIGN WITH IDEALIZED DESIGN:

(a) ACTUAL (4 COL.)   (b) IDEALIZED (2 COL.)

COLUMN MATERIAL WILL BE DISTRIBUTED ACCORDING TO RELATIVE LOAD AREAS

MATERIAL DISTRIBUTED TO 2-COLUMN PAIRS } { INTERIOR $\longrightarrow$ $\frac{2}{3}$

EXTERIOR $\longrightarrow$ $\frac{1}{3}$

ALL MATERIAL CONCENTRATED IN EXTERIOR COLUMNS

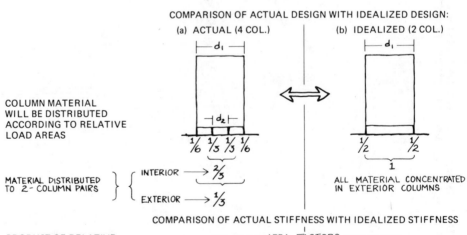

COMPARISON OF ACTUAL STIFFNESS WITH IDEALIZED STIFFNESS

PRODUCT OF RELATIVE AREA AND DEPTH FACTORS FOR COLUMN PAIRS INDICATES OVERALL SHAPE STIFFNESS FACTORS*

EXTERIOR: $\frac{1}{3} \times (1)^2 = \frac{1}{3}$
INTERIOR: $\frac{2}{3} \times (\frac{1}{3})^2 = \frac{2}{27}$ } $\frac{11}{27}$

EXTERIOR: $(1)^2 \times 1 = 1$

OVERALL SHAPE FACTORS

(* NOTE: SINCE THE COMPONENTS ARE COLUMN PAIRS, THEIR SHAPE FACTORS ARE (1) & COMPONENT STIFFNESS IS INDICATED BY THE PRODUCT OF THE RELATIVE AREA AND DEPTH FACTORS ONLY.)

differ in terms of the depth and material quantity factors but not in terms of shape (Figure 2-21a). Hence, if the depth of the interior column pair (d_2) is expressed in terms of (d_1), the relative stiffness factor of the overall system is given by

$$\left[\frac{1}{3}(1)^2 + \frac{2}{3}\left(\frac{1}{3}\right)^2 = \frac{11}{27}\right]$$

This compares with the maximum factor of (1) for an idealized case (Figure 2-21b), which could be achieved only if all of the material were concentrated in the two exterior column locations. Since, in Figure 2-21a, the material is less than half as effective as in the idealized case, the reader may find it instructive to explore what options to follow in improving the footprint design while maintaining the four-column elevation.

One's knowledge of overall stiffness control factors can be combined with that of a maximum balanced design eccentricity to achieve efficient design of any type of support-system footprint. For example, Figure 2-22 illustrates that a maximum balanced design eccentricity factor (k) exists for any type of footprint such that the balanced eccentricity is given by $e_b = kd$). Further, these (k) values can be used to compare the stiffness of one support system layout relative to another. This becomes evident when one recognizes that, for balanced design, the maximum amount of bending rotation is determined by the available DL resistance stress (f_{DL}) in the exterior columns and the overall depth, while the corresponding moment capacity is determined by the eccentricity. Assuming equal (f_{DL}) and depth (d), the balanced design rotation will be the same for (Figures 2-22b and 2-22c), and a comparison of eccentricity factors indicates relative stiffness. Thus, if k_b or k_c is compared with the (maximum) balanced design eccentricity factor of an idealized two-column elevation of equal depth (i.e., $k = \frac{1}{2}$), the overall stiffness factors are indicated as $(\frac{5}{9}$ and $\frac{11}{27})$ as shown. Note that if $d_b = d_c$, the relative stiffness of Case b : Case c is $(\sim\frac{3}{2})$, as given by comparing either the stiffness factors shown above or by direct comparison of eccentricity factors. But, if $d_b \neq d_c$, the comparison yields only a shape factor ratio, and the combined effect of the depth ratio will also have to be considered as discussed previously.

The reader should also note that Figure 2-22 illustrates that the balanced design eccentricity factor (k) is relatively easy to approximate for any type of footprint. This is accomplished by understanding that the balanced design stress pattern and the distribution of material (area) over the footprint depth determine the resistance force pattern. From the force pattern, the resultant of DL resistance can be found, and its eccentricity with the resultant of DL is revealed as shown.

As it turns out, the location of the resultant of DL resistance also determines the overall moment-resisting arm for a given footprint. For example, the overall resisting arm of a bearing-wall system such as illustrated in Figure 2-19 can be determined directly from the triangular

FIGURE 2-22
BALANCED DESIGN ECCENTRICITY RATIO $k_1d_1 : k_2d_2$ INDICATES RELATIVE STIFFNESS.

(a)

BALANCED DESIGN OF A SYMMETRIC 4 COLUMN ELEVATION

AXIAL RESISTANCE TO DL   MOMENT RESISTANCE

COLUMN LOCATIONS
AND DISTRIBUTION OF
AXIAL AND OVERTURN
RESISTING STRESSES
OVER (d)

AXIAL LOAD STRESS PATTERN

ROTATIONAL STRESS PATTERN

AXIAL STRESSES ARE UNIFORM
FOR ANY LOCATION

ROTATIONAL STRESSES VARY
WITH COLUMN LOCATION

REQUIRED TO BALANCE
DL AND OVERTURN
RESISTANCE STRESSES

NET BALANCED DESIGN
STRESSES VARY WITH
COLUMN LOCATION
RELATIVE TO ('0')

BALANCED STRESS PATTERN

THE RESISTANCE FORCES IN COLUMNS WILL VARY WITH
THE PRODUCT OF THE NET BALANCED DESIGN STRESS
FACTOR AND THE DISTRIBUTION OF COLUMN AREAS

(b) ←——— COMPARE CASES ———→ (c)

AXIAL DL FORCES (F_{DL})
VARY WITH DISTRIBUTION
OF MATERIAL TO COLUMNS

BALANCED DESIGN
DISTRIBUTION OF DL
RESISTANCE FORCES
(IN F_{DL} UNITS)

BALANCED DESIGN
ECCENTRICITY (kd)
INDICATED BY RESULTANT
OF DL RESISTANCE

OVERALL
M-RESISTING ARM

OVERALL STIFFNESS
RELATIVE TO IDEALIZED
2 COLUMN ELEVATION

$$\frac{\frac{5}{18}(d)}{\frac{1}{2}(d)} = \frac{5}{9} \rightarrow \left(\begin{array}{c} \text{NOTE THAT CASE (b) IS} \\ \sim \frac{3}{2} \text{ STIFFER THAN (c)} \end{array} \right) \leftarrow \frac{\frac{11}{54}(d)}{\frac{1}{2}(d)} = \frac{11}{27}$$

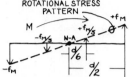
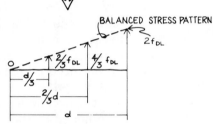

stress distribution pattern to be $(\frac{2}{3})$ the moment-resisting depth. Thus, $(k = \frac{1}{6})$ since the balanced design eccentricity for the *DL* resultant is $\frac{2}{3}(d) - \frac{1}{2}(d) = \frac{1}{6}(d)$. When this is compared with the idealized eccentricity factor $(k = \frac{1}{2})$ of Figure 2-19a, the ratio (k_r) is only $\frac{1}{3}$, indicating the poor shape stiffness factor of the wall system as compared with an ideal two-column system. Or, for a given depth, the stiffness of the wall system relative to the four-column system which is illustrated in Figure 2-22b is indicated by the eccentricity factor ratio $(\frac{1}{6} : \frac{11}{54} = \frac{9}{11})$.

Figure 2-23 shows that this approach is also useful when it is desired to determine and control the combined forces in a building's footprint. When values for (H), (c), (h/d), and the overall $(\Sigma DL = W)$ are approximated for a building design option, the ratio of actual to balanced design eccentricity (zero stress in an exterior column) is given by

$$\left[e_r = \frac{Hch}{Wkd} = \frac{\beta c}{k} \cdot \frac{h}{d} \right],$$

where $(k = \frac{1}{2})$ for a two-column elevation. An (e_r) value of, say, 2.0, would mean that the overturn resistance stress, and thus the forces, would be twice that allowed for balanced design. It also means that twice the rotation allowed for balanced design is required to develop those stresses. Therefore, the design must be improved by a factor of two in order to balance the overturn forces with the *DL* forces.

When $(e_r > 1)$, a designer can use his knowledge of relative stiffness to improve a footprint layout. But it should be noted that a design cannot be balanced by simply adding more material while keeping the same shape and depth. Although that would reduce the amount of rotation and thus the overturn stresses, it would also reduce *DL* resistance stresses equally, and the same unbalanced condition would exist. But it can be balanced without changing the depth by changing the distribution of the material, and thus the shape factor, so that the available eccentricity factor (k) will be two times larger. This means both the rotation and the forces would vary inversely with a change in shape factor.

Balanced design can also be accomplished by increasing the overall depth of the footprint by a factor of two times while holding the shape factor constant. In this case, the eccentricity *factor* will not change, but the required forces will vary inversely with an increase in depth to achieve the balance desired. But since stiffness varies with the square of a change in depth, the rotation will be $\frac{1}{4}$ as large. Of course, a combination of these methods could be employed to control the force and rotation picture.

The relative stiffness approach is also useful when some tensile force in the support system is allowed, because there is always a practical limit as to how much. Therefore if, in the example, the overturn resistance forces could be $\frac{3}{2}$ larger than the balanced design value, but not two times, a stiffness improvement in the shape and/or depth factors need total only $\frac{2}{3} \times 2 = \frac{4}{3}$.

FIGURE 2-23

**THE COMBINED AXIAL AND OVERTURN RESISTANCE FORCES CAN BE
DETERMINED BY REFERRING TO THE BALANCED DESIGN ECCENTRICITY
RATIO ($e_r = Hch/Wkd = M/M_b$).**

SCHEMATIC
DESIGN AND RELATIVE
COLUMN AREAS

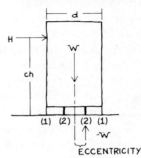

RELATIVE STRESSES
AT COLUMN LOCATIONS
DUE TO AXIAL LOAD

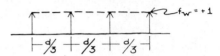

@ BAL. DESIGN M ⟵——— COMPARE ———⟶ @ 2x BAL. DESIGN M
    ($e_r = 1$)   ($e_r = 2$)

RELATIVE STRESSES
AT COLUMN LOCATIONS
DUE TO OVERTURN M

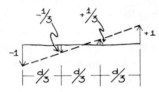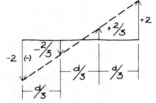

COMBINED AXIAL AND
OVERTURN STRESSES
(RELATIVE TO $f_w = +1$)

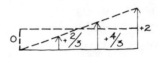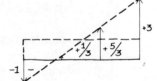

RELATIVE FORCES AT
COLUMN LOCATIONS
(STRESS X AREA FACTORS)

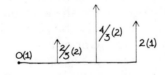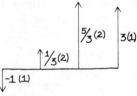

THE RATIO OF ACTUAL
TO BALANCED DESIGN
ECCENTRICITY INDICATES
THE RATIO OF ACTUAL
TO BALANCED DESIGN
OVERTURN STRESSES

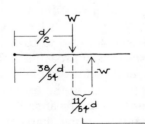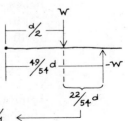

SECTION 5: Symmetry and Asymmetry in Building Forms

In general, dead loads will not produce overall horizontal bending in a symmetric building elevation. That is so because normally the load is axially supported by the overall system. However, when the building elevation is asymmetric, or when the support system resultant is not axial with the building mass, overall bending results (Figure 2-24).

When asymmetry results in bending, the forces in the resisting system may be analyzed in a manner similar to that of a horizontal load. The ratio of actual eccentricity between the resultants of dead load and resistance may be compared with the balanced design eccentricity, and the force distribution determined accordingly.

But in most cases the eccentricity of vertical load will not be as major a design problem, as is the case for wind or earthquake loads. This is so because the column and wall areas will vary with the vertical load distribution and mitigate the eccentricity. However, in combination with earthquake and wind loads, it is possible for eccentric dead load conditions to become important. For example, action by earthquake or wind against rotation due to *DL* eccentricity can be beneficial, but action in the same direction would make the overturn problem worse. Since wind or earthquake can act in any direction, it is useful to analyze the vertical load eccentricity in the beginning and to combine the vertical and horizontal effects of asymmetry.

Asymmetry between overall building form and the support system can also result in horizontal asymmetry between the resultant of wind and earthquake loads and that of shear resistance. This can produce horizontal twisting (torsion) as illustrated in Figure 2-25. Figure 2-25a illustrates that asymmetry of building form can result in twisting, whereas in Figure 2-25b

FIGURE 2-24
ECCENTRICITY BETWEEN THE RESULTANT OF THE *DL* AND THE E OF THE SUPPORT SYSTEM WILL PRODUCE AN OVERTURN MOMENT.

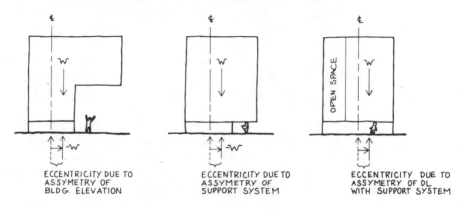

ECCENTRICITY DUE TO ASSYMETRY OF BLDG. ELEVATION

ECCENTRICITY DUE TO ASSYMETRY OF SUPPORT SYSTEM

ECCENTRICITY DUE TO ASSYMETRY OF DL WITH SUPPORT SYSTEM

FIGURE 2-25

HORIZONTAL ASYMMETRY BETWEEN A BUILDING FORM AND THE SUPPORT SYSTEM FOOTPRINT WILL PRODUCE TWISTING.

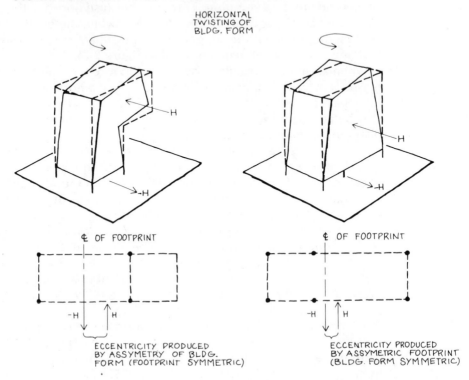

the problem is due to asymmetry of the support system footprint. Analysis of this situation will be dealt with separately in Chapter 7.

In many cases of design, a complex building form will emerge that has noticeably asymmetric form and mass distribution properties as measured in the vertical and/or horizontal dimension. But Figure 2-26 illustrates that the concepts of overall analysis may be applied to such complex forms by dealing with major parts separately. This is possible because the idea of a "foundation" support system can be taken to represent any level of a structure, the parts above a given level being supported by the level immediately below.

By assuming integrity in each major form component, it is possible to concentrate on the overall requirements for design of each and for interface of the components. In fact, the interface represents the critical area of schematic design, since it will determine the options for actually achieving integrity in the form components being joined.

FIGURE 2-26
COMPLEX BUILDING FORMS CAN BE FACTORED WHEN USING AN OVERALL APPROACH TO SCHEMATIC STRUCTURAL DESIGN.

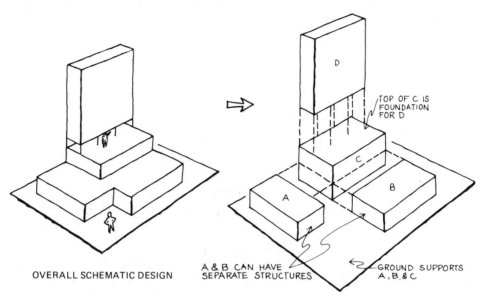

OVERALL SCHEMATIC DESIGN

A & B CAN HAVE
SEPARATE STRUCTURES

GROUND SUPPORTS
A, B, & C

If the idea of form subdivision is carried to the extreme, each level of a building, from top to bottom, could be so analyzed to determine optimum layout and size requirements for effective design of the support systems. In fact, this is exactly what is done at detailed design levels (known as *taking free bodies* in analytical mechanics). The degree of analysis is different, but not the basic concept.

Thus, by assuming overall integrity, a designer can first recognize and mitigate against the overall problems of a structure-form relationship at schematic levels, when modification of form and/or structural scheme is easiest. The smaller subproblems can be dealt with at a later stage when refinement is necessary. Of course, while trying to solve these big problems in a more-or-less conceptual way, the economic and construction feasibility of the subsystems must be kept in mind. But these issues will be treated separately in Chapters 13 and 14.

In Chapter 3 we will introduce the basic structural subsystems that are required to give substance to the assumption of overall form integrity. We will also present simple methods for approximating the vertical and horizontal load-resistance effect of basic types of subsystems on overall structural behavior. Later chapters will deal with the specifics of subsystem design and touch on the basic factors of economic feasibility.

3
Overall Integrity and Major Subsystem Interaction

SECTION 1 The Hierarchy of Structural Action in Building Forms

At some point, when designing a building form, a designer must consider the basic subsystems that are required to substantiate his assumption of total-system integrity. Fortunately, the overall approach facilitates creativity in this regard by establishing a means for applying total-system thinking to subsystem design. As we shall now learn, one can examine total-system behavior in terms of the basic requirements for schematic interaction and design of its key subsystems.

An assumption of overall integrity implies that one can understand what basic types of subsystems and interactions are required to actually achieve the integrity assumed. At the schematic level this means that one should be able to conceptualize fundamental design options for laying out the interaction of key subsystems. Then, at the preliminary level, the objective is to work hierarchically to prove the feasibility of such subsystems by determining their key properties. Here, approximation is sufficient and consideration of details can be postponed until the final level of the design process (Figure 3-1). The idea is to promote maximum flexibility at early stages of architectural design, when it is most desirable, by allowing a simplified exploration of total-system structural design options.

The inherent value of such a hierarchic approach becomes evident when one recognizes the spatial nature of the interaction between total and subsystem design thinking. First, the designer is concerned with establishing a total-system understanding of the three-dimensional (3-D) implications of architectural space-form options. And, since the total structure will have to be constructed of more-or-less planar components, he will then wish to extend this understanding to include the basic requirements for preliminary design of major two-dimensional (2-D) subsystems. As shown in (Figure 3-1), we can consider schematic (3-D) thinking as Level I of the design process and preliminary design of (2-D) subsystems as Level II. The final level, III, is associated with one-dimensional (1-D) design, where the object is to elaborate and refine the more generalized decisions made at Level II in terms of individual elements and connection details.

Although the three levels are distinguishable, note that feedback is necessary to optimize the interaction between total and subsystem thinking. At the first two levels the emphasis is on overall thinking and analysis of fundamental properties by quick approximation. Thus, when the more basic

FIGURE 3-1
DESIGN LEVELS AND THE HIERARCHY OF STRUCTURAL THINKING.

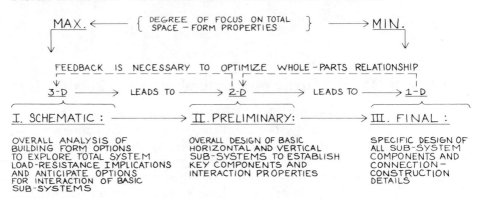

components of schematic and preliminary designs are optimized, they become a total-system context for dealing in-depth with the final design of all elements and details.

Figure 3-2 diagramatically illustrates how such a hierarchy of structural thinking might evolve by taking an overall-to-specific look at an architectural space-form scheme that is similar to an actual building designed by the architectural firm HOK (Photo 3-1). Note that the whole space-form scheme is first conceptualized in terms of layout, as a total space-structure system, and then elaborated in two successive stages to provide for the more specific and interactive properties of subsystem design. Being of broader scope, decisions made at Level I or II serve to establish a necessary context for dealing with the more specific decisions made at the next lower level. Viewed in terms of feedback, this hierarchic approach also allows consideration of elemental options (at Level III) to suggest changes in subsystem options (at Level II), which may in turn suggest changes in the total-system concept. Or, if contradiction or confusion of the whole-parts relationship occurs at Level II or III, one can easily refer to the next bigger level to regain (or change) the broader perspective of overall design purpose.

Such an approach reflects the organic concept that the whole (of a design scheme) should give rise to the need for details and not vice versa. It also reflects the natural hierarchy of an architectural design problem, that is, 1-D structural elements will become architecturally relevant only when it is understood how they can work together to contribute to the fulfillment of a broader need for 2-D space-enclosing, organizing, and building subsystems. Similarly, these 2-D subsystems become architecturally relevant as they are understood to contribute to the overall effectiveness of a 3-D space-form scheme as a total environmental system. Thus, elements are essential; but it is the more comprehensive conceptualization of overall space-form and subsystem interactions that should determine how and when elements are

FIGURE 3-2
BUILDING FORMS AND THE HIERARCHY OF STRUCTURAL THINKING.

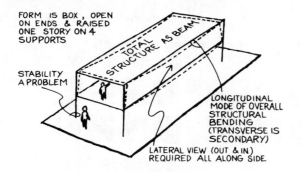

FORM IS BOX, OPEN ON ENDS & RAISED ONE STORY ON 4 SUPPORTS

TOTAL STRUCTURE AS BEAM

STABILITY A PROBLEM

LONGITUDINAL MODE OF OVERALL STRUCTURAL BENDING (TRANSVERSE IS SECONDARY)

LATERAL VIEW (OUT & IN) REQUIRED ALL ALONG SIDE

3-D (SCHEMATIC LEVEL)

STRUCTURE-FORM RELATIONSHIPS ANALYZED AS TOTAL SYSTEM BY ASSUMING OVERALL INTEGRITY AND BEHAVIOR

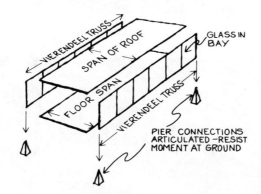

VIERENDEEL TRUSS

SPAN OF ROOF

GLASS IN BAY

FLOOR SPAN

VIERENDEEL TRUSS

PIER CONNECTIONS ARTICULATED —RESIST MOMENT AT GROUND

2-D (PRELIMINARY LEVEL)

BASIC HORIZONTAL AND VERTICAL SUB-SYSTEMS IDENTIFIED AND KEY COMPONENT PROPERTIES AND INTERACTIONS ESTABLISHED

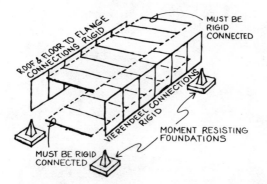

MUST BE RIGID CONNECTED

ROOF & FLOOR TO FLANGE CONNECTIONS RIGID

VIERENDEEL CONNECTIONS RIGID

MOMENT RESISTING FOUNDATIONS

MUST BE RIGID CONNECTED

1-D (FINAL LEVEL)

ALL LINEAR ELEMENTS AND CONNECTION DETAILS SPECIFIED SUFFICIENT FOR PREPARATION OF ENGINEERING AND CONSTRUCTION DOCUMENTS

dealt with in design.

Now that the concept of total-to-subsystem thinking has been introduced, the remainder of this chapter will focus on identifying basic types of horizontal and vertical 2-D subsystems and exploring their interaction requirements.

PHOTO 3-1*a* and *b*
LABARGE, INC. (FORMERLY THE AMERICAN ZINC BUILDING). TWO STEEL VIERENDEEL TRUSSES CARRY 50-FT-CLEAR SPANS. (HOK, ST. LOUIS, ARCHITECTS; THE ENGINEERS COLLABORATIVE, STRUCTURAL ENGINEERS).

SECTION 2: Building Forms Conceived as Solid Structures

Buildings are seldom designed to be primarily solid structures. A few historical examples are found in such buildings as Egyptian pyramids and Aztec and Cambodian temples. Since there is little usable space enclosed, such structures served more as pure monuments than as space enclosures— life swirled around, but not within, the forms. Contemporary examples are even more rare, and solid architecture is only approximated in the form of the Washington Monument or technological structures such as tall industrial masonry smokestacks and some concrete television transmission towers.

Nevertheless, the basic issues of providing overall structural integrity through key subsystem interactions can be introduced in their simplest form by conceiving of building forms as solids. In Section 3 we will see that the basic *issues* of subsystem design for solid structures are essentially the same for buildings designed as space-enclosing structures.

When a building is conceived as a solid, it is easy to visualize that, as a whole, the form will act as both a column, for transfer of vertical loads to the ground, and a cantilever, for transfer of horizontal loads. That is, axial and compressive capabilities are required for transferring vertical loads to the ground, but shearing and bending capabilities are required to transfer horizontal loads and to resist overturning moments. We will now examine the implications of these overall requirements for design of subsystems by imagining the form to consist of stacked slices. These slices represent basic planar (2-D) subsystems that must interact both horizontally and vertically to provide the overall integrity of column and cantilever action.

Figure 3-3 illustrates the axial-compressive load-transfer requirements for an imaginary subsystem slice taken from a simple rectangular building form. In Figure 3-3a, the *idea* of axially symmetric load distribution across a slice of the form is illustrated for a symbolic external load (P). The compressive stress ($f_c = P/A$) would be the same for any such slice cut along the height, since the full load must be vertically transferred from slice to slice.

In contrast, Figure 3-3b illustrates that the actual self-weight (dead load) on a particular slice would increase linearly from zero at the top to a maximum at the bottom as shown in the triangular loading diagram. Thus, for a symmetric external load (P) and dead load per slice (DL), a thin 'slice' of the form at any point along its height would be under uniform axial compressive stress ($f_c = P/A + \Sigma_{h_i} DL/A$). Each imaginary slice must be capable of picking up the load accumulated from the slices above and transferring that load, plus its own weight, to slices below.

When the material in a solid form is actually continuous, the slice subsystems are only imagined and can occur at any point and be of any thickness. But it should also be recognized that slice subsystems could be literal in that a solid form can be constructed of discontinuous blocks of material laid one on top of the other. In fact, columns and walls of most primitive masonry buildings were constructed in this manner and behaved well under axially symmetric loadings.

FIGURE 3-3
**VERTICAL LOADS MUST BE AXIALLY TRANSFERRED FROM ONE SUBSYSTEM
SLICE TO THE NEXT.**

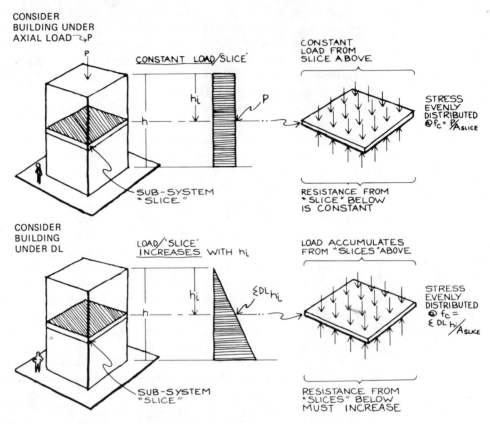

For vertical load transfer, the axial interaction between horizontal slice subsystems is direct and simple enough to visualize. But when the issue is overall resistance to horizontal load, the same (2-D) subsystems must provide an indirect kind of interaction if they are to transfer horizontal loads to the ground and resist overturn of the building.

A horizontal load cannot be axial with a vertical cantilever. Therefore, a symbolic horizontal load (H) applied at the top of a building must be transferred to the ground by indirect means of horizontal resistance to slippage between slices (Figure 3-4a). This kind of action between subsystem slices is called shear resistance. For a symbolic load (H) concentrated at the top of a building, the total shear resistance (V) required from top to bottom would be the same between each slice (i.e., $V = H$). Similarly the average shear stress is ($\bar{v} = V/\text{area}$).[1]

---

[1]Throughout the remainder of the text, (V) will stand for total shear, (v) for shear stress.

FIGURE 3-4
HORIZONTAL LOADS MUST BE TRANSFERRED FROM ONE SUBSYSTEM SLICE TO THE NEXT BY SHEAR RESISTANCE BETWEEN SLICES.

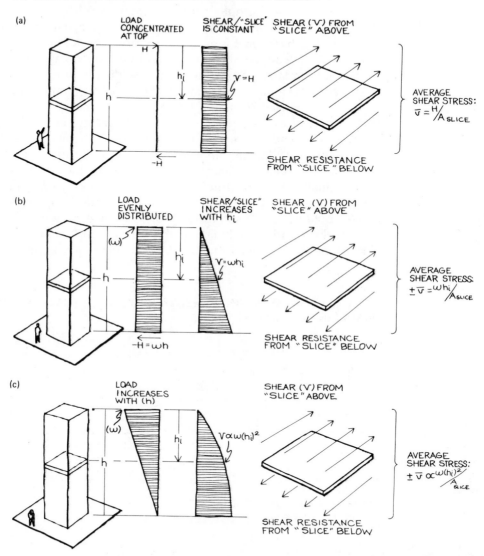

On the other hand, Figure 3-4*b* illustrates that if the total horizontal load (H) were to be distributed evenly along the height of a building ($w = H/h$), the shear at any point (h_i) would accumulate from zero at the top to ($V_i = wh_i$). The shear diagram would be triangular with the maximum ($V = H$) at the foundation and ($\bar{v}$) will vary with slice location.

Distributed loads can arise from the surface of a building because it offers resistance to wind. Thus, for symmetric elevations, a wind load is often assumed to be fairly evenly distributed over a building's height and, in design, would be assumed to yield a triangular shear diagram.

Unevenly distributed horizontal loads may result from foundation movement due to earthquake or nonsymmetric elevations under wind load. Figure 3-4c illustrates a common type of earthquake load distribution, wherein the shear force is assumed to vary over the height of a rectangular building. The load per slice varies according to its mass and its distance from the foundation. Thus, for even mass distribution, the load diagram is assumed to be an inverted triangle, the shear diagram parabolic.

Thus far we have illustrated that horizontal slices represent basic subsystems that must interact to resist and transfer not only axial compression under vertical loading, but also horizontal shearing forces under horizontal loading. In a continuously solid form, such a structural capability between imaginary slices is inherent. However, if a solid form were actually to be constructed by stacking blocks, it would not necessarily tend to satisfy the requirement for shear resistance. Figure 3-5 illustrates what would happen if sufficient shear resistance were not achieved between the blocks. In primitive buildings this was accomplished by either relying on friction between heavy stone blocks or by physically keying the joint

FIGURE 3-5
FAILURE TO TRANSFER SHEAR RESISTANCE BETWEEN SUBSYSTEM BLOCKS WILL RESULT IN COLLAPSE OF FORM.

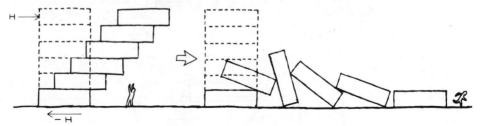

FIGURE 3-6
PRIMITIVE MEANS OF TRANSFERRING SHEAR RESISTANCE BETWEEN SUBSYSTEM BLOCKS.

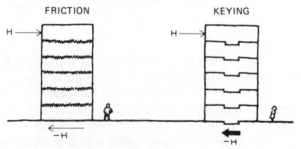

between the blocks (Figure 3-6). In brick building, morter plays this role.

Shear resistance is essential for horizontal load transfer. But it also results in a corollary requirement for rotational interaction between slices to resist overturn. That is, resistance to transverse slippage will produce an overturning moment (M) that must be resisted by rotational integrity between the slices. (Figure 3-7). Note that the overturning moment is generated because there is a lever arm (a) between the resultant of horizontal load (H) and any lower horizontal plane of shear resistance. Thus, where there is shear, there will be moment and the slices must also transfer a compressive and tensile resisting force couple (C-T) as shown in (Figure 3-7b).[2] To develop the resisting force couple, each imaginary slice must resist being pulled apart on the tensile side of a neutral axis of rotation and compressed on the opposite side. This, in turn, causes the slices to be elongated (in tension) on one side and shortened (in compression) on the other (Figure 3-7c). The overall result is a curving of the structural form, which is termed "bending."

Figure 3-8 illustrates that the distribution of horizontal load determines the distribution of shear, which in turn controls both the distribution and the magnitude of moment along the height of a rectangular form.[3] Note

[2]The symbols (C and T) will now replace the symbol ($\pm V$) used in Chapter 2.
[3]The reader can refer to Figure 2-9 for an indication of moment diagrams for other types of forms.

FIGURE 3-7
TO DEVELOP OVERTURN RESISTING FORCES (C and T) THERE MUST BE A ROTATIONAL INTERACTION BETWEEN THE SUBSYSTEM SLICES.

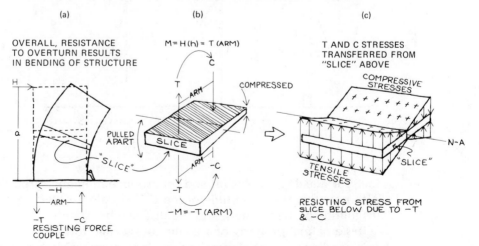

FIGURE 3-8
**HORIZONTAL LOAD DISTRIBUTION DETERMINES SHEAR (*V*) AND MOMENT (*M*);
SECTIONAL SHAPE AND DEPTH DETERMINES THE REQUIRED *C-T* FORCES.**

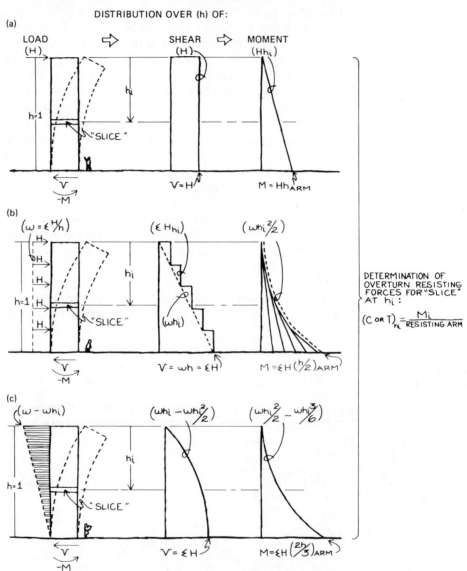

that, for equal (*V*), cases (*a*) and (*c*) produce moments larger than case (*b*).
Also note that the magnitude of the *C-T* couple is determined by the
magnitude of the moment and the distance between *C* and *T*.

It is a natural property of a continuously solid form to resist both shear
and bending to some degree, although most materials will exhibit different

FIGURE 3-9
**STACKED BLOCKS MAY NOT PROVIDE SUFFICIENT TENSILE RESISTANCE TO
OVERTURN MOMENTS.**

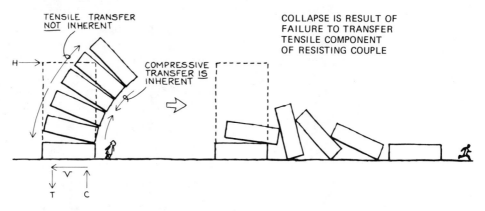

resistance to tension and compression. But with a discontinuous stack of
blocks there is a generic lack of tensile transfer capability (Figure 3-9). Thus,
a simple stack of blocks can resist a compressive axial load and possibly
provide some frictional shear resistance to horizontal loads. But, except for
the stabilizing effect of a vertical load, such a structure could literally
overturn, because tensile forces cannot be transferred to resist rotation
between the blocks.

As discussed in Chapter 2, one could count on the dead load
compression to provide for some overturn stability, even if a building of
stacked blocks were not physically tied together to resist tension at the
interface. Nominally this means that an overturn moment can be resisted to
the extent that the compressive axial stress (due to vertical load) balances
the requirement for tensile resistance. However, unless provision is made to
assure a capability for tensile interaction between the blocks, they will
separate on the tensile side as the overturn tension exceeds the *DL*
compression. Therefore, it is often desirable to design to avoid this
possibility. Figure 3-10 summarizes the basic requirements for resistance to
shear, compressive, and tensile stresses on a slice subsystem such that the
dead load will balance the need for tensile resistance.

Assuming the capability for developing shear and tensile resistance is
adequate, we can now turn our attention to the problem of controlling the
horizontal deflection (Δ) that results from the overall bending action of a
building form under horizontal load. These deflections can become
excessive, particularly at the top, and must be restricted in such a way that
the horizontal displacement of a building goes unnoticed to its inhabitants.
A building must not only be strong enough to avoid failure, but it must also
be "stiff" enough to prevent excessive deflections under horizontal load.

Chapter 2 also introduced the concept that resistance of a structural form
to deflection under transverse load is determined by sectional shape and

FIGURE 3-10
**DESIGN OF A RECTANGULAR SECTION FOR SHEAR, AND BALANCE OF *DL*
AND *C-T* STRESSES.**

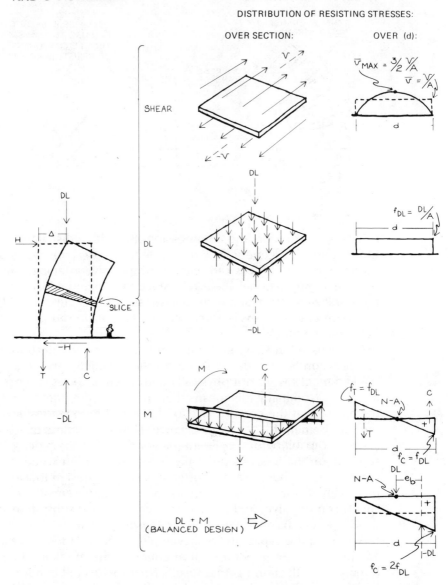

depth more than by material strength. In overall terms this means that
bending stiffness in any structural form is achieved by gaining the greatest
separation between the resultants of tensile and compressive stresses (*C*
and *T*), which form the overall resisting couple. We shall now take a more
penetrating look at the implications of this for design of subsystem slices.

The influence of shape can be explained by examining a rectangular slice under bending (Figure 3-11a). The resisting stresses are generated when each subsystem slice is forced to rotate about its neutral axis (*N-A*) by its interaction with the slice above and restrained by the slice below. The symmetry of the section determines the location of the *N-A* and, for elastic action, the stresses vary linearly across its rotational depth (*d*). As a result, the stress effectiveness of the tensile and compressive areas will also vary with their distribution about the *N-A*. For a rectangular shape, this means

FIGURE 3-11
STIFFNESS AND MOMENT RESISTING CAPACITY.

(a) THE OVERALL RESISTANCE ARM DETERMINES THE MOMENT RESISTING FORCES AND STRESSES

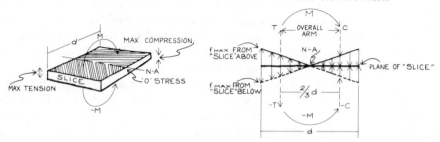

(b) THE M-RESISTING CAPACITY OF A SECTION IS DETERMINED BY ITS DEPTH AND DISTRIBUTION

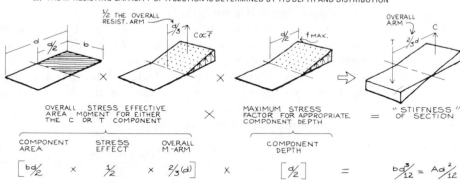

(c) THE RELATIVE STIFFNESS OF EQUIVALENT SECTIONAL TYPES IS GIVEN BY THE Ad^2 RATIO

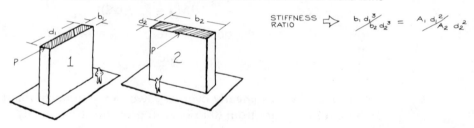

that the resultant forces (C and T) act over a resisting arm of 2/3d and the *average* stress [$\bar{f}_{C\,or\,T} = (C$ or $T)/(A_C$ or $A_T)$] is only one half (f_{max}). Similarly, the stress effectiveness of the tensile and compressive areas is only one-half the maximum possible.

Note that for a given moment (M), the amount of rotation required to generate (f_{max}): will vary with the shape, depth, and modulus of elasticity (E) of a slice. But, whatever the modulus, it is the shape and depth of a slice that determines its overall moment-resisting *effectiveness* and, thus, how large (f_{max}) must be. The lower the effectiveness, the larger (f_{max}) and the more rotation will occur. Sectional shape and depth are therefore important geometric properties that act independently of material properties to influence the bending stiffness of a structure.

The engineering term for the geometric stiffness property of a slice is *moment of inertia* (*I*). *I* is technically defined in terms of sectional area and depth as $\Sigma A_d x_d^2$, where A_d represents each element of tensile and compressive area and x_d^2 is the product of the distance from each element to the "neutral" axis of rotation and its stress effectiveness at that distance. Thus, for any type of section, *I* represents the summation of *stress-effective area moments* about the *N-A* of the whole section.

The foregoing defines *I*, but it is the *meaning* of stiffness that is most important to our purposes. Fortunately, the task of illustrating the meaning of *I* is simplified by the fact that (C = T) in the case of pure bending. Therefore the overall stiffness of a section can be expressed as the product of the *overall* stress effective *area moment* of either the C or T area component and its maximum stress factor as determined by the appropriate component depth (Figure 3-11*b*).

Figure 3-11*b* illustrates how shape and depth determine the average stress effectiveness and overall arm for A_C or A_T. But the *usefulness* of the expression becomes more evident when it is understood that the overall arm and component depth (stress factor) determine the maximum stress (f_{max}) required for A_C or A_T *to produce* C and T and resist a given moment.

Thus, if we divide a force moment (M) by the product of a component area moment and its depth, we will have

$$\left(\frac{\text{force} \times \text{arm}}{\text{eff. area} \times \text{arm} \times \text{comp. depth}} = \frac{\text{force}}{\text{eff. area} \times \text{depth}}\right)$$

which represents the maximum stress per unit of C or T area depth, *i.e.*,

$$[M/I = f_{max}/(d/2)].$$

Now, if we express the depth of the C or T area as ($d/2 = c_{max}$), we can write

$$\left(\frac{Mc_{max}}{I} = f_{max}\right)$$

And, since *f* varies linearly over c_{max}, we can generalize that ($f = Mc/I$), where (c) may vary from (0 to c_{max}) to give the stress at any point along the

C or T area depth. We can also generalize that f_{max} varies directly with the load-moment and inversely with the *overall* stress-effective area moment of either A_C or A_T:

$$f_{max} = \frac{M}{(A_{C\ or\ T} \times \text{effectiveness factor} \times \textit{overall} \text{ arm})}$$

Applied to the rectangular case, the validity of the preceding generalization is made obvious by recalling that moment equilibrium requires that

1. C or $T = \dfrac{M}{\text{overall arm}} = \dfrac{M}{\frac{2}{3}d}$;

2. $f_{max} = \dfrac{C \text{ or } T}{(A_{C\ or\ T}) \times \text{effectiveness factor}}$; and therefore,

3. $f_{max} = \dfrac{M}{(\text{effective area}) \times (\text{overall arm})} = \dfrac{M}{\frac{A}{4} \times \frac{2}{3}d} \times \dfrac{d/2}{d/2} = \dfrac{M(d/2)}{(Ad^2/12)} = \dfrac{Mc_{max}}{I}$

We can now generalize even further. Assuming a slice to be a solid, I can be expressed (Ad^2/z), where (z) is constant for any given sectional type. This means that $(1/z)$ is an overall stiffness factor. Thus, for sections of the same type, I will vary directly with a change in Ad^2 (Figure 3-11c). It also means that, for a given load-moment, the required resisting stress (f_{max}) will vary inversely with a change in Ad, the required resultant force (C or T) varying inversely with a change in (d).

The actual I (or z) values need be determined only when a designer wants to specify the actual stress value for a given section or to compare different types of sections under a given (M). To compare sections of equal area and depth, only the (z) values are needed, whereas the relative stiffness of *any* two sections is indicated by I_1/I_2; relative C or T stress by the inverse ratio of overall stress effective area moments (i.e., $f_1/f_2 = I_2/I_1 \times c_1/c_2$); and the resultant forces (C or T) will be indicated by the inverse ratio of *overall* moment arms (Arm$_2$/Arm$_1$).

In terms of slice subsystems, the usefulness of the above is that the effectiveness of modifications in shape *can* be compared. However, it is obvious that the practical applicability of this would be increased if it were possible to develop a simplified means for (roughly) approximating the (z) values of most sectional shapes. We shall now see that this can be done if it is recognized that the stiffness of any section can be given by summing:

1. The product of the compressive and tensile area components times the square of their centroidal distance from the N-A.
2. The independent rotational stiffness value for each component.

For a rectangular section this yields $Ad^2[1/16 + 1/48] = Ad^2/12$ (Figure 3-12a). The centroid distance squared factor (1/16) represents the portion

FIGURE 3-12
STIFFNESS IS LARGELY DETERMINED BY THE DISTANCE FROM THE CENTROID OF THE TENSILE AND COMPRESSIVE AREAS TO THE *N-A*.

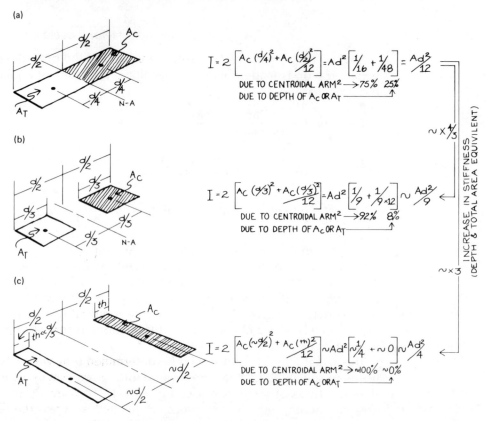

of total stiffness contributed by the two area components (A_C and A_T) as though they were concentrated at the centroid point. The rotational factor (1/48) represents the *additional* contribution for the components owing to the fact that they are not concentrated. This is significant because it allows one to compare the proportion of the overall stiffness contributed by the centroid arm squared factor with that of the rotational factor. Note that the rotational factor is significantly smaller.

The possibility for simple approximation becomes evident when the stiffness, as computed for a rectangular section (Figure 3-12a), is compared with that of two idealized sections where all of the material is placed further away from the *N-A* (Figures 3-12b and 3-12c). Note that, in the rectangular Case *a*, the centroid arm is $d/4$ and the centroidal arm squared factor $[(\frac{1}{4})^2 = \frac{1}{16}]$ accounts for 75 percent of the total stiffness factor ($\frac{1}{z} = \frac{1}{16} + \frac{1}{48} = \frac{1}{12}$). But when the arm is $\geq d/3$ (Cases *b* and *c*), the arm squared factor accounts for nearly all of the stiffness.

This means that when the component area centroid arm is $\geq d/3$, the overall stiffness of any sectional shape can be approximated by summing the products of the component areas times their centroidal arms squared. Thus, Figure 3-12c represents the idealized limiting case where the centroid arm is nearly $d/2$, and the overall stiffness factor ($\frac{1}{z} \approx \frac{1}{4}$) approaches a value three times larger than that of Case a.

It is now possible to utilize the idealized ratio of Figures 3-12a : 3-12c (i.e. 3 : 1) to approximate the relative change in stiffness that would result from a redistribution of the material in a rectangular section to form an (H) shape. That is, any material that is placed to form the flange of an (H) shape can be assumed to be about three times as effective as the web material. In fact, when the flange thickness is $\leq\frac{1}{4}$, the component depth of A_C or A_T, the approximation is very close. This is illustrated in Figure 3-13, where the

FIGURE 3-13
THE RELATIVE STIFFNESS OF H SECTIONS CAN BE COMPARED IN TERMS OF THE RELATIVE EFFECTIVENESS OF MATERIAL PLACED IN THE FLANGE AND WEB.*

RELATIVE STIFFNESS FACTORS*:

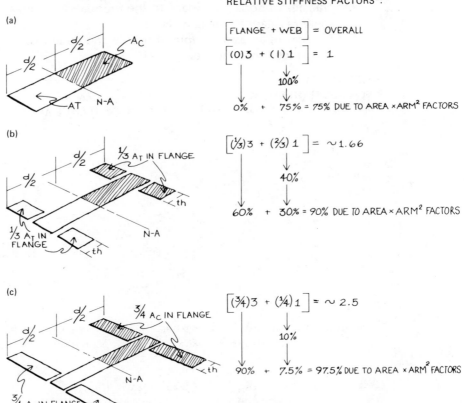

(a)

$$\left[\text{FLANGE} + \text{WEB} \right] = \text{OVERALL}$$

$$\left[(0)3 + (1)1 \right] = 1$$

$\downarrow$ 100%

0% + 75% = 75% DUE TO AREA × ARM² FACTORS

(b)

⅓ A_T IN FLANGE

⅓ A_T IN FLANGE

$$\left[(\tfrac{1}{3})3 + (\tfrac{2}{3})1 \right] = {\sim}1.66$$

$\downarrow$ 40%

60% + 30% = 90% DUE TO AREA × ARM² FACTORS

(c)

¾ A_C IN FLANGE

¾ A_T IN FLANGE

$$\left[(\tfrac{3}{4})3 + (\tfrac{1}{4})1 \right] = {\sim}2.5$$

$\downarrow$ 10%

90% + 7.5% = 97.5% DUE TO AREA × ARM² FACTORS

* 3:1 RATIO APPLIES WHEN th $\leq$ ¼ (⅟₂) = (⅜) FOR SYMMETRIC SECTIONS

idealized ratio is used to compare a rectangular section with three *H* sections of equivalent depth but varied proportions of total area in the flanges. The relative contribution of the flange indicates roughly how much of a moment on such a section will be carried by the more or less uniform stresses in the flange only.

The reader should note from the above that it is possible to simplify even further. As it turns out, if the flange thickness is $\leq\frac{1}{2}$ the tensile or compressive depth of an *H* section, and the *gross* flange : web area is $\geq 2:1$, a reasonable approximation of the overall stiffness value can be made by dealing with the gross flange area only, the web contribution being small (Figure 3-14). This will include most *H* section designs.

We can now summarize that the assumption of overall integrity in a solid building form is possible only when its subsystem slices are capable of providing three basic actions. They are:

1. Axial transfer of vertical loads from one horizontal slice to another and ultimately to the foundation.
2. Transfer of horizontal loads to the foundation by means of horizontal shear resistance between slices.
3. Transfer of an *axial* couple of tensile and compressive overturn resistance forces between slices.

FIGURE 3-14
FOR MOST (H) SECTIONS, *C* OR *T*, *I*, AND σ_{max} CAN BE APPROXIMATED BY DEALING WITH THE GROSS FLANGE AREA.

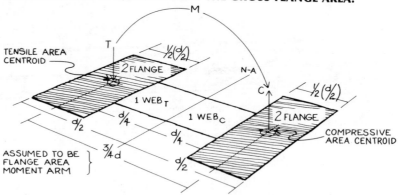

WHEN THE FLANGE DEPTH IS $\leq (9/4)$ AND THE FLANGE TO WEB RATIO $\geq 2:1$:

1. ASSUME C & T ACT AT FLANGE CENTROIDS
  ∴ C OR T = $M/\frac{3}{4}d$

2. BY SYMMETRY ASSUME,
$$I \sim 2\left[A_{FLANGE} \times \left(\tfrac{3}{8}d\right)^2\right] = \{A_F d^2 \left(\tfrac{9}{64}\right)$$
(NOTE : THE READER CAN VERIFY THAT SUCH AN APPROXIMATION RESULTS IN A CONSERVATIVE ERROR OF LESS THAN 7%)

3. APPROXIMATE STRESS BY $6 = Mc/I$ APPROXIMATE
  ∴ $6_{MAX} = \dfrac{M(d/2)}{A_F d^2 (9/64)}$

*GROSS FLANGE AREA GIVEN BY FLANGE THICKNESS X ITS WIDTH

When such integrity can be assumed, it is possible to approximate the amount and distribution of internal forces required to resist vertical and horizontal loads. Generally, axial vertical loads will be distributed evenly across the sectional area of the slice. Horizontal loads will result in horizontal shear forces that will be distributed across the section in some fashion, frequently assumed to be uniform for approximate designs. We also know that the overturn resisting couple (C or T) varies with the overall resisting arm, and the maximum stress $f_{max} = Mc_{max}/I$ of the slice. For most H sections, I can be reasonably approximated by dealing with the flange area only because the web becomes significant only as the flange-to-web area ratio becomes less than $2:1$. Then it will be more accurate to add for web stiffness by assuming that a unit of web area will be roughly $\frac{1}{3}$ as stiff as a unit of flange area.

SECTION 3: Building Forms Conceived as Space-Structures

In basic terms, a space-enclosing form differs from a solid form only insofar as its structural subsystems are distributed in space to enclose rather than fill up forms. That is, overall integrity in a space-structure requires that provision be made for the same three types of basic subsystem interactions as for a solid structure.

This can be simply illustrated by conceiving of a rectangular building form as a (3-D) space-structure that has been defined and enclosed by four thin vertical planes (2-D subsystems) joined to form a rectangular tube. The top is open and the bottom is fastened to the earth by some foundation system (Figure 3-15a). If the planes are imagined to be very thin relative to their

FIGURE 3-15
A MINIMUM PHYSICAL MODEL OF A BUILDING AS A SPACE STRUCTURE.

(a)

4 THIN VERTICAL PLANES
JOINED TO FORM
OPEN TOP TUBE

(b)

IF PLANES ARE TOO THIN,
THEY WILL BUCKLE AND
THE FORM WILL COLLAPSE

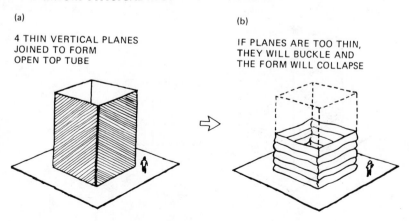

width and height, we can say that this conceptualization represents a physically minimum model of a building form as a space-structure. But the planes cannot be too thin. For instance, if the planes were constructed of Saran Wrap, they would tend to buckle under their own weight and the form would collapse (Figure 3-15b). Thus the type of material would have to be changed, or the planes thickened, so that the 2-D subsystems could be made sufficiently stiff (in the tube form) to act as rigid diaphragms that could at least support their own weight without buckling.

A self-supporting model of a similar rectangular-plan building as an open-top tube form space-structure could be constructed of writing paper. And, such a model can also be imagined to consist of imaginary horizontal slice subsystems (Figure 3-16). But, in contrast to a solid structural form, horizontal slices through this building will look like a rectangular ring. Even so, the *DL* of the vertical subsystems will be axial and thus evenly distributed across its surface. The principle of transmitting axial vertical loads from one slice to another is the same as for a solid structure. The difference is only geometric in that the material in a tube form is concentrated at the periphery of a section rather than being distributed across the entire section.

The similarity of basic structural interaction will also hold for transfer of horizontal forces from slice to slice. However, as Figure 3-17 illustrates, a horizontal plane of resistance called diaphragm action must be added to pick up and distribute horizontal load (*H*) to the *two vertical planes* that can resist shear. These will be the longitudinal planes, because they are much stiffer than the transverse planes. Without the top diaphragm, a load applied normal to the top edge of one of the vertical planes (Figure 3-17a) will meet with little resistance, since the thin edge is transversely flexible and can easily be imagined to bend inward. Such local bending will in turn cause the

FIGURE 3-16
_DL_ IS CARRIED BY RINGLIKE SLICES.

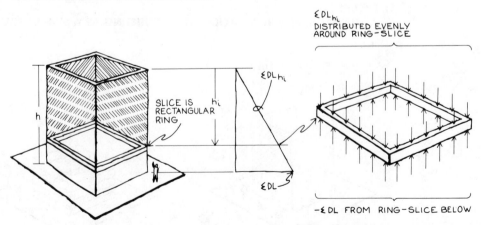

FIGURE 3-17
HORIZONTAL LOADS ON A RECTANGULAR TUBE SPACE-STRUCTURE.

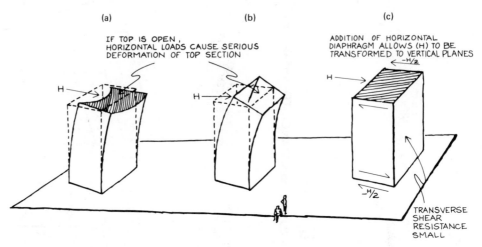

(a)

IF TOP IS OPEN,
HORIZONTAL LOADS CAUSE SERIOUS
DEFORMATION OF TOP SECTION

(b)

(c)

ADDITION OF HORIZONTAL
DIAPHRAGM ALLOWS (H) TO BE
TRANSFORMED TO VERTICAL PLANES

TRANSVERSE
SHEAR
RESISTANCE
SMALL

two shear-effective planes to twist inward. The vertical plane on the opposite side of the force-receiving plane will bend less, but it is clear that the top section will change its shape.

If the horizontal force is applied at a corner (Figure 3-17b), the top section of each vertical plane will also tend to deform, because there is little resistance to horizontal rotation at the tops of all four thin planes. In either Case a or b, the building form as a whole will deform seriously (or even collapse) if the top section is allowed to deform. Thus the top section must be made rigid by addition of a 2-D horizontal subsystem to pick up the horizontal load (H) and distribute it to the two shear-resisting vertical subsystems, as in Figure 3-17c. In the case of a diagonally applied horizontal load, the diaphragm action will make all four of the vertical subsystems act in transferring shear to the foundation (since they are not allowed to twist).

The above examples illustrate that a space-structure *must* be made to maintain its overall (and thus sectional) shape under load. This is inherent to a solid structure, because all of the horizontal subsystem slices are horizontally rigid. But in a space-structure, overall integrity of form and structural action requires that *separate* horizontal and vertical subsystems be provided and made to interact to resist shear. We shall now see how this also applies to bending resistance.

You should be able to verify from experience that a concentrated horizontal load applied anywhere at the top of a closed rectangular tube form will be transferred to the foundation, with a minimum of overall or local deformation. And, since the vertical and horizontal subsystems can be thin, one can sense that such a space-structure is very efficient. On the other hand, if the same amount of material used to construct a tube form were compacted into a solid form of the same height (but much smaller

74

FIGURE 3-18

INTERACTION OF VERTICAL AND HORIZONTAL SUBSYSTEMS TO PROVIDE RESISTANCE TO SHEAR AND
BENDING.

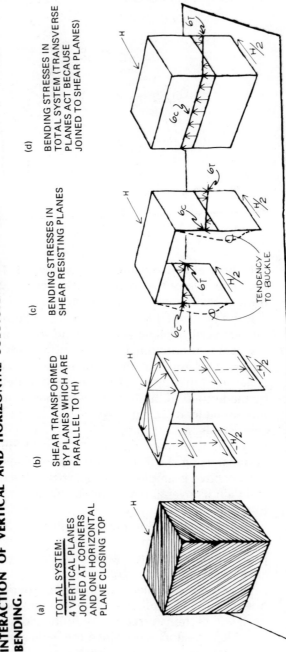

(a)

TOTAL SYSTEM:
4 VERTICAL PLANES
JOINED AT CORNERS
AND ONE HORIZONTAL
PLANE CLOSING TOP

(b)

SHEAR TRANSFORMED
BY PLANES WHICH ARE
PARALLEL TO (H)

(c)

BENDING STRESSES IN
SHEAR RESISTING PLANES

(d)

BENDING STRESSES IN
TOTAL SYSTEM (TRANSVERSE
PLANES ACT BECAUSE
JOINED TO SHEAR PLANES)

rectangular section), an equivalent horizontal load would result in very large bending deflections.

We may know intuitively that, pound for pound, a space-structure can be much stiffer than a solid structure. But it is worth exploring why this is so to develop a few more basic ideas about the fundamental horizontal and vertical structural interactions required of 2-D subsystems in space-structures.

Consider the four vertical planes in Figure 3-18a separately: only two offer significant shear resistance potential for transferring symbolic load (H) to the foundation (Figure 3-18b). As suggested in Figure 3-17, this is so because relative shear resistance is determined by the relative stiffness of the vertical planes *in the direction of* bending resistance. Thus the horizontal diaphragm picks up and distributes the horizontal load to the two vertical planes that are parallel to the load. If the load (H) is symmetric with these two planes, the shearing forces will be (H/2) for each plane.

Of course, (H) will produce an overturn moment, and resisting stresses will develop in the two shear-resisting planes (Figure 3-18c). Note that this could be a problem for Case c. The stresses could produce buckling along the compressive edge of the two planes because they are very thin. However, the fact that there are actually four vertical planes that are connected at their edges provides resistance to such buckling and causes all four planes to act together to resist bending (Figure 3-18d).

The overall action and stress pattern illustrated in Figure 3-18d is significant. It suggests a similarity with the discussion on solid structure stiffness. That is, the distribution of sectional area about the neutral axis will determine the proportion of the overall moment resistance that is provided by each of the vertical subsystems. In fact, the bending resistance of a rectangular tube structure is analogous to an (H) section solid structure (Figures 3-19a and 3-19b). The transverse subsystems (flange areas) can be assumed to be approximately three times as stiff as the two shear-resisting subsystems (web areas). This means that if the area components are equal, 3/4 of the total overturn moment will be resisted by the transverse subsystem (flange) areas. The appropriate C or T forces will be approximately

$$\left(C \text{ or } T = \frac{M_F}{d} = \frac{3/4 \times M}{d} \right)$$

where M_F is the amount of M resisted by the flange only. Similarly, if the flange-to-web area ratio is $\geq 2:1$, the web effect can be neglected. In any case, the flange forces will be distributed evenly over each transverse edge (@ M_F/db), and f_{flange} is given by M_F/dA_{flange} (Figure 3-19c). Note that f_{web} begins with this same value and decreases to 0 at the N-A. The flange provides most of the moment resistance, while shear is almost entirely resisted by the webs.

The basic requirements for overall integrity of space structures under horizontal load may now be generalized:

FIGURE 3-19
**THE RELATIVE MOMENT RESISTING EFFECTIVENESS OF THE VERTICAL SUBSYSTEMS
DETERMINES THE OVERTURN FORCES ON THE SUBSYSTEMS.**

(a)

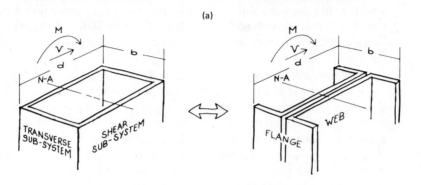

THE SECTION OF A RECTANGULAR TUBE SPACE STRUCTURE
IS ANALOGOUS TO AN 'H' SECTION FOR A SOLID STRUCTURE

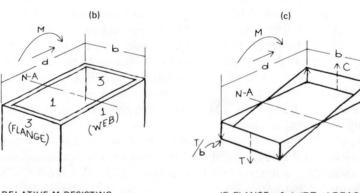

(b)

RELATIVE M RESISTING
EFFECTIVENESS OF VERTICAL
SUB-SYSTEM AREAS
(ABOUT N-A)

(c)

IF FLANGE & WEB AREAS EQUAL,
¾ M IS RESISTED BY FLANGE & :

1. $C \text{ OR } T = \dfrac{\frac{3}{4}M}{d}$

2. $C \text{ OR } T \text{ OVER } (b) = \dfrac{\frac{3}{4}M}{db}$

3. $\overline{\sigma}_{FLANGE} = \dfrac{\frac{3}{4}M}{dA_{FLANGE}}$

1. Horizontal subsystems are required to pick up and transfer horizontal
    loads to the vertical subsystems and to maintain sectional geometry.
    (They will also carry local vertical loads, as will be discussed in
    Chapter 6.)
2. Vertical subsystems are required to transfer both overall dead load
    and horizontal shear to the foundation.
3. Vertical subsystems must also be tied together (by horizontal sub-
    systems, etc.) to optimize overall resistance to bending and buckling.

FIGURE 3-20
**ADDITIONAL HORIZONTAL SUBSYSTEMS MUST BE PROVIDED TO
PICK UP HORIZONTAL LOADS APPLIED OVER *h*.**

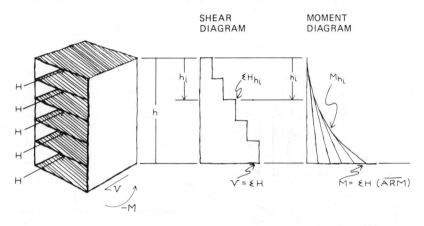

From the information given above, it is easy to add that horizontal loads that are applied at several points along the height of a tube structure will require more than one horizontal-acting subsystem (Figure 3-20). For most buildings this action is supplied by the diaphragm effect of roof and floors. Such distributed diaphragms will not only pick up horizontal loads; they will also act to resist localized compressive buckling of the vertical subsystems by tying them together. Because the *local bending length* of the vertical subsystems is reduced, their overall load capacity is increased (Figure 3-21). Thus, additional horizontal subsystems make it possible for a very thin-walled space-tube to pick up horizontal loads at any point along its height (and to carry heavy vertical loads) without significant local bending deformation (Figure 3-22). Of course, roof and floor subsystems will also pick up and transfer their own dead load, plus whatever live load they carry, to the vertical subsystems.

Note that, in a building conceived as a space-structure, the vertical subsystems need only be locally stiff to enable transfer of horizontal forces applied to surface areas between the horizontal subsystems and to withstand axial compression from floor to floor without buckling. This means that the thickness of vertical subsystems is determined by the distance between horizontal subsystems rather than the overall height of a building.

Thus far, we have illustrated space-structures as simple rectangular tube forms in order to simplify our explanation of how overall loads and moments are carried by the interaction of enclosing vertical and horizontal planes of action. But it should be noted that tube-structure action is also possible for building forms of other sectional shapes (Figures 3-23*a* and 3-23*b*). In Case *a*, the difference is only that the shear and overturn portions

FIGURE 3-21
HORIZONTAL SUBSYSTEMS INCREASE THE LOCAL STIFFNESS OF VERTICAL SUBSYSTEMS.

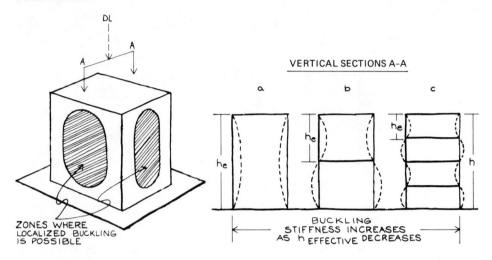

are less distinct for rounded forms. But one can approximate this by quartering the tube surface. In Case *b*, the only difference is that there are two flanges that will act in proportion to the square of their centroidal distance from the *N-A*.

One can also envision designs in which the horizontal planes either encompass or extend through and beyond an interior tube-structure to provide a core-tube design (Figure 3-23c). However, the principles of tube action are the same whether the structure is conceived as a total space enclosure or as a core within an enclosed space. Figure 3-23d illustrates that it is even possible to explode the core tube. In any case, a core tube allows the periphery of a building to be enclosed by a very light secondary vertical subsystem. The core tube(s) can carry most of the vertical load and provide all of the shear and overturn resistance, while the *enclosing subsystems* need only carry a small part of the total vertical load and to pick up and transfer local increments of the horizontal load to the horizontal subsystems. Of course, if the space-forms illustrated in Figure 3-23 are imagined to be of the same overall dimensions, there would obviously be a significant difference in overall stiffness. A rough approximation of the relative stiffness of two tube-structures can be approximated by the ratio:

$$\frac{A_{F_1}d_1^2}{A_{F_2}d_2^2}$$

where A_F is the flange area.

FIGURE 3-22
**VERTICAL SUBSYSTEMS MUST BE ABLE TO PICK UP AND TRANSFER WIND LOAD
FROM TRIBUTARY AREAS TO APPROPRIATE HORIZONTAL SUBSYSTEMS.**

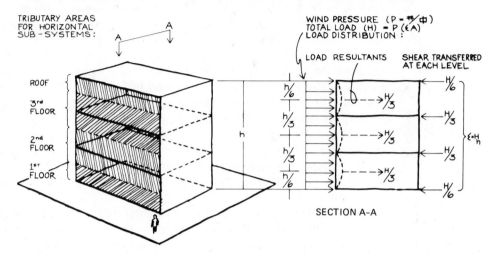

FIGURE 3-23
**TUBE ACTION CAN BE ACHIEVED FOR A VARIETY OF SECTIONAL SHAPES AND BY
MEANS OF STRUCTURAL CORE DESIGNS.**

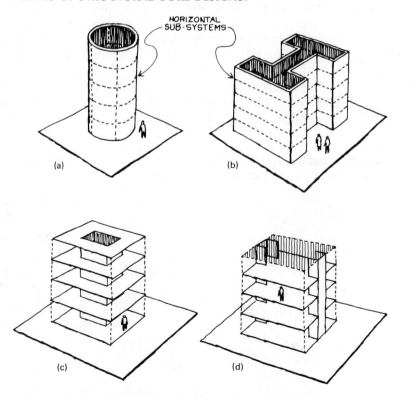

SECTION 4: Space-Structures of Columns and Frames

We will begin our discussion of buildings conceived as columnar space-structures by considering a simple rectangular space-form enclosed by four slender columns (Figure 3-24). Since each column acts independently as a cantilever supported at its base, it will have to be sufficiently stiff to at least maintain its verticality without buckling under self-weight. And, because the columns do act independently, they will have to be thicker than the thin planes of the tube structures discussed in Section 3. Even so, such columns will offer little bending resistance to, and will result in large deflections under, horizontal load (Figure 3-25a).

It should also be noted that the deflection inadequacy will not be improved by the addition of a pin-connected horizontal subsystem (Figure 3-25b). Such a subsystem would cause the columns to deflect together, but would have no effect on the independent cantilever bending stiffness of each column.

One could think of improving horizontal deflection resistance by increasing the sectional dimensions of each of the four slender columns (Figure 3-25c). But that would be a brute-force approach, and is frequently not the best strategy for efficient design.

On the other hand, the elevations in Figure 3-26 illustrate that if there is fixity between the horizontal subsystem and the columns, there will be a rotational interaction that improves overall efficiency. With such fixity, the horizontal subsystem can act as a connector that resists independent cantilever rotation of the column tops and causes them to bend in opposite directions from top to bottom (Figures 3-26a and 3-26b). This is called frame action, and it improves efficiency because the bending reversal (called contraflexure) causes each column to act as two short components, the point of inflection (*PI*) determining their relative length. As a result, overall deflections due to horizontal load are reduced, and resistance to buckling

FIGURE 3-24
FOUR INDEPENDENT CANTILEVERED COLUMNS CAN SYMBOLIZE A BUILDING AS A SPACE-FORM.

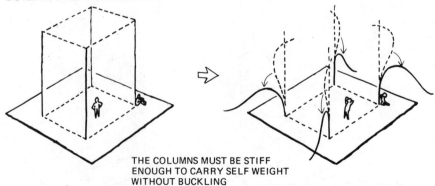

THE COLUMNS MUST BE STIFF
ENOUGH TO CARRY SELF WEIGHT
WITHOUT BUCKLING

FIGURE 3-25
INDEPENDENT CANTILEVER ACTION PROVIDES INEFFICIENT RESISTANCE TO HORIZONTAL LOADS.

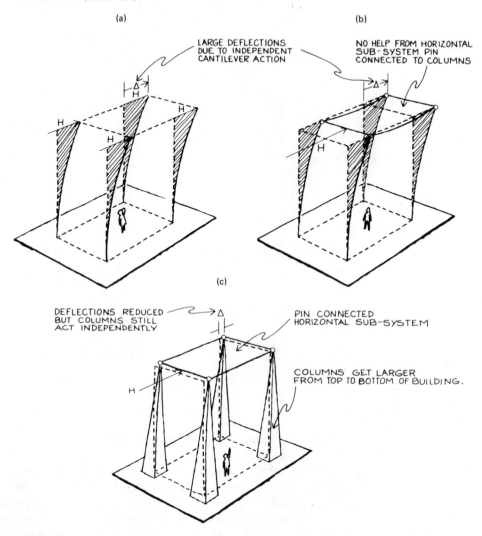

(a)

LARGE DEFLECTIONS
DUE TO INDEPENDENT
CANTILEVER ACTION

(b)

NO HELP FROM HORIZONTAL
SUB-SYSTEM PIN
CONNECTED TO COLUMNS

(c)

DEFLECTIONS REDUCED
BUT COLUMNS STILL
ACT INDEPENDENTLY

PIN CONNECTED
HORIZONTAL SUB-SYSTEM

COLUMNS GET LARGER
FROM TOP TO BOTTOM OF BUILDING.

under axial load is improved (Figures 3-26c and 3-26d).

When two columns are rigidly connected by a horizontal component, the whole is a structural subsystem called a frame. But it should be noted that the action is two-dimensional because forces are most effectively resisted when they are in the plane of frame action. As a result, we can say that a frame subsystem can play a role in columnar building design that is somewhat analogous to that of the vertical planes discussed earlier. That is, more than one plane of stiffness must be used.

FIGURE 3-26
BY FIXING HORIZONTAL SUBSYSTEM TO COLUMNS, ROTATIONAL INTERACTION IMPROVES OVERALL STIFFNESS.

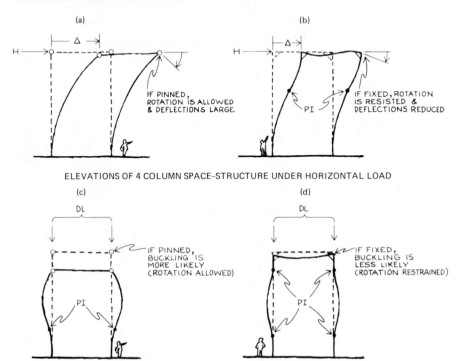

ELEVATIONS OF 4 COLUMN SPACE-STRUCTURE UNDER HORIZONTAL LOAD

ELEVATIONS OF 4 COLUMN SPACE-STRUCTURE UNDER VERTICAL LOAD

   The degree of frame versus independent cantilever action that may be achieved by a fixed connection between columns is determined by the degree to which the connector can resist independent cantilever rotation of the column tops (Figures 3-27a and 3-27b). If we idealize that the column bases are fully fixed and a fixed connector is stiff enough to resist all rotation at the top, the point of inflection (PI) would be at column midheight (Figure 3-27b). In this idealized case, we have total-frame action, and only one-half of the overall moment ($M = Hh$) is resisted by half-length column bending [i.e., $M/2 = H(h/2)$], the remainder being provided by *axial* forces in the columns (Figure 3-27c). And, since bending deflections are proportional to (ML^2), this means that the overall deflection of columns would be one-fourth that for full-column cantilever action (i.e., $2(\frac{1}{2})^3 = \frac{1}{4}$).[4] Thus total frame action is roughly 4 times stiffer than total cantilever action.

[4]L represents the effective bending length of columns. In this example $M = \frac{1}{2}$ and $L = h/2 = \frac{1}{2}$.

FIGURE 3-27
THEORY OF TOTAL FRAME VERSUS CANTILEVER ACTION.

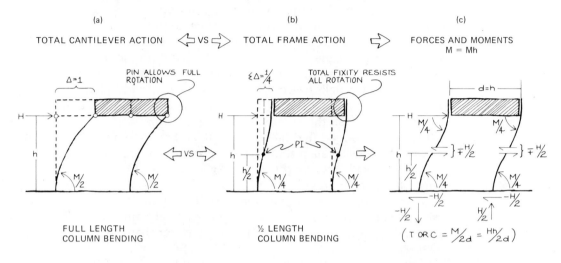

(a)

TOTAL CANTILEVER ACTION ⟨⟩ VS ⟩ TOTAL FRAME ACTION ⟩

FULL LENGTH
COLUMN BENDING

(b)

½ LENGTH
COLUMN BENDING

(c)

FORCES AND MOMENTS
$M = Mh$

$\left(T \text{ OR } C = \dfrac{M}{2d} = \dfrac{Hh}{2d} \right)$

The above example *is* idealized. However, it does illustrate the general principal that, with frames, the amount of axial moment and the amount of overall deflection are largely determined by the proportion of overall moment taken by column bending. It also illustrates that, for a given axial moment, the magnitude of axial forces varies directly with the aspect ratio (Figure 3-28). But note that, in real frames, the bending moment in the columns will vary with the *actual* degree of frame action achieved.

FIGURE 3-28
THE ASPECT RATIO (*h/d*) AND AXIAL FORCES, ASSUMING TOTAL FRAME ACTION.

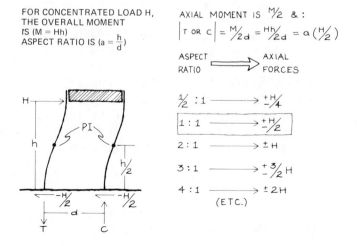

FOR CONCENTRATED LOAD H,
THE OVERALL MOMENT
IS (M = Hh)
ASPECT RATIO IS $(a = \dfrac{h}{d})$

AXIAL MOMENT IS $\dfrac{M}{2}$ & :

$\left| T \text{ OR } C \right| = \dfrac{M}{2d} = \dfrac{Hh}{2d} = a\left(\dfrac{H}{2}\right)$

ASPECT ⟹ AXIAL
RATIO ⟹ FORCES

$\dfrac{1}{2} : 1 \longrightarrow {}^{+}_{-}\dfrac{H}{4}$

$\boxed{1 : 1 \longrightarrow {}^{+}_{-}\dfrac{H}{2}}$

$2 : 1 \longrightarrow \pm H$

$3 : 1 \longrightarrow {}^{+}_{-}\dfrac{3}{2}H$

$4 : 1 \longrightarrow \pm 2H$

(ETC.)

In real frames, the degree of frame action achieved is primarily determined by the connector-to-column stiffness ratio. If the columns are individually stiffer than the connector, more of the overturn moment will be resisted by bending in each column. If the connector is stiffer, column bending moments are reduced and the coupled axial forces in the columns will carry a significant part (up to $\frac{1}{2}$) of the overturn moment. Thus, moment resistance is transferred from independent column bending, where the resisting arm is small, to the frame as a whole, where the arm is large between the coupled columns. However, this does require a relatively stiff connector between the columns.

It is acknowledged that the elongation and shortening of the columns due to axial forces will cause some rotation of the connector. But this is normally a small factor and is generally neglected. On the other hand, the connector-to-column stiffness ratio is very significant and can be considered as a rough indicator of the degree of frame action achieved. For purposes of approximation, one can assume that nearly total frame action is achieved when the connector is at least *four* times stiffer than the columns (Figure 3-29a).[5]

As the stiffness ratio decreases, there will be significant rotation of the column tops, and the overall action will degenerate from that of a total frame assumption (*PI* at midheight) to that of nearly full cantilever action with *PI* at column top. For example, a stiffness ratio of 1:1 is a crude indicator that the degree of frame action would be on the order of one-half that of the pure-frame assumption, and the *PI* could be assumed to be at roughly the $\frac{3}{4}$ point (Figure 3-29b).[6] This would mean that overall deflections would be roughly one-half that for full cantilever action and the axial moment would be one-half as large as for an assumption of total frame action (i.e., *M*/4). Again, for approximation purposes, one can assume that little frame action will result from ratios of less than 1:1.

In design of buildings, the floor and roof subsystems are often used to act as the horizontal connectors of columns. But often their stiffness will not be sufficient to cause optimum frame action. Thus Figure 3-29c illustrates that, if floor and roof subsystems are too thin to provide enough rotational resistance, it can be economical to stiffen their edges to improve frame action.

For clarity, a simple two-column frame elevation has been used in the above examples to illustrate the basic difference between overall frame and independent cantilever action of a columnar structure under horizontal load. It will now be useful to look at three strategies that can be employed to increase the overall stiffness, and thus improve the load-carrying capacity,

---

[5]Relative stiffness will vary directly with $I(E)$ and inversely with length (i.e. $\propto EI/L$).
[6]At 2:1 the action is $\sim\frac{2}{3}$ that for a pure frame, at 4:1 the action is $\sim\frac{4}{5}$, etc.

FIGURE 3-29
THE CONNECTOR TO COLUMN STIFFNESS RATIO CAN BE USED AS A ROUGH INDICATOR OF THE DEGREE OF FRAME ACTION ACHIEVED.

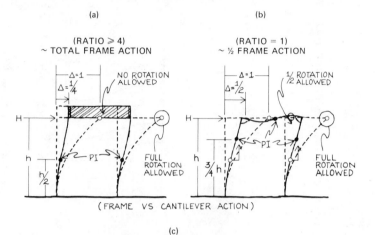

(a)

(RATIO ≥ 4)
~ TOTAL FRAME ACTION

(b)

(RATIO = 1)
~ ½ FRAME ACTION

(FRAME VS CANTILEVER ACTION)

(c)

THIN HORIZONTAL SUB-SYSTEM WITH STIFF
EDGES FOR IMPROVED FRAME ACTION

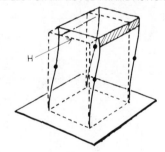

of a basic two-column frame subsystem:

1. Add interior columns within a basic two-column frame subsystem
2. Increase the depth of the two columns in the plane of bending action
3. Stiffen the overall frame vertically by addition of more and/or stiffer horizontal connectors

Figure 3-30*a* represents a two-column frame elevation of a rectangular building conceived to withstand symbolic horizontal load (*H*). Assume that the connector and columns are of equal stiffness. If four interior columns of equal stiffness to the exterior columns are added and fixed to the connector (Figures 3-30*b* and 3-30*c*), two benefits are gained: First, frame action will be improved because *connector* stiffness is increased by a factor of five, since the effective length for each two-column pair is one-fifth that of Case *a*. Second, horizontal deflections will be about one-third that of the basic

FIGURE 3-30
THE OVERALL FRAME ACTION AND STIFFNESS OF A BASIC TWO-COLUMN SUBSYSTEM CAN BE IMPROVED BY THE ADDITION OF INTERIOR COLUMNS.

(a)

BASIC 2-COLUMN FRAME,
PARTIAL FRAME ACTION,

AXIAL MOMENT $< \dfrac{M}{2}$

$(M = Hh)$

(b)

ADD (4) COLUMNS
OF EQUAL STIFFNESS
TO THOSE OF CASE (a)

(c)

6-COLUMN FRAME
AXIAL MOMENT $\sim \dfrac{M}{2}$
$(M = Hh)$

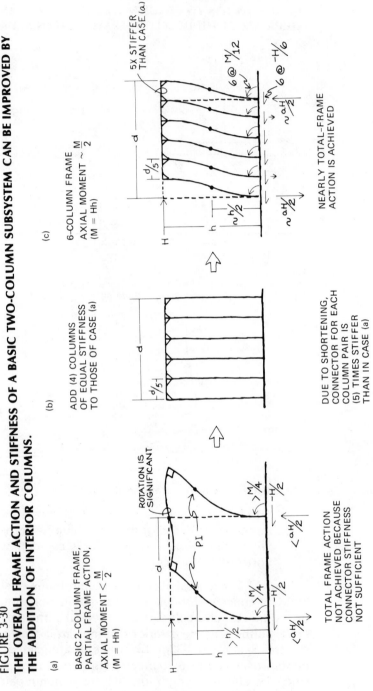

TOTAL FRAME ACTION
NOT ACHIEVED BECAUSE
CONNECTOR STIFFNESS
NOT SUFFICIENT

DUE TO SHORTENING,
CONNECTOR FOR EACH
COLUMN PAIR IS
(5) TIMES STIFFER
THAN IN CASE (a)

NEARLY TOTAL-FRAME
ACTION IS ACHIEVED

two-column case because the average load, and thus bending moment, on each of the six columns will be about $\frac{2}{6}$ that of Case a. Overall, this means that total frame action will be more closely approximated, and overall stiffness would be increased by about three times.

In the examples above, it should be noted that the *exterior* columns would supply the larger part of axial forces to resist the overturn moment. There could be small axial forces in the interior columns because, with equal stiffness, there would be somewhat unequal location of the *PI*. However, in a usual building layout, the interior columns would be designed to carry a larger tributary portion of a building's overall vertical loads than the exterior columns (Figure 3-31a). As a result, interior columns will be inherently stiffer than the exterior columns. If the difference is on the order of 2:1 (as is common), the overall frame can be assumed to act as a composite of independent (and equal) two-column frame components (Figure 3-31b). Thus, the *PI* location will be equal, and each component frame can be visualized to carry an equal part of the horizontal force. In this case the axial forces on the interior columns would cancel out, and, assuming nearly total frame action, the axial forces in the exterior columns will be the same as in a basic two-column subsystem (Figure 3-31c).

We have now observed that two benefits are gained by the addition of interior columns to stiffen a basic two-column frame:

1. The relative stiffness of the horizontal connectors is increased because the distance between two column pairs is decreased. This means that the degree of overall frame action is increased.

FIGURE 3-31
BALANCED FRAME ACTION REQUIRES THAT INTERIOR COLUMNS BE ABOUT TWO TIMES STIFFER THAN EXTERIOR COLUMNS.

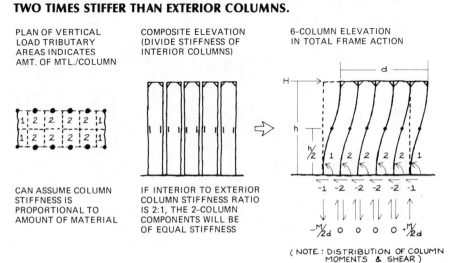

PLAN OF VERTICAL LOAD TRIBUTARY AREAS INDICATES AMT. OF MTL./COLUMN

COMPOSITE ELEVATION (DIVIDE STIFFNESS OF INTERIOR COLUMNS)

6-COLUMN ELEVATION IN TOTAL FRAME ACTION

CAN ASSUME COLUMN STIFFNESS IS PROPORTIONAL TO AMOUNT OF MATERIAL

IF INTERIOR TO EXTERIOR COLUMN STIFFNESS RATIO IS 2:1, THE 2-COLUMN COMPONENTS WILL BE OF EQUAL STIFFNESS

(NOTE : DISTRIBUTION OF COLUMN MOMENTS & SHEAR)

2. Each column carries less horizontal load and therefore bends less. This means that overall deflections will decrease.

The second fundamental strategy for improving the overall stiffness of a basic two-column frame is to simply make the two columns deeper in the plane of action (Figures 3-32a and 3-32b). Providing sufficient connector-to-column stiffness ratio is preserved, horizontal deflections can be expected to vary inversely with $(Ad^2_{col.})$, where $d_{col.}$ is the bending depth of the columns. Of course, if the overall frame width (d) is held constant, any increase in $(d_{col.})$ will also increase the axial forces somewhat, because the centroidal moment arm will be decreased (Figure 3-32c).

In fact, if the six columns in Figure 3-31 were combined to form two columns that are three times deeper, their stiffness would be 27 times larger. The overall deflection would be much less even though each column would carry three times the load because the net increase is nine. Note that Figure 3-32c shows that if the column depth approached the limiting value of $(d_{col.} \sim d/2)$, the centroidal lever arm for the axial force couple would be approximately one-half the width of the building, and the axial forces will be twice as large as for Figure 3-31.

If the elevation of a four-column building is as shown in Figure 3-32b, the columns would be (L) shaped and the building would appear as in Figure

FIGURE 3-32
OVERALL STIFFNESS OF TWO-COLUMN FRAME CAN BE INCREASED BY INCREASING THE DEPTH OF COLUMNS.

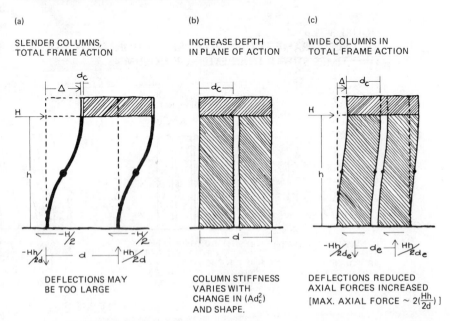

(a)

SLENDER COLUMNS,
TOTAL FRAME ACTION

(b)

INCREASE DEPTH
IN PLANE OF ACTION

(c)

WIDE COLUMNS IN
TOTAL FRAME ACTION

DEFLECTIONS MAY
BE TOO LARGE

COLUMN STIFFNESS
VARIES WITH
CHANGE IN (Ad^2_c)
AND SHAPE.

DEFLECTIONS REDUCED
AXIAL FORCES INCREASED
[MAX. AXIAL FORCE $\sim 2(\frac{Hh}{2d})$]

3-33*a*. In overall appearance it would be similar to that of a rectangular tube structure (Figure 3-33*b*). But full tube action would not be achieved because of the vertical discontinuity between the two columns at the center of each building elevation.

Assuming equivalent material and overall dimensions, a tube design

FIGURE 3-33
ASSUMING EQUIVALENT MATERIAL, THE STIFFNESS OF A WIDE-COLUMN FRAME WILL ALWAYS BE LESS THAN THAT OF A TUBE DESIGN.

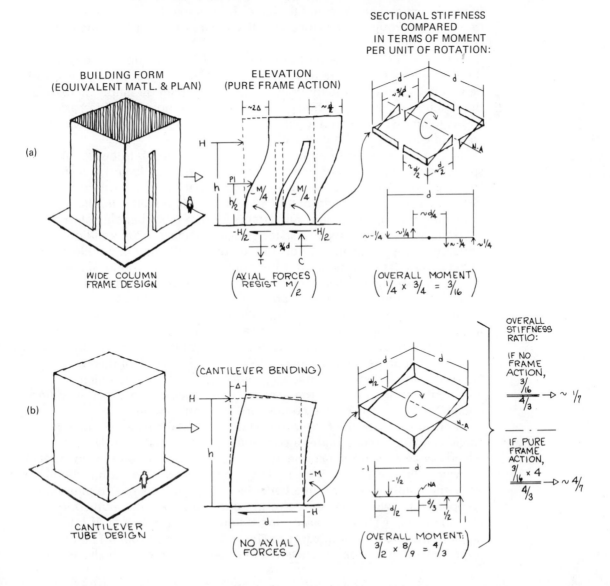

would be stiffer by several times. Figure 3-33 and the following discussion offer a crude but instructive explanation.

If, in Figure 3-33a, $d_{col.} \sim d/2$ and there were no frame interaction, the total stiffness of the four independent (L) section columns would be on the order of $\frac{1}{7}$ that for a tube design, and overall deflections would be about seven times larger. But if the connector shown is assumed stiff enough for total frame action, the overall deflection would be reduced to $\frac{7}{4}$, or roughly two times that of cantilever tube action. However, since neither of the above would be the actual case, the comparative stiffness of Case *a* would fall somewhere in between. Further, the stiffness of an (L) section column frame design can be expected to diminish from that for the maximum case ($d_{col.} \sim d/2$) in rough accordance with the square of the ratio of the actual column depth to $d/2$.

Of course, a combination of both more *and* wider columns is possible. But the general point to remember is that overall frame action stiffness can be improved *more by widening* column sections than by simply increasing the number of slender columns. In any case, as compared with a basic two-column system, the effect of an increase in both the number and the width of columns will be cumulative.

Thus far we have explored the column-frame approach to design of a space-structure in terms of a single symbolic horizontal load (H) at the top. However, as we shall see, the fact that real horizontal loads will be distributed along the height of a building form will not alter the basic principles of designing frame subsystems for maximum bending stiffness.

Just as with the tube type of space-structure (Figure 3-20), additional horizontal subsystems may be distributed over the height of a building designed as a frame structure for the purpose of picking up local areas of horizontal loading and distributing them to the resisting frame subsystems. And, if these added subsystems are rigidly connected (Photo 3-2), it is reasonable to expect that the overall stiffness of the frame would be improved. The improvement is due to a rather complicated interaction of the columns with the added connectors. As a result, a precise analysis of how much improvement is achieved is beyond the scope of this text. However, we can offer the following simplified explanation, which is admittedly crude, but gives the reader a rough idea of the effect of adding connectors to a basic (single-connector) frame subsystem.

To begin with, consider the addition of horizontal connectors to a basic single-connector frame under a concentrated horizontal load (H). If the added connectors are pin-connected, there would be no improvement at all in overall stiffness (Figure 3-34a). On the other hand, if the connections are rigid, the multiconnector frame becomes a stack of (n) short column frame components (Figure 3-34b). As a result, one can expect the multiconnector case to be stiffer than the basic, single-connector case. In fact, if total frame action is assumed for both cases in Figure 3-34b, the amount of stiffness improvement would be proportional to the deflection of each short-column

PHOTO 3-2
**COLUMNS AND RIGIDLY CONNECTED GIRDERS FORMING A STRUCTURAL
FRAME.**

component frame ($1/n^3$). Thus, relative to the deflection of a basic frame,
the overall deflection with (n) connectors would be ($n/n^3 = 1/n^2$). In our
idealized example we could say the four-connector frame would be 16 times
stiffer, or that a load 16 times greater would result in the same deflection as
the basic case.

Now, if the load in Figure 3-34b were evenly distributed, as in Figure
3-34c, the overall moment would be one-half as large. And, assuming total
frame action, it turns out that there is a rough correspondence between the
overall moment and the average deflection of the frame components. Thus
a comparison of Figures 3-34b and 3-34c illustrates that with total frame

FIGURE 3-34

OVERALL FRAME ACTION STIFFNESS MAY BE INCREASED BY THE USE OF MORE THAN ONE CONNECTOR.

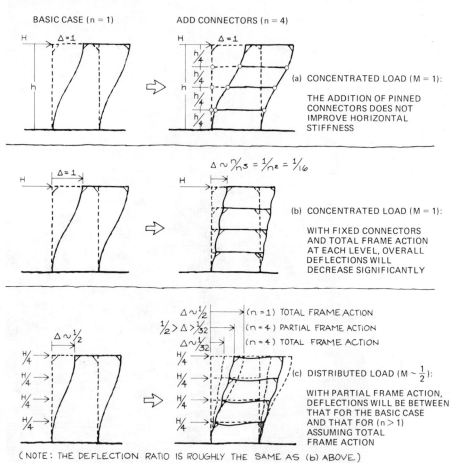

BASIC CASE (n = 1)   ADD CONNECTORS (n = 4)

(a) CONCENTRATED LOAD (M = 1):

THE ADDITION OF PINNED CONNECTORS DOES NOT IMPROVE HORIZONTAL STIFFNESS

$\Delta \sim \frac{n}{n^3} = \frac{1}{n^2} = \frac{1}{16}$

(b) CONCENTRATED LOAD (M = 1):

WITH FIXED CONNECTORS AND TOTAL FRAME ACTION AT EACH LEVEL, OVERALL DEFLECTIONS WILL DECREASE SIGNIFICANTLY

$\Delta \sim \frac{1}{2}$ (n = 1) TOTAL FRAME ACTION

$\frac{1}{2} > \Delta > \frac{1}{32}$ (n = 4) PARTIAL FRAME ACTION

$\Delta \sim \frac{1}{32}$ (n = 4) TOTAL FRAME ACTION

(c) DISTRIBUTED LOAD ($M \sim \frac{1}{2}$):

WITH PARTIAL FRAME ACTION, DEFLECTIONS WILL BE BETWEEN THAT FOR THE BASIC CASE AND THAT FOR (n > 1) ASSUMING TOTAL FRAME ACTION

(NOTE: THE DEFLECTION RATIO IS ROUGHLY THE SAME AS (b) ABOVE.)

action, the *degree of improvement* in overall stiffness is roughly the same for distributed as for concentrated loads.

We can generalize that the addition of connectors will theoretically improve the overall stiffness of a basic frame (n = 1). And, the effectiveness of adding connectors will be approximately the same whether one is dealing with concentrated or distributed loads. But for real cases, the amount of improvement will lie somewhere between the actual stiffness indicated for single-connector design and that indicated for a multiconnector design assuming total frame action (Figure 3-34c). This is so because, in real cases, total frame action at each level will *not* be automatically achieved.

Even if a single top connector is stiff enough to approximate total frame

FIGURE 3-35

A DIAGRAMMATIC EXPLANATION OF THE DECREASING EFFECTIVENESS OF THE CONNECTORS BELOW THE TOP IS GIVEN BY CONSIDERING THAT EACH LOAD COMPONENT ACTS ON A BASIC FRAME AND RELATIVE COLUMN STIFFNESS VARYS INVERSELY WITH RELATIVE LENGTH AT EACH LEVEL.

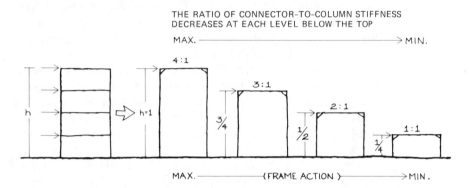

action in a basic frame design, the addition of more connectors of equal stiffness will not assure total frame action at lower levels. In reality, the effective stiffness of the added connectors will diminish at each level below the top because the remaining column length will decrease while the connector length will not (Figure 3-35). Therefore, at each level below the top, the ratio of connector stiffness to remaining column stiffness steadily decreases, and the connector stiffness at lower levels may not be sufficient to support an assumption of total frame action.

Roughly speaking, the amount of improvement can be approximated by averaging the declining connector-to-column stiffness ratios as shown in Figure 3-35. If the average ratio is 4:1, one can assume that total frame action will be roughly approximated. If the average ratio is 1:1, the overall improvement in stiffness will be on the order of one-half that indicated by a total frame action assumption. This is reasonable because an average ratio of 1:1 would indicate significant rotation at the base of a frame where it contributes most to overall deflection. If the average ratio is much less than 1:1, there will be little benefit gained from vertical subdivision of basic frame design ($n = 1$). On the other hand, the stiffness of added connectors can also be increased from top to bottom to maintain their effectiveness at lower levels.

In addition to reducing overall deflection, the use of more than one connector has the effect of reducing the proportion of overall moment that must be resisted by column bending and *increasing* the proportion to be supplied by coupled axial forces in the columns (refer to Figures 3-34*b* and 3-34*c*). Thus, in the case of a concentrated load, and assuming total frame action (*Pl* at the midpoint of short column components), bending at the bottom level would account for resistance to 1/2n of the overall moment (i.e., $M/2n$), and moment resistance due to column axial forces would account for the balance $\{M[1-(1/2n)]\}$. If the load were evenly distributed, the overall moment M would be about one-half that for a concentrated load, and for $n \geq 2$, the column proportion would be roughly M/n, the axial proportion being $\{M[1-(1/n)]\}$.[7] For less than total frame action, at the base, the column moment will vary between the idealized proportion given above and twice those values as determined by the actual location of the *PI* at the bottom frame component.

We can now summarize that simple addition of either more columns or more horizontal connectors will improve the overall stiffness, but not necessarily the efficiency, of a frame subsystem. However, it is possible to combine both approaches to obtain a favorable subdivided column-to-connector stiffness ratio and thus increase both stiffness and efficiency (Figure 3-36).

Obviously, if such horizontal and vertical subdivision is carried to an extreme, the proportional depth of the column and connector components could become so great that a building designed as a frame system would, in effect, evolve into a perforated tube system (Figure 3-36c and Figure 3-37a). To the degree that this occurs, the sum and the distribution of axial forces

---

[7]For evenly distributed load on a single-connector frame, the column moment would be $\frac{3}{4}M$ and the axial load $\frac{1}{4}M$. For uneven load distribution and $n \geq 2$, assume the column portion varies with the ratio $[(1/2n)(h):ch = (1/2n):c]$, the axial portion being $M[1-(\frac{1}{2}nc)]$.

FIGURE 3-36
THE OVERALL STIFFNESS AND EFFICIENCY OF A BASIC FRAME IS IMPROVED BY A COMBINATION OF MORE COLUMNS AND CONNECTORS.

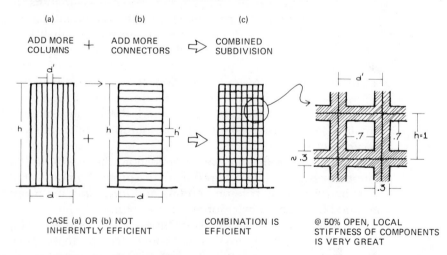

(a) ADD MORE COLUMNS + (b) ADD MORE CONNECTORS ⇨ (c) COMBINED SUBDIVISION

CASE (a) OR (b) NOT INHERENTLY EFFICIENT

COMBINATION IS EFFICIENT

@ 50% OPEN, LOCAL STIFFNESS OF COMPONENTS IS VERY GREAT

FIGURE 3-37
AS HORIZONTAL AND VERTICAL SUBDIVISION INCREASES, A FRAME DESIGN EVOLVES INTO A PERFORATED TUBE STRUCTURE

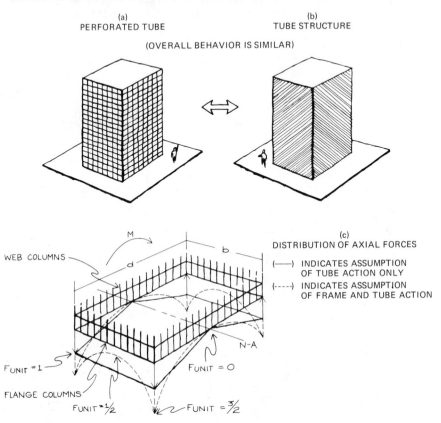

(a)
PERFORATED TUBE

(b)
TUBE STRUCTURE

(OVERALL BEHAVIOR IS SIMILAR)

(c)
DISTRIBUTION OF AXIAL FORCES

(——) INDICATES ASSUMPTION
OF TUBE ACTION ONLY

(-----) INDICATES ASSUMPTION
OF FRAME AND TUBE ACTION

WEB COLUMNS

$F_{UNIT} = 1$

FLANGE COLUMNS

$F_{UNIT} = 1/2$

$F_{UNIT} = 0$

$F_{UNIT} = 3/2$

in the columns will approximate that illustrated previously for a space-tube design (Figure 3-37*b* and solid line force diagram in Figure 3-37*c*). When horizontal and vertical subdivision is not so extreme, action will be somewhere between that for a pure frame assumption and that for a pure tube assumption. The dotted force diagram in Figure 3-37*c* represents the type of axial force distribution that can be expected of a system acting one-half as a frame and one-half as a tube design (due to connector shear lag).

For purposes of roughly approximating the forces in a perforated tube structure, one can simply assume the simple triangular distribution of Figure 3-37*c*. The web columns can be assumed to be $\frac{1}{3}$ as effective as the columns in the flanges, and the overall resisting forces can be approximated as discussed in Section 3. A more exact approximation would adjust the force values obtained by the tube structure assumption to reflect the dotted pattern as shown in Figure 3-37*c*. In any case, nearly all of the *overturn*

FIGURE 3-38
VERTICAL AND HORIZONTAL SUBSYSTEMS MAY BE COMBINED IN MANY WAYS TO PROVIDE OVERALL STRUCTURAL INTEGRITY.

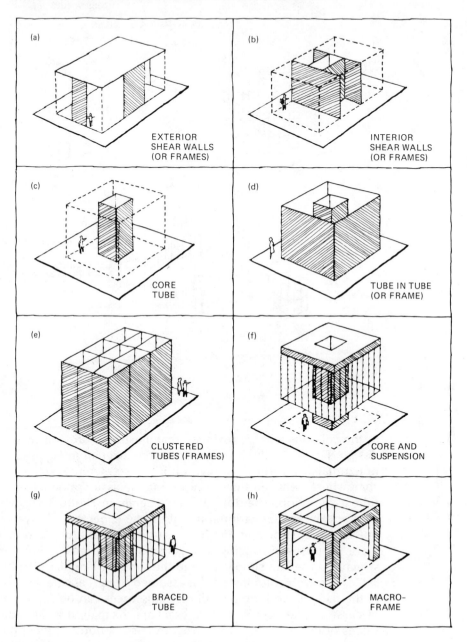

(a) EXTERIOR SHEAR WALLS (OR FRAMES)

(b) INTERIOR SHEAR WALLS (OR FRAMES)

(c) CORE TUBE

(d) TUBE IN TUBE (OR FRAME)

(e) CLUSTERED TUBES (FRAMES)

(f) CORE AND SUSPENSION

(g) BRACED TUBE

(h) MACRO-FRAME

moment will be resisted by axial forces. Only the shear force in each story will be resisted by *local* bending in the columns and distributed according to their stiffness.

We will now conclude our introduction to the basic subsystem interactions that are required in the design of space-structures by pointing out that tube and/or frame subsystems can be combined and laid out in many ways. For example, components of a tube may be exploded (or imploded) to produce a variety of tube and/or shear wall designs (Figures 3-38*a*, 3-38*b*, and 3-38*c*). In either case the shear-resisting walls may be designed as solid planes or very stiff frames. But it should be recognized that in all these layouts there must also be horizontal subsystems (or diaphragms) to pick up and transfer horizontal loads to the vertical shear-resisting subsystems. Such combination layouts are also useful because it will often be necessary to provide interior vertical support of the horizontal subsystems.

It is also possible that buildings may be designed as tubes within tubes (Figure 3-38*d*), a cluster of tubes, or a grid of frames (Figure 3-38*e*). A core tube layout and a very stiff horizontal subsystem may be combined to allow a suspension design of floors or to cause columns to interact with the core to resist overall bending (Figure 3-38*f* and 3-38*g*). It is also possible to use tubes as vertical and horizontal components in the design of a very large-scale macro-frame (Figure 3-38*h*). Within the macro-frame, any number of very lightweight supplementary column or frame designs can be utilized.

We will discuss the design of some of the above options in forthcoming chapters. But here we want the reader to recognize that the creative opportunities are many. At conceptual stages, the designer need only keep in mind the four basic structural subsystem interactions that must be provided in order to achieve overall integrity in the structural action of a building form:

1. Horizontal subsystems must pick up and transfer vertical loads to the vertical subsystems.
2. Horizontal subsystems must also pick up horizontal loads accumulated along the height of a building and distribute them to the vertical shear-resisting subsystems.
3. All of the vertical subsystems must carry the accumulated dead and live loads, and some must be capable of transferring shear from the upper portions of a building to the foundation.
4. Key vertical subsystems that can resist bending and/or axial forces due to overturning moments must be provided. Where possible, they should be interacted by horizontal subsystems.

4
Schematic Analysis of Buildings as Total Structural Systems

SECTION 1: Space-Organizing Components as Major Structural Subsystems

We have said that, at schematic stages of design, the architect tends to think of space-form options in overall terms. He does this to control complexity and to generate an overall design concept that represents a space-form response to a variety of basic activity, environmental, and experiential performance needs. This problem presents itself in terms of both space-to-site and space-to-space differentiation and interaction objectives. The designer responds by working to conceptualize an organizational scheme that optimizes the interaction of major space-enclosure, articulation, and service components, such as walls, floors, roofs, and large circulation and service arteries.

For example, most buildings must satisfy a variety of often conflicting activity needs. The architect must design for a variety of more-or-less enclosed spatial components and optimize their modes of interaction. Thus, one of his first tasks is to organize the various space components in terms of their relative character, configuration, size, service needs, and site context. To do this, he has to manipulate their interaction by controlling the overall properties of spatial adjacency, accessibility, interface, and service distribution. It may also be necessary to consider requirements for subdivision within major space components at schematic levels.

Since an architect's basic organizational decisions will ultimately be manifest as enclosing walls, floors, ceiling, and roofs, they must also deal with the climatic and energy-related aspects of their overall space-form scheme. And, in order to allow for the movement of people and objects as well as energy and services into, throughout, and out of the building, large arterial access, egress, and service routes must be provided. Such routes will often take the form of large horizontally and vertically continuous space components, such as halls, elevator and stair shafts, and trunk and branch shafts and ducts that penetrate enclosing surfaces. Doors and other types of interface portals must also be provided.

The overall issues of spatial differentiation and organization are obviously major form determinants and can be complicated. Therefore, it is reasonable that the architect will tend to focus on them at schematic stages of design. But it is also reasonable to expect that it would be very helpful to consider the overall structural implications of one's space-form decisions at schematic levels, when flexibility is maximum.

Most architects do recognize that an important contribution to creative design thinking *could* be realized if they could evaluate the structural implications of basic spatial organization schemes in an overall way. But it is commonly difficult for them to do so because the context of their knowledge of structure and/or the type of input from their consultants is often much more specific than is needed and therefore limited in usefulness at this early stage. Thus, the object of this chapter is to demonstrate how an overall approach can be used to deal with the issues of structural system design in a hierarchic fashion.

Recognition of a hierarchy is important because it allows designers to focus on the more basic structural implications of space-form schemes before becoming entangled in premature consideration of consequential details. It allows them to insure compatibility between the overall nature of structural and space-form thinking, and that a total structural system scheme will become a context for thinking about the more local issues of subsystem interaction and design.

For example, there may be dozens of ways of framing a floor using many types of materials. But the floor, as a whole, must effectively contribute to an efficient overall structural-architectural system before it makes sense to evaluate the framing options. It is simply a recognition of the need for a *big-things-first* approach that will lead to total solutions to environmental design problems (Photo 4-1 and Figure 4-1).

Thus, an overall approach to structural design of buildings calls for recognition of the potential for utilizing large space-defining, enclosing, and servicing components as basic structural subsystems. Exterior and interior wall, roof and floor components should be looked at to determine their potential for performing both as space-organizing and as structural subsystems.

Figure 4-1 illustrates that this can be done at schematic levels by viewing major space-form components of overall architectural schemes as basic components of a total structural system. Viewed as wholes, the boxlike properties of the major space-form components in Figure 4-1c suggest three-dimensional structural subsystem potentials. This may also be true when one takes an overall look at folded-roof or wall subsystems (Figures 4-2a and 4-2b). In other cases, an overall approach will suggest that large vertical elevator and stair shafts may be used as core-tube structural subsystems (Figure 4-2c). Similarly, it becomes possible to explore the overall possibilities for tying smaller vertical shaft and chase-space components together to act as a total system mega-frame. It is even possible that hall or major horizontal duct spaces can be utilized as horizontal connector components, and walls as vertical load-carrying subsystems (Figure 4-2d).

PHOTO 4-1
MILWAUKEE COUNTY WAR MEMORIAL, EERO SAARINEN ASSOCIATES, ARCHITECT; WING COMPONENTS CANTILEVER 34 FT, 6 IN. EAST-WEST AND 28 FT, 6 IN. NORTH-SOUTH.

(Courtesy of Maynard W. Meyer and Associates, Milwaukee).

SECTION 2: Overall Versus Local Issues of Total-System Analysis

A total-to-subsystem hierarchy is the key to successful application of structural design ideas at schematic stages of architectural design. When basic structural subsystems are conceived as schematic components of a total system, individual elements need be considered only to the extent that their layout may be *critical* to the assumption of overall integrity. In fact, at this level the really basic issue is how to *visualize* major components of a space-form scheme as being able to play basic structural subsystem roles. Thus, a spatial component should first be thought of in terms of its overall structural role in the total scheme of things, and secondarily in terms of specific options for laying out and sizing the individual elements or specifying connection details.

Figure 4-1 illustrated that initially an architect can think of tying together the four major space-form components as cantilevered structural box subsystems. Subsequently one can explore the more local requirements for design of each box as a subsystem. Finally, when overall feasibility is established, one can deal with determination of specific element and construction properties. In a nutshell, the idea is to establish an overall structural scheme to serve as a context for dealing with the more localized issues of subsystem design. Again, the total system space-structure concept should control the need for details and not vice versa.

FIGURE 4-1
MAJOR SPACE-ORGANIZING COMPONENTS CAN BE USED AS BASIC STRUCTURAL SUBSYSTEM COMPONENTS.

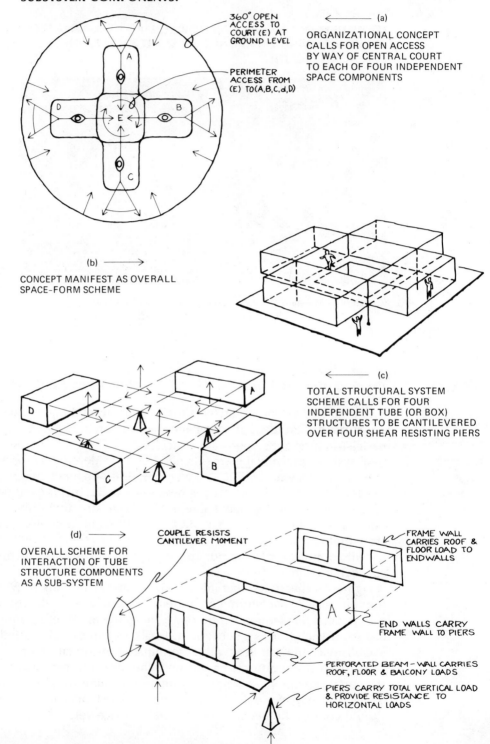

360° OPEN ACCESS TO COURT (E) AT GROUND LEVEL

PERIMETER ACCESS FROM (E) TO (A,B,C,d,D)

←———— (a)
ORGANIZATIONAL CONCEPT CALLS FOR OPEN ACCESS BY WAY OF CENTRAL COURT TO EACH OF FOUR INDEPENDENT SPACE COMPONENTS

(b) ———→
CONCEPT MANIFEST AS OVERALL SPACE-FORM SCHEME

←———— (c)
TOTAL STRUCTURAL SYSTEM SCHEME CALLS FOR FOUR INDEPENDENT TUBE (OR BOX) STRUCTURES TO BE CANTILEVERED OVER FOUR SHEAR RESISTING PIERS

(d) ———→
OVERALL SCHEME FOR INTERACTION OF TUBE STRUCTURE COMPONENTS AS A SUB-SYSTEM

COUPLE RESISTS CANTILEVER MOMENT

FRAME WALL CARRIES ROOF & FLOOR LOAD TO END WALLS

END WALLS CARRY FRAME WALL TO PIERS

PERFORATED BEAM-WALL CARRIES ROOF, FLOOR & BALCONY LOADS

PIERS CARRY TOTAL VERTICAL LOAD & PROVIDE RESISTANCE TO HORIZONTAL LOADS

FIGURE 4-2
OVERALL PROPERTIES OF MAJOR SPACE-FORM COMPONENTS CAN SUGGEST MAJOR STRUCTURAL SUBSYSTEM DESIGN OPTIONS.

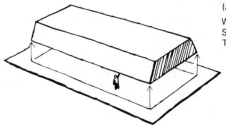

(a)

WHEN VIEWED AS A WHOLE, A FOLDED ROOF SUGGESTS THE POSSIBILITY OF BEING DESIGNED TO ACT AS A FOLDED PLATE SUB-SYSTEM

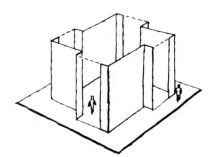

(b)

WHEN DESIGNED AS WHOLES, THE FLANGE-WEB INTERACTION OF THE FOLDED WALLS CAN BE MUCH STIFFER THAN FOR FLAT DESIGN ONLY

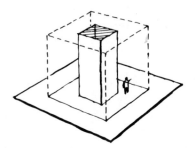

(c)

A LARGE AND CENTRALIZED VERTICAL CIRCULATION SPACE MAY BE DESIGNED TO ACT AS A CORE TUBE SUB-SYSTEM

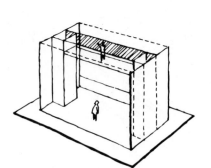

(d)

A COMBINATION OF SHEAR WALL AND MEGA-FRAME ACTION CAN BE VISUALIZED BY CONSIDERING THAT THE UPPERMOST HALL SPACE CAN BE DESIGNED TO CONNECT THE SLENDER SHAFTS

At schematic levels, when one is attempting to conceive of an optimum spatial and structural scheme for a building as a whole, the value of overall thinking may now seem obvious. But note that overall thinking also applies to design of individual subsystems.

Figures 4-1 and 4-2 show that you may first assume overall integrity of subsystems and analyze their force and behavior implications as 2-D or 3-D wholes. Second, you may proceed to substantiate that assumption by establishing the basic force and geometric properties of key components. Figure 4-3 illustrates how an overall approach may be applied to analyze the basic properties of a folded-roof subsystem that must span between four corner supports. We see that the two folded-plate schemes can be compared in terms of their overall versus local structural design implications

FIGURE 4-3
AN OVERALL COMPARISON OF TWO FOLDED ROOF SUBSYSTEM SCHEMES.

SCHEME A: WITH HIGH RISE ARCH ACTION

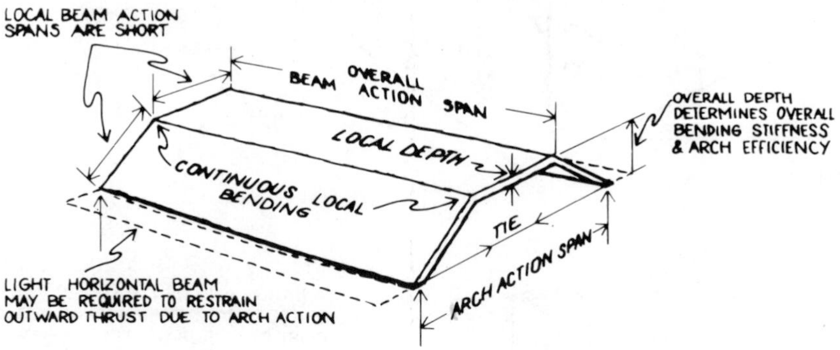

LOCAL BEAM ACTION
SPANS ARE SHORT

BEAM OVERALL ACTION SPAN

LOCAL DEPTH

OVERALL DEPTH DETERMINES OVERALL BENDING STIFFNESS & ARCH EFFICIENCY

CONTINUOUS LOCAL BENDING

TIE

ARCH ACTION SPAN

LIGHT HORIZONTAL BEAM MAY BE REQUIRED TO RESTRAIN OUTWARD THRUST DUE TO ARCH ACTION

SCHEME B: WITH LOW RISE ARCH ACTION

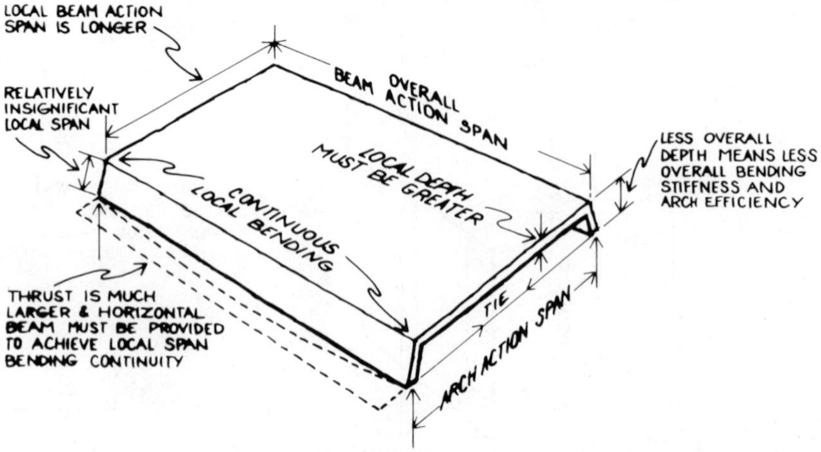

LOCAL BEAM ACTION SPAN IS LONGER

BEAM OVERALL ACTION SPAN

RELATIVELY INSIGNIFICANT LOCAL SPAN

LOCAL DEPTH MUST BE GREATER

CONTINUOUS LOCAL BENDING

LESS OVERALL DEPTH MEANS LESS OVERALL BENDING STIFFNESS AND ARCH EFFICIENCY

THRUST IS MUCH LARGER & HORIZONTAL BEAM MUST BE PROVIDED TO ACHIEVE LOCAL SPAN BENDING CONTINUITY

TIE

ARCH ACTION SPAN

without actually dealing with specification of elemental properties. Subsequent to this kind of schematic analysis of basic component interactions, it is reasonable to work out specific properties of the key elements required to actually achieve the overall integrity assumed in schematic analysis. Of course, one's approach to design of structural subsystems will be influenced by one's *intentions* for their interaction as a total space-form scheme (Figure 4-4).

FIGURE 4-4
THE TOTAL SYSTEM SCHEME SERVES AS A CONTEXT FOR OVERALL CONSIDERATION OF ROOF SUBSYSTEM DESIGN OPTIONS.

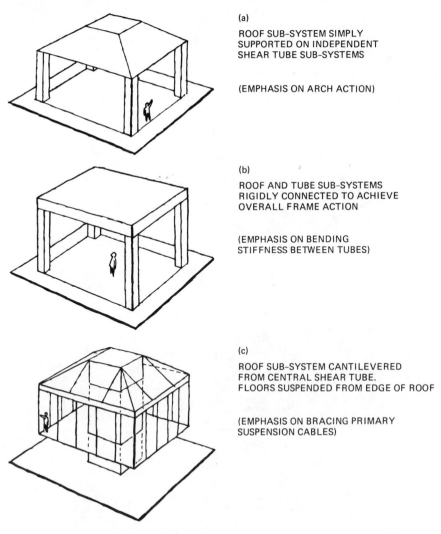

(a)

ROOF SUB-SYSTEM SIMPLY SUPPORTED ON INDEPENDENT SHEAR TUBE SUB-SYSTEMS

(EMPHASIS ON ARCH ACTION)

(b)

ROOF AND TUBE SUB-SYSTEMS RIGIDLY CONNECTED TO ACHIEVE OVERALL FRAME ACTION

(EMPHASIS ON BENDING STIFFNESS BETWEEN TUBES)

(c)

ROOF SUB-SYSTEM CANTILEVERED FROM CENTRAL SHEAR TUBE. FLOORS SUSPENDED FROM EDGE OF ROOF

(EMPHASIS ON BRACING PRIMARY SUSPENSION CABLES)

At the total system level, the issue is to identify and evaluate options for achieving overall interaction between subsystems that satisfy both spatial and structural needs. As a result, some feedback between total and subsystem levels of design is required for optimum design. But eventually, the more key properties of total and subsystem schemes will be optimized, and one's attention can turn to element design. And, at some point, the service of a structural specialist may be required to verify and refine, by detailed analysis, what is basically a sound total scheme. Undoubtedly, some further trade-offs between spatial and physical requirements will be required even at these final stages. But, if the schematic thinking is sound, these will be both expected and minimized.

Figure 4-5 offers us another example of how the overall approach allows a

FIGURE 4-5
THE OVERALL APPROACH IS USEFUL IN DEALING WITH UNUSUAL SPACE-FORM DESIGNS.

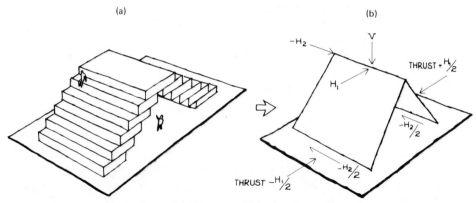

(a) (b)

OVERALL SPACE-FORM SCHEME SUGGESTS OVERALL STRUCTURAL SYSTEM CONCEPT

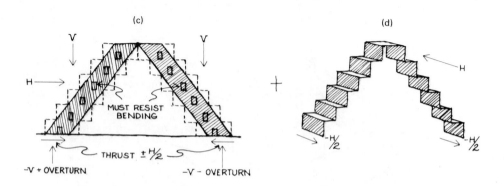

(c) (d)

TRANSVERSE ROOM DIVIDING WALLS CAN SERVE AS ARCH SUB-SYSTEMS

LONGITUDINAL WALLS AND FLOORS CAN SERVE AS STAGGERED SHEAR TRANSFER SUB-SYSTEM

PHOTO 4-2
**WALT DISNEY WORLD CONTEMPORARY RESORT HOTEL,
WELTON BECKET ASSOCIATES, ARCHITECTS.
(*a*) (TOP) A VIEW OF THE ATRIUM SPACE.
(*b*) (BOTTOM) UNDER CONSTRUCTION.**

designer to take a hierarchic look at the structural implications of a space-form scheme similar to that of the Disney World Hotel (Photo 4-2). When viewed as a whole, it is evident that the form suggests a continuous arch total system scheme. Further, the possibility exists for alignment of the hotel room spaces such that the transverse subdividing walls are continuous and can be used as two-dimensional arch subsystems. Structural continuity between the arch subsystems can be achieved by interacting the staggered longitudinal walls with floors as diaphragms.

Such schematic conceptualizations can be made prior to consideration of the specifics of subsystem design. There need be no preconceptions because overall feasibility is not changed by the fact that the arch subsystems can be designed in a number of ways, employing either concrete or steel. They could be designed as perforated arched beams, or as stepped trusses with openings arranged to coincide with hallway spaces and adjoining doorways. Similarly, the longitudinal arch connectors can be designed as either trusses or shear walls.

With an overall approach, creativity at either schematic or specific levels of design is limited only by the imagination and insight of the designer. For example, a high-rise building, designed by T. Y. Lin International (Photo 4-3a) incorporated what would normally be thought of as a nonstructural enclosing component to fill both overall structural and specific constructive needs simultaneously. In this case, the enclosing components (Photo 4-3b) were prefabricated to act as spatial enclosures, window boxes, spandrel elements, and column forms, and to help resist seismic forces. When erected, the whole became a very efficient frame-wall space structure.

The overall approach can also be applied to complex building forms as well as to the simpler symmetric and/or single-form buildings illustrated thus far. The problem of asymmetry may be obvious as in a multi-form design. But it can also be evident in single-forms that have internal asymmetries. The basic overall approach would be to zone, or factor, a complex space-form scheme into simpler components that can then be dealt with in a semiindependent fashion. For example, Figure 4-6 illustrates how a multiple-form building can be simplified by factoring the overall form into tall-to-short building components, long-to-short-span, or other simple space-structure components. Since the effect of horizontal forces may be very different for the tall versus the short portions of buidlings, it is logical to treat each part as a separate structure. Similarly, a building may be factored into varied structural zones according to their basic span or occupancy properties. Thus the building in Figure 4-6 could call for each zone to be designed differently. The only source of real complexity is in the need for achieving compatibility (of displacements and force transmission, etc.) at the interface between different structural schemes.

Having discussed and illustrated the idea and logic of applying an overall approach to structural analysis, we will devote the remaining sections of this chapter to four examples of its application at the schematic level. The forms

PHOTO 4-3
**PRECAST OFFICE BUILDING, SAN JOSE, CALIFORNIA, ALLAN WALTER,
ARCHITECT; T. Y. LIN INTERNATIONAL, STRUCTURAL ENGINEERS.**
(a) BUILDING UNDER CONSTRUCTION
**(b) WALL PANEL BEING ERECTED BETWEEN COLUMN STEEL CAGES WHICH WILL
BE CONCRETED TO INTEGRATE THE PANELS.**

FIGURE 4-6
COMPLEX FORMS AND OCCUPANCY REQUIREMENTS MAY REQUIRE FACTORING TO SIMPLIFY.

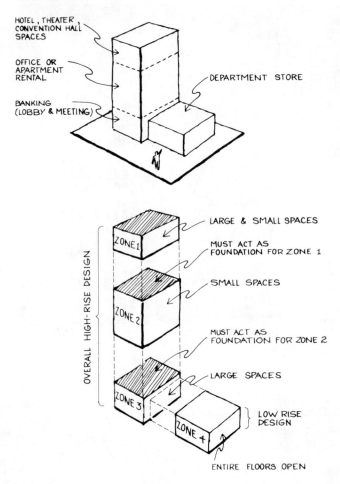

will be simple, but they will vary from single-story to 2-, 12-, and 15-story building designs. The types of usage will range from large clear-span single-story space to subdivided high-rise office and apartment designs, and include a two-story garage. Thus, several types of subsystem interaction schemes will be illustrated and analyzed to show how an assumption of overall integrity can be actually substantiated.

The examples will not deal with questions of preliminary design for particular materials or elemental details, as this will be treated in later chapters. Instead, they are primarily intended to show how a building can be viewed as a total structural system and how the overall vertical and horizontal forces can be transmitted into the ground support of a building through the interaction of various types of major structural subsystems. In all examples, efficient designs in either steel or reinforced or prestressed concrete are possible. But, in any case, in following through these examples you should come to realize that major structural decisions about total system schemes can often be made with insightful assumptions and only a minimum of calculations. Thus, the calculations you will see are conveniently rounded, rather than precise, to emphasize the overall, versus detailed, picture.

SECTION 3: A Single-Story Open-Space Building

Example 4-1

A single-story open-space building with a plan area of 100 ft × 200 ft is to be covered with a hipped-frame roof (Figure 4-7). It will be schematically analyzed to determine the overall requirements for resistance of dead load, live load, and wind load. The specifics of estimating actual loading conditions will be treated in Chapter 5, but for now loading will be assumed as follows.

FIGURE 4-7
OVERALL DETERMINATION OF LOADING CONDITIONS (ALL LOADS ARE ASSUMED TO BE EVENLY DISTRIBUTED OVER PROJECTED AREAS).

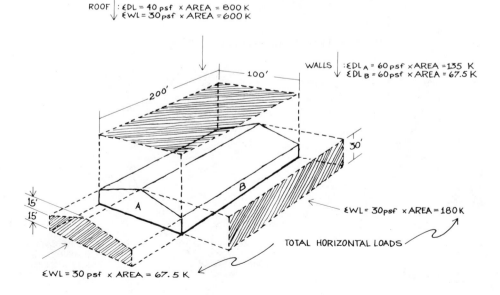

The dead load (*DL*) of the roof is assumed to be 40 psf (pounds per square foot) of horizontal projection. This weight includes the framing, the roofing, and any mechanical duct or other service equipment to be supported on or hanging from the roof. The *DL* of the side walls is assumed at 60 psf of wall area. This will include the weight of the columns, the sidings, and any other material incorporated into the walls. It is further assumed that the horizontal wind load (*WL*) acting on the windward side of the building will be 30 psf of projected area, which has a height of 30 ft and a width of 200 ft. Suction on the leeward side is neglected here, but will be discussed in Chapter 5. A vertical wind load of 30 psf will act downward (pressure) on the roof and will be computed on a horizontally projected area of 100 ft × 200 ft, again neglecting any suction force.

Live load on the roof will be 20 psf of horizontal projection. This is smaller than the wind load, and would not seem to control the design. However, the effect of live load may be more serious than the wind load, even though the latter has a higher intensity. This can be so because building codes permit all parts of a structure to be worked to a higher stress level (generally $\frac{1}{3}$ higher) under wind load than under live load.

Initially, the enclosing surfaces can be assumed to act as rigidly connected plates for determination of overall requirements for resistance. Actual structural integrity is pictured as being provided by a series of frames, each consisting of two end columns and two inclined roof girders, rigidly connected to form a primary hipped-frame type of structural subsystem that picks up and transmits the roof load onto the foundations (Figure 4-8). Secondary roofing and wall siding components will be provided as necessary between the frames to hold them together and to enclose the

FIGURE 4-8
OVERALL STRUCTURAL SYSTEM SCHEME.

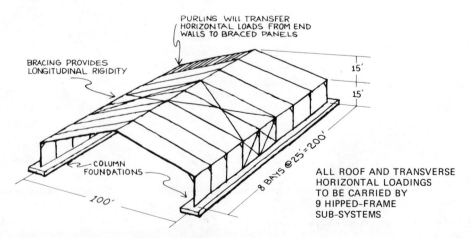

PURLINS WILL TRANSFER HORIZONTAL LOADS FROM END WALLS TO BRACED PANELS

BRACING PROVIDES LONGITUDINAL RIGIDITY

15'

15'

COLUMN FOUNDATIONS

8 BAYS @ 25' = 200'

100'

ALL ROOF AND TRANSVERSE HORIZONTAL LOADINGS TO BE CARRIED BY 9 HIPPED-FRAME SUB-SYSTEMS

space. End walls at two ends of the building are also to be provided with the required door openings and windows.

Horizontal forces in the transverse direction to the form can be carried by the rigid frames. But, to transmit horizontal forces applied in the longitudinal direction to the foundation, two center panels of the roof are braced with *X*-braces. These braced panels will stiffen the roof planes in the longitudinal direction to allow them to pick up and transmit horizontal forces to the two similarly braced wall panels on the two sides. The braced wall panels are required to transmit the longitudinal forces received from the roof into the ground.

Overall, the loads shown in Figure 4-7 must be transmitted to the foundations. At the foundations, reactions to both vertical and horizontal loads will be developed in order to maintain static equilibrium of the building as a whole under dead load and wind load. Owing to the vertical (dead and wind) loads, an additional inward horizontal reaction must be developed, since a hipped-frame type of structure will produce an outward horizontal kick at the bottom of the columns as shown in Figure 4-9. The actual computation of such kickout forces would have to allow for a combination of frame and arch action (which will vary with the sizes of the members), as will be explained in Chapter 11. But for schematic purposes,

FIGURE 4-9

AN APPROXIMATION OF TOTAL HORIZONTAL KICK PRODUCED BY OVERALL FRAME ACTION UNDER TOTAL ROOF LOAD.

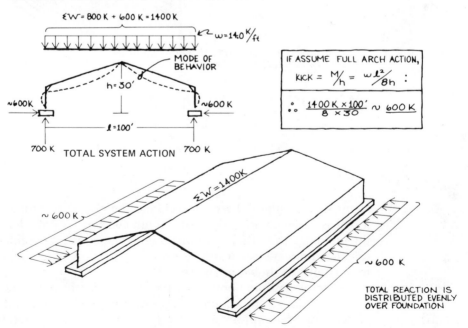

FIGURE 4-10
REACTIONS AT FOUNDATIONS DUE TO VERTICAL AND TRANSVERSE HORIZONTAL LOADS.

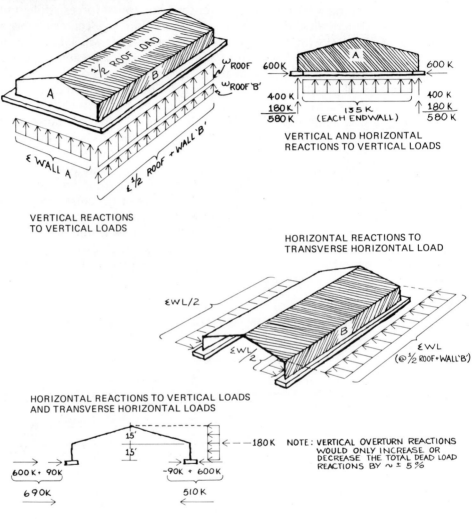

VERTICAL AND HORIZONTAL
REACTIONS TO VERTICAL LOADS

VERTICAL REACTIONS
TO VERTICAL LOADS

HORIZONTAL REACTIONS TO
TRANSVERSE HORIZONTAL LOAD

HORIZONTAL REACTIONS TO VERTICAL LOADS
AND TRANSVERSE HORIZONTAL LOADS

NOTE: VERTICAL OVERTURN REACTIONS
WOULD ONLY INCREASE OR
DECREASE THE TOTAL DEAD LOAD
REACTIONS BY ~ ± 5%

Figure 4-9 assumes full arch action, and the overall magnitude of these forces is easily computed to total about 600 kips for each longitudinal side of the building. The other reactions due to static loading conditions are calculated directly according to projected areas (Figure 4-7), and act on the foundations as shown in Figure 4-10.

Now let us examine how each set of these forces is separately transmitted through various components of the building into the ground. Take the case of dead load or wind pressure on the roof. They are carried from the

FIGURE 4-11
TRANSMISSION OF HORIZONTAL FORCES FROM END WALLS TO THE FOUNDATION.

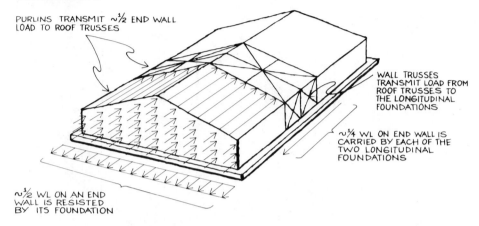

PURLINS TRANSMIT ~½ END WALL LOAD TO ROOF TRUSSES

WALL TRUSSES TRANSMIT LOAD FROM ROOF TRUSSES TO THE LONGITUDINAL FOUNDATIONS

~¼ WL ON END WALL IS CARRIED BY EACH OF THE TWO LONGITUDINAL FOUNDATIONS

~½ WL ON AN END WALL IS RESISTED BY ITS FOUNDATION

secondary roof members (purlins) to the primary rigid-frame girders and then down the frame columns into the foundation. The purlins also act to transmit about one-half of the wind loads acting on the end walls of the building to the trussed panels (Figure 4-11). The roof truss picks up and transmits the purlin forces into the braced column panels and finally into the longitudinal foundation. The other one-half of the end loads will be transmitted directly to the transverse foundation.

Wind loads that act horizontally on both the side walls and on the roof must be resisted either by the two transverse end walls of the building or by the frames. If it were desired to carry all the transverse wind load by the end walls, it would be necessary to design the entire roof as a rigid diaphragm strong enough to carry transverse wind load from the roof and longitudinal walls onto the end walls. But our example exploits the possibility of carrying the wind load falling within the tributary area of each frame by the frames themselves (Figure 4-12a). This is reasonable in this case, since the transverse wind load is not high. Figure 4-12 illustrates the transmission of these loads by frame action.

The analysis of the frame under vertical dead load and wind load can now be approximated (Figure 4-12b). The maximum horizontal reaction for each frame is about 75 k, which must be resisted by the support to the footing in some manner. It is also noted that the knee of the frame will be subjected to a bending moment of $\sim 75\,k \times 15\,ft \approx 1100\,k\,ft$ due to *DL* only. One can determine the change in these values due to *WL*. Thus the major forces and moments on a typical interior frame can be determined so as to serve as a guide for an approximate sizing or feasibility and economic study of the structural layout.

FIGURE 4-12
PROJECTED TRIBUTARY AREA DETERMINES LOAD ON EACH FRAME.

(a)

TRIBUTARY AREAS FOR INTERIOR AND END FRAMES

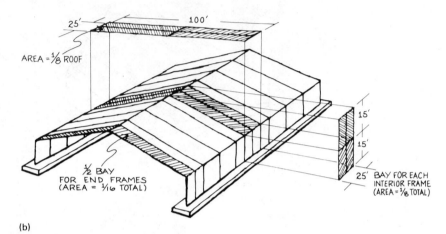

(b)

LOADS AND REACTIONS ON A TYPICAL INTERIOR FRAME
(OVERTURN REACTIONS NOT SHOWN)

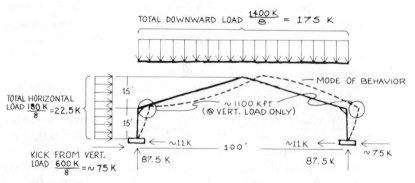

SECTION 4: A Two-Story Parking Garage

Example 4-2

A two-story parking structure is shown in Figure 4-13. The rather long spans (60 ft) are desirable to facilitate auto movement and right-angle parking. To save the cost of joining the girders and the columns into rigid frames, this building is designed with the girders being simply supported on the columns. But the main concern in this example is not the girder or column

FIGURE 4-13
OVERALL SCHEME FOR DESIGN OF A TWO-STORY PARKING STRUCTURE.

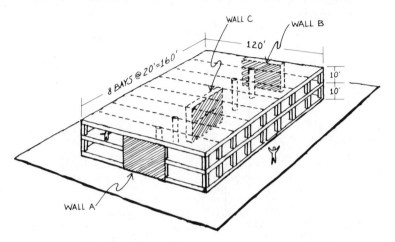

design (to be treated in Chapters 6 and 7). Here we analyze the overall requirements for design of shear walls that must resist the lateral forces (wind and/or earthquake) acting on this structure. To provide resistance to horizontal forces in the longitudinal direction, a shear wall C is located near the center of the structure. To resist forces in the transverse direction, two shear walls (A and B) are located at each end. It is assumed that both the roof and the floor will be capable of acting as rigid horizontal diaphragms that are stiff enough to transmit the horizontal load from various parts of this building into the appropriate shear walls without local buckling. Of course, the floors must also carry vertical loads between girders in bending, but this is a local design problem to be discussed in Chapter 8.

The first step is to determine the total vertical and horizontal forces acting on each of these shear walls (and their foundations) so that their overall strength and stability design requirements can be noted. Observe that the tributary loading areas for the vertical loads versus the horizontal loads are totally different (Figure 4-14). For the horizontal load applied in the longitudinal direction, the entire force must be resisted by wall C, which carries only a small tributary component of the total vertical load. Similarly, shear walls A and B will share the entire transverse horizontal force on the structure, but each wall will carry an even smaller tributary area of vertical load. It is possible that the vertical loads will not be sufficient to stabilize the shear walls against the overturning action of the horizontal forces.

The following simple circulations will allow a designer to quickly determine the adequacy of shear wall C to resist earthquake forces (Figure 4-15). For wall C: Begin by calculating the total weight of each floor or roof @ 80 psf, (80 psf × 120 ft × 160 ft ≈ 1550 k). If the total earthquake force is

FIGURE 4-14
**HORIZONTAL AND VERTICAL LOADS ESTIMATED BY TRIBUTARY AREA
ASSUMPTION.**

VERTICAL LOAD TRIBUTARY AREAS

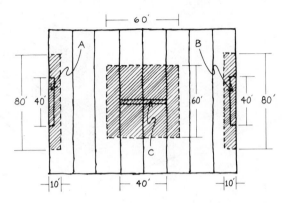

WIND LOAD TRIBUTARY AREAS

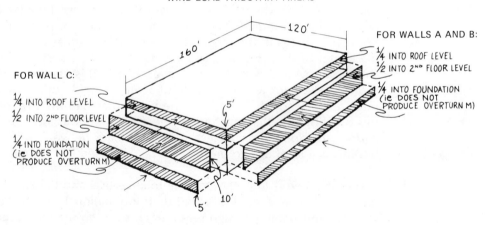

FOR WALL C:

¼ INTO ROOF LEVEL
½ INTO 2ND FLOOR LEVEL

¼ INTO FOUNDATION
(ie DOES NOT
PRODUCE OVERTURN M)

FOR WALLS A AND B:

¼ INTO ROOF LEVEL
½ INTO 2ND FLOOR LEVEL
¼ INTO FOUNDATION
(ie DOES NOT
PRODUCE OVERTURN M)

assumed to be about $0.1W$, there will be forces of 155 k applied at each of
the two levels (Figure 4-15). This will produce an overturning moment that
ideally should be resisted by the vertical load on wall C plus its own weight:

$$\text{Two floors} = 80 \text{ psf} \times 60 \text{ ft} \times 60 \text{ ft} \times 2 = 576 \text{ k on wall C}$$
$$\text{Wall C} = 100 \text{ psf} \times 20 \times 40 = \underline{80} \text{ k in wall C}$$
$$\text{Total} = 432 + 80 = \overline{656 \text{ k}} \text{ at foundation.}$$

From the ratio of the overturn moment and the available vertical force, it
can be seen that the eccentricity ($e = 7.1$ ft) will be outside the middle third

FIGURE 4-15

OVERALL REQUIREMENTS FOR BALANCING OVERTURN *M* AND VERTICAL LOAD STRESSES FOR WALL C.

E. Q. FORCES AND OVERTURN 'M' ON WALL 'C':

$\overline{W}$/FLOOR = 80 psf × 120' × 160' = 1550 K/FLOOR
ASSUME : F_{EQ}/FLOOR = 0.1$\overline{W}$ = 155 K/FLOOR

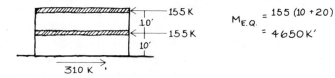

$$M_{E.Q.} = 155 (10 + 20)$$
$$= 4650 K'$$

VERTICAL LOAD SUPPORTED ON WALL 'C':

FROM 2 FLOORS : 80 psf × 60' × 60' × 2 = 576 K
FROM WALL C : 100 psf × 40' × 20' = 80 K
V_C = 656 K

FOR BALANCED DESIGN OF FOUNDATION:

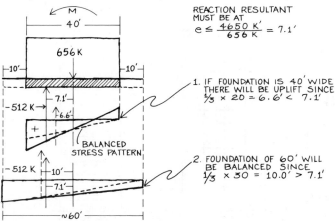

REACTION RESULTANT MUST BE AT
$$e \leq \frac{4650 K'}{656 K} = 7.1'$$

1. IF FOUNDATION IS 40' WIDE THERE WILL BE UPLIFT SINCE
⅓ × 20 = 6.6' < 7.1'

2. FOUNDATION OF 60' WILL BE BALANCED SINCE
⅓ × 30 = 10.0' > 7.1'

point (6.6 ft) for the wall section, although well within the toe of the wall. Therefore, if the foundation were designed to be larger, say 60 ft, the eccentric load would be well inside the third point ($\frac{1}{3}$ × 30 ft = 10 ft) and the available vertical load would easily balance the overturn moment without uplift in the foundation.

Readers should use the same approach to check shear walls *A* and *B* and propose what to do for them to resist seismic overturning moment.

SECTION 5: A 12-Story Office Building

Example 4-3

Typical plan, elevations, and floor sections of a 12-story office building are shown in Figure 4-16. This building is supported by nine columns on each side and a 20 ft × 40 ft elevator and utility shaft in the center of the building. The vertical loads (dead and live load) acting on each floor level and the roof are transmitted primarily through the columns and partly through the central core. Lateral or horizontal load transmission can be accomplished in several ways. In our example, the central core will act as a tube subsystem that can resist lateral forces in any direction caused by wind or earthquake (Figure 4-17a). The 18 exterior columns will carry most of the vertical load, but they will not resist any lateral forces.

Figure 4-17b shows an alternate scheme, which would call for the exterior columns and their connecting girders to act as rigid frames in longitudinal

FIGURE 4-16
A 12-STORY OFFICE BUILDING DESIGN SCHEME.

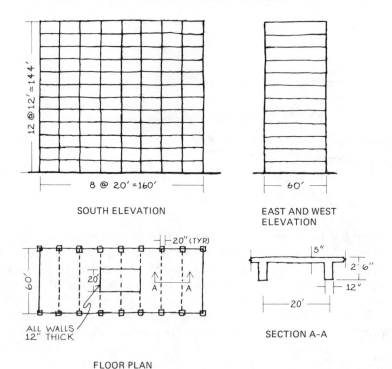

FIGURE 4-17
TWO SCHEMES FOR LATERAL FORCE RESISTANCE.

(a) LATERAL FORCES RESISTED BY CORE TUBE

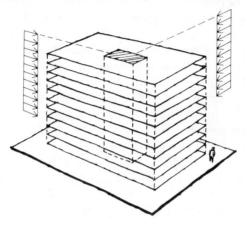

(b) LATERAL FORCES RESISTED BY COLUMN FRAMES

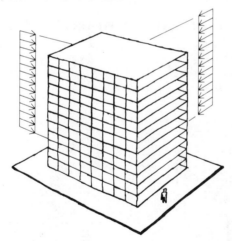

and transverse directions. Essentially, longitudinal forces will be resisted by two rigid-frame subsystems that have 7 interior columns and 12 horizontal connecting girders, one corresponding to each story. In the transverse direction, lateral forces will be resisted by 9 frames, each of which would have 2 exterior columns connected by longer girders at each story. Of course, it is also possible that the building could be designed in such a way that both the central core and the frames could act together to resist lateral force in any direction.

In our example, the lateral forces are to be transmitted through the building with the floors acting as horizontal diaphragms, which also stiffen the columns and connect them to the central shaft (Figure 4-18). The specific role played by horizontal diaphragm action will differ depending on how the lateral-force-resisting components are assumed or arranged. The floor diaphragms may either transmit lateral forces to the shaft as is assumed for our example, to the frames, or to both. In any case, the floors must be more-or-less rigid in a horizontal plane in order to act as a diaphragm and force the vertical elements to act together.

The vertical loads on each vertical-acting component of the building can be estimated by using the tributary area method (Figure 4-19). This would divide each floor or roof area into sections tributary to each vertical load-carrying element (the shaft or each column). Then the live and dead loads of each element can be computed. For our example, we will assume a *DL* for roof and floors at 120 psf, and a horizontal wind load of 30 psf on wall surfaces. This means that the shaft will carry a total *DL* from the 11 floors above ground plus the roof load of 3450 k.

FIGURE 4-18
SCHEMATIC INTERACTION OF CORE TUBE, FLOOR DIAPHRAGM AND COLUMN SUBSYSTEMS.

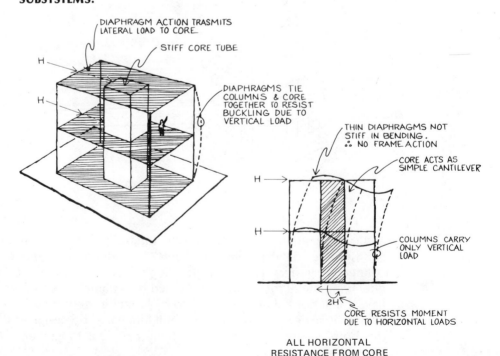

DIAPHRAGM ACTION TRASMITS LATERAL LOAD TO CORE

STIFF CORE TUBE

H

H

DIAPHRAGMS TIE COLUMNS & CORE TOGETHER TO RESIST BUCKLING DUE TO VERTICAL LOAD

THIN DIAPHRAGMS NOT STIFF IN BENDING.
∴ NO FRAME ACTION

CORE ACTS AS SIMPLE CANTILEVER

H →

H →

COLUMNS CARRY ONLY VERTICAL LOAD

2H

CORE RESISTS MOMENT DUE TO HORIZONTAL LOADS

ALL HORIZONTAL
RESISTANCE FROM CORE

FIGURE 4-19
APPROXIMATION OF COMPONENT LOADING BY TRIBUTARY AREA ANALYSIS (*NOTE*: USED $\frac{1}{2}$ SPACE ASSUMPTION).

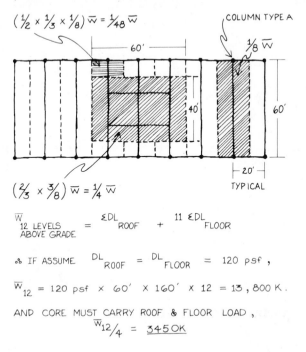

$\left(\frac{1}{2} \times \frac{1}{3} \times \frac{1}{8} \right) \overline{W} = \frac{1}{48} \overline{W}$

COLUMN TYPE A

60'

$\frac{1}{8} \overline{W}$

40'

60'

$\left(\frac{2}{3} \times \frac{3}{8} \right) \overline{W} = \frac{1}{4} \overline{W}$

20'

TYPICAL

$\dfrac{\overline{W}}{\underset{\text{ABOVE GRADE}}{\text{12 LEVELS}}} = \Sigma DL_{ROOF} + 11 \, \Sigma DL_{FLOOR}$

$\therefore$ IF ASSUME $DL_{ROOF} = DL_{FLOOR} = 120 \text{ psf}$,

$\overline{W}_{12} = 120 \text{ psf} \times 60' \times 160' \times 12 = 13,800 \text{ K}.$

AND CORE MUST CARRY ROOF & FLOOR LOAD,

$\overline{W}_{12}/4 = \underline{345 \text{ OK}}$

In the example in Figure 4-20, we will only calculate a rough moment for wind load of 30 psf in the transverse direction. It is more critical because of the larger exposed area and the lesser shaft dimension. Assuming the floor and roof *DL* on the shaft to be 3450 k, and that the shaft walls represent a *DL* of 100 psf, the total load available to resist overturn is 5170 k. As a result, Figure 4-20 illustrates that the 9.6 ft eccentricity of the resultant force is very near the edge of the shaft, which appears to be unstable. These calculations indicate that the shaft must be made quite strong and the foundation must be widened and perhaps anchored down to avoid uplifting.

In contrast, if we could assume that the 60 ft-wide frame was capable of acting to resist transverse forces without help from the shaft, then we have an overall *T* or *C* force in the columns of approximately 49,600/60 ft = 830 k, which would be more-or-less equally shared between the nine columns on each side: 830/9 = 92 k per column. This is small compared to the dead load on each column. Hence, a frame design appears to be a better option to follow up than a core design. A more detailed comparison may also show that a combination of frame and core would be the most economical, especially when considering the ability of the shaft to resist local story-shear and to reduce story drift.

FIGURE 4-20
OVERALL ANALYSIS OF CORE-TUBE SCHEME FOR TRANSVERSE LOAD.

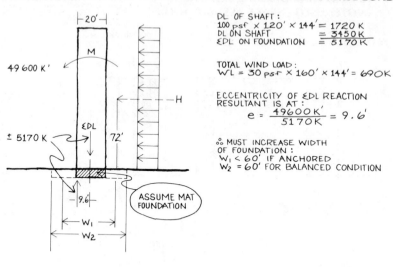

DL OF SHAFT :
100 psf $\times$ 120' $\times$ 144' = 1720 K
DL ON SHAFT   = 3450 K
ΣDL ON FOUNDATION   = 5170 K

TOTAL WIND LOAD :
WL = 30 psf $\times$ 160' $\times$ 144' = 690K

ECCENTRICITY OF ΣDL REACTION
RESULTANT IS AT :
$$e = \frac{49600 K'}{5170 K} = 9.6'$$

$\therefore$ MUST INCREASE WIDTH
OF FOUNDATION :
W_1 < 60' IF ANCHORED
W_2 = 60' FOR BALANCED CONDITION

ASSUME MAT FOUNDATION

SECTION 6: A 15-Story Apartment Mega-Structure

Example 4-4

An unusual 15-story apartment project consists of a row of large buildings. The buildings are connected at the upper floors, and step back at the tenth and fifth floors to provide more space between building units at ground level (Figures 4-21 and 4-22). Each building unit is 90 ft $\times$ 160 ft in plan at the top five floors, and the total vertical load is to be carried by only four circulation and service shafts spaced 70 ft clear in both directions. The scheme also calls for three heavy (mega-) structural floors, A, B, and C, to rigidly connect the shafts and support the five thin conventional apartment floors above. As a result, seismic forces are to be resisted by the shafts acting together with the structural floors A, B, and C, as overall rigid-frame subsystems. For purposes of schematic analysis, openings in the hollow shafts can be ignored, and their overall action can be assumed to be that of a structural tube. At this stage, we will also neglect any frame action participation from the intermediate thin floors.

Assume the total dead load of each building unit (90 ft width $\times$ 130 ft *average* length $\times$ 15 stories high) has been computed to be 15,600 k, and the total live load to be 3140 k. Make preliminary calculations to determine the feasibility of the system, relative to their seismic resistance, as controlled by

FIGURE 4-21
A MEGA-FRAME DESIGN SCHEME.

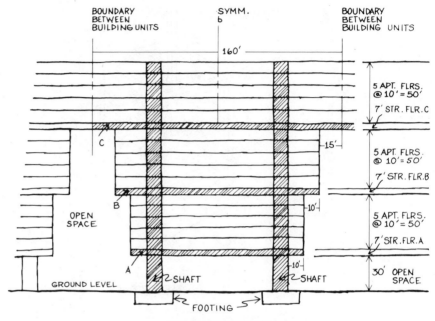

BUILDING ELEVATION

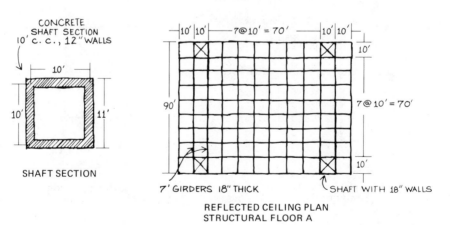

SHAFT SECTION

REFLECTED CEILING PLAN
STRUCTURAL FLOOR A

the lower 30 ft height of the shafts:

1. Assume seismic shear (F_{EQ}) = 6 percent of the dead load weight (i.e., $0.06W$): $0.06 \times 15,600$ k = 940 k/bldg. = 470 k/bent, and from Figure 2-9, assume F_{EQ} acting at $\frac{3}{4}$ of 200 ft = 150 ft and, 470 k × 150 ft = 70,500 k'.
2. For resisting lever arm = 70 ft + 10 ft = 80 ft, the force (C or T) induced in the shaft is 70,500 k'/80 ft = 880 k up or down.

FIGURE 4-22
THE MEGA-FRAME STRUCTURAL SCHEME.

PERSPECTIVE:

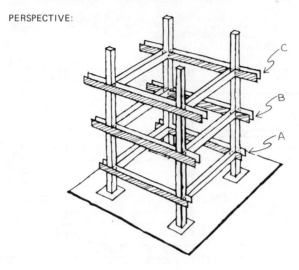

ELEVATIONS:

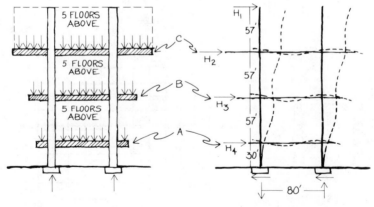

GIRDERS CARRY VERTICAL
LOAD FROM 5 FLOORS ABOVE

MEGA-FRAME RESISTS HORIZONTAL
FORCES AND OVERTURN

3. Since the *DL* on each shaft is 15,600/4 = 3900 k, the overturning effect of ±880 k is small.

4. Now, the seismic shear per column is 940/4 = 235 k. Thus, Figure 4-23 illustrates that, for a story height of 30 ft, and assuming end moment in the shaft is equally shared by top and bottom of column, 235 k × 30 ft/2 = 3520 k'.

5. Therefore, eccentricity produced by this moment, *e* = 3520 k'/3900 k = 0.9 ft, is small compared with 10 ft width of shaft.

FIGURE 4-23
PURE FRAME ACTION ASSUMPTION IN BOTTOM 30 FT OF EACH SHAFT.

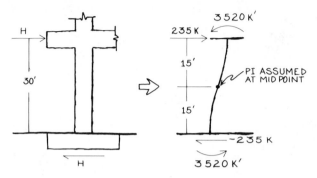

It should be noted that step 5 assumed equal moments in the top and bottom 30 ft of each bent. But, even if all the 235×30 ft moment were carried at the bottom, the eccentricity is only 1.8 ft, slightly beyond the one-third point of the 10 ft column. Now, if the shaft is of reinforced concrete with 12 in.-thick walls, the area of concrete is

$$A_c = 36 \text{ ft} \times 12 \text{ in.} \times 12 \text{ in.} = 5184 \text{ in}^2.$$

Therefore, the average compressive stress under *DL* and *LL* $(3140/4 \simeq 800 \text{ k})$ is

$$3900 + 800 = 4700 \text{ k}/5184 = 0.9 \text{ ksi},$$

which may be high, because to this must be added the axial load of 880 k due to overturn and the moment of 3520 k ft (or more) due to seismic shear. It would be desirable to thicken the wall, say to 18 in., or to increase the size of the shaft, say to 12 ft × 12 ft. Thus the structural feasibility of this unusual layout can be checked by relatively simple though quite approximate means.

5
Structural Loads and Responses

SECTION 1: Introduction

The actual forces acting on a building structure are seldom accurately known. In fact, they are pretty hard to predict with certainty even when the most complete information is available. This problem is particularly apparent at the beginning of the design process when practically no physical data are specified. Even in final design, precision is a problematic issue. But it is possible to make some reasonable assumptions at schematic stages, which will not be seriously inaccurate, in order to get started.

In the examples in Chapter 4, the various external and *DL* forces acting on a structure were assumed more-or-less arbitrarily in order to illustrate how overall design loads can be quickly estimated. In this chapter we describe the various types of forces and how to estimate design loads with sufficient accuracy for schematic and preliminary levels of design. We will also introduce the response of structures to these loads.

For simplicity, it is usually assumed that forces acting on building structures can be reduced to static (unchanging) loads, in pounds per square foot (psf or Kg/m^2). In fact, these loads are not always static. Sometimes they are dynamic or changing in nature. Live loads, seismic disturbances, gusting of wind, movement of machinery, or any other source of fairly rapid load variations will produce dynamic loads. At other times they are produced by the strain and movement in the structure caused by temperature and shrinkage. Additional strains, and thus forces, may also be produced by the uneven settling of foundations, even though dead-load conditions are static.

Nonetheless, it is generally possible to express the effects of these more-or-less changing forces in terms of equivalent static load in psf. Thus, basic live loads are often expressed in terms of a uniform "design load" in psf, whether the area involved is of a roof or a floor.

An even more truly dynamic source of additional live load can result from the sudden application (or movement) of live load in the form of lifting equipment, oscillating machinery, cars in a garage, and so on. These produce impact forces that act in addition to the gravity weight of the basic live load itself. But, because the use of a building can be anticipated, these impact loads can also be expressed by an "impact factor" that yields an additional statical load as a percentage of the basic live load. When applicable, the factor may vary from 5 percent or 10 percent up to 100 percent or more, but generally it will not be found to be in excess of 20 percent or 30 percent.

Wind load on a building is dynamic, but it is conveniently expressed as an equivalent statical load in psf of exposed surface area. The area may be the exposed wall area of a building; or it may be the roof surface on which either a wind pressure and/or a wind suction may be acting. As we shall see, wind loads vary with wind speed, surface shape, and exposed area.

Lateral loads produced by earthquakes are also dynamic but are usually expressed as a percentage of the overall mass or gravity load (*W*) of a building. When considering a building as a whole, that percentage may vary from as little as 2 percent to 5 percent (of *W*) for tall buildings in moderate seismic zones, to as much as 10 percent to 20 percent for short stiff buildings in active seismic zones.

Internal forces may also be produced in a building as a result of a temperature differential between various parts of the building. In this situation, one part of a structure will tend to resist the expansion or contraction movement of another part. Relative shrinkage of the materials, or uneven settling of foundations, produces internal forces similar to those of self-stressing produced by a temperature gradient within the structure itself. Each condition can cause unequal movement across a structure that can produce significant forces in various parts of a building.

The magnitude of the various types of forces and their approximate equivalent design loads will be discussed in the following sections.

SECTION 2:   Dead Loads

When considering a building as a whole at the preliminary design stage, it is best and easiest to estimate the total weight of the building in terms of average psf of floor area. As an approximation, such a gross *DL* for ordinary timber buildings may be assumed at 40 to 50 psf. For steel buildings, the gross *DL* average will be on the order of 50 to 80 psf. For ordinary reinforced concrete buildings, it will likely be between 100 and 150 psf. For prestressed concrete buildings, a value of 70–80 percent of that suggested for ordinary reinforced concrete buildings may be used.

Although expressed in terms of floor area, these overall dead-load values attempt to include the weight of the structural floor itself, the roof, walls, shafts, columns, and possibly the floor surfacing, ceilings, and partitions. But, it should be clear that they are only *rough* approximations. They can be varied greatly depending upon the designer's skill and his estimate of the materials to be used for the basic structural system, the floors themselves, the walls, and other permanent portions.

For example, as shown in Figure 5-1, in steel buildings the weight of the

FIGURE 5-1
STRUCTURAL WEIGHT AND FORM ARE RELATED.

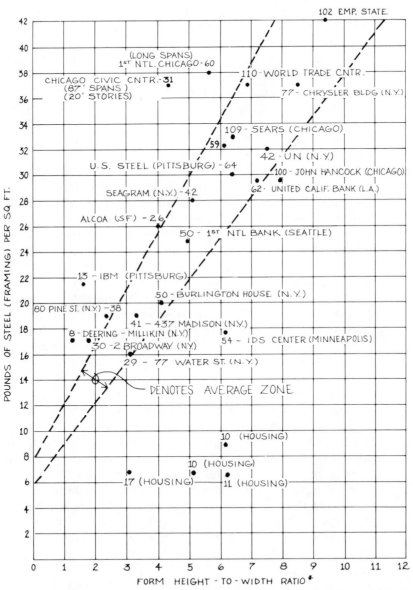

* THE NUMBER BESIDE EACH ENTRY INDICATES BUILDING HEIGHT IN STORIES.

structural framing members may be only a small part of the total dead load and will vary from as low as 5–10 lb psf to 40 lb or more for high-rise buildings with long spans. Therefore, the rough starting values suggested above should eventually be verified or modified when a preliminary design is being refined.

However, despite the roughness, the ability to assume a reasonable, if approximate, dead load for a building structure is useful in schematic design of major subsystems. Provision for the overall interaction of walls, shafts, and columns as major structural subsystems can be facilitated by approximation of their dead-load values to provide a basis for initiating and quickly comparing alternate overall structural schemes. Such overall estimates can be particularly useful when dealing with the lower stories of a building where the total structure and the foundation interface and the overall layout and component dimensions are critical. As discussed in previous chapters, overall estimates enable the designer to consider the overturning effect of wind or earthquake forces, since the gross weight of a building will have to act to help stabilize the building. Approximate knowledge of the distribution of the dead-load values can also be used to indicate whether the structural proportioning of the building is tending to be effective insofar as lateral force resistance is concerned.

When one is analyzing the dead load acting on a *specific floor* or *roof*, loading assumptions based on an overall average of the dead load may not be accurate enough. But greater accuracy calls for an estimate of a particular type of floor structural subsystem to get a start. As will be discussed in Chapter 6, this will involve making preliminary decisions about the design of particular floor or roof framing and covering systems in order to compute their actual dead load. Assumptions will have to be made regarding the likely thickness of materials involved to obtain a reasonable picture of the load condition. Then the unit weight of the material can be obtained from handbooks or tables like those shown in Tables 5-1a and 5-1b and used to estimate the actual dead load of the subsystem itself.

Walls and partitions can be expressed in psf of their own projected area and applied to the floor along their plan lines. However, when walls are more-or-less distributed, it is more convenient to reduce such dead load to pounds psf distributed over the entire floor area. This means that one will need to have an idea of the ratio of the wall or partition area per floor compared to the floor area. For example, if there are 3000 sq ft of wall surfaces per floor, weighing 60 lb psf, and 6000 sq ft of floor area per floor, the ratio is one wall unit to two floor units, and it can be estimated that the dead load of the walls will be equivalent to 30 lb psf of floor area. This value must be added to the weight of the floor subsystem. Ceilings hung below floors as well as floor coverings will also add to the load on a specific floor subsystem.

Table 5-1a

Approximate Weights of Building Materials (lb/sq ft)

MATERIALS	WEIGHT LB/SQ FT	MATERIALS	WEIGHT LB/SQ FT
Ceilings		*Partitions* (cont.)	
Channel suspended system	1	4 in.	18
Lathing and plastering	See Partitions	6 in.	28
Acoustical fiber tile	1	8 in.	34
		10 in.	40
Floors		Gypsum block	
Steel deck	2–10	2 in.	$9\frac{1}{2}$
Concrete-reinforced 1 in.		3 in.	$10\frac{1}{2}$
Stone	$12\frac{1}{2}$	4 in.	$12\frac{1}{2}$
Slag	$11\frac{1}{2}$	5 in.	14
Lightweight	6–10	6 in.	$18\frac{1}{2}$
Concrete-plain 1 in.		Wood studs 2×4	
Stone	12	(12–16 in. o.c.)	2
Slag	11	Steel partitions	4
Lightweight	3–9	Plaster 1 in.	
Fills 1 in.		Cement	10
Gypsum	6	Gypsum	5
Sand	8	Lathing	
Cinders	4	Metal	$\frac{1}{2}$
Finishes		Gypsum board $\frac{1}{2}$ in.	2
Terrazzo 1 in.	13		
Ceramic or quarry tile $\frac{3}{4}$ in.	10	*Walls*	
Linoleum $\frac{1}{4}$ in.	1	Brick	
Mastic $\frac{3}{4}$ in.	9	4 in.	40
Hardwood $\frac{7}{8}$ in.	4	8 in.	80
Softwood $\frac{3}{4}$ in.	$2\frac{1}{2}$	12 in.	120
		Hollow concrete block (heavy aggregate)	
Roofs		4 in.	30
Copper or tin	1–5	6 in.	43
Corrugated steel	1–5	8 in.	55
3 ply ready roofing	1	12 in.	80
3 ply felt and gravel	$5\frac{1}{2}$	Hollow concrete block (light aggregate)	
5 ply felt and gravel	6	4 in.	21
Shingles		6 in.	30
Wood	2	8 in.	38
Asphalt	3	12 in.	55
Clay tile	9–14	Clay tile (load bearing)	
Slate $\frac{1}{4}$	10	4 in.	25
Sheathing		6 in.	30
Wood $\frac{3}{4}$ in.	3	8 in.	33
Gypsum 1 in.	4	12 in.	45
Insulation 1 in.		Stone 4 in.	55
Loose	$\frac{1}{2}$	Glass block 4 in.	18
Poured in place	2	Windows (glass, frame and sash)	8
Rigid	$1\frac{1}{2}$	Structural glass 1 in.	15
		Corrugated cement asbestos	
Partitions		$\frac{1}{4}$ in.	3
Clay tile			
3 in.	17		

Table 5-1b
Approximate Weights of Materials (lb/cu ft)

Masonry		*Water, Snow*	
Marble, granite	140–165	Water	62.5
Brick masonry	100–150	Snow, fresh	5
Concrete, normal	150	Snow, packed	10 and up
Concrete, lightweight	90–120	Snow, wet	40–50
Metals		*Miscellaneous*	
Steel	480	Sand	100–120
Aluminum	165	Glass	160
Brass	530	Asphalt	80–100
		Mortar	100
	Timber		
Redwood	26	Pine	35–40
Douglas fir	32	Oak	54

SECTION 3: Live Loads

The actual live load on the floor of a building varies greatly. It is possible to concentrate a heavy loading such as a safe box or moving equipment over a rather small area (2–6 sq ft) amounting to, let us say, 500 or 1000 psf of that small area. Then it is necessary to design that small area for a heavy concentration. On the other hand, when we are talking of a large tributary area (over 100 or 150 sq ft) supported by a primary structural component, the significance of that concentration as compared with the overall load will be reduced correspondingly.

Since the actual live load over the floor area of a building can vary greatly with time of day and occupancy type, it becomes quite difficult to determine a precise live load in pounds per square foot to a specific floor design. Nevertheless, an average design load value can be assigned when the actual or probable type of building occupancy is known. This would be classified as the basic live load for application when considering the larger tributary areas.

When smaller areas (less than 30 to 50 sq ft) are considered, the effect of a concentrated live load should be considered as a special case and may be expressed as a single concentrated load or as a uniform load in psf, but of a magnitude much higher than the specified basic live load. Such concentrations are often stated or suggested in building code specifications. For example, Table 5-2 [from Chapter 23 of the *Uniform Building Code UBC*] gives the uniform load as a basic value to use for floor design; but at the same time it also gives concentrated values of live load to be applied in case of smaller areas.

Although the *basic* live load does not apply to smaller areas, it also will not be accurate to apply it without some modification to larger tributary areas. Consider that, in most buildings, it would be unusual to load every square foot of a large tributary area completely, as could be the case for a limited area. Therefore, the size of a tributary area determines the probability of loading every square foot to basic live load levels. To simulate this effect, the *UBC* specifies that the basic live load can be reduced when considering large tributary areas (over 150 sq ft). The usual rules for such reduction of basic uniform live load for large tributary areas are noted in the following excerpt from the *UBC* ("Reduction of Live Loads," Section 2306, p. 123):

Except for places of public assembly, and except for live loads greater than 100 pounds per square foot, the design live load on any member supporting 150 square feet or more may be reduced at the rate of 0.08 percent per square foot of area supported by the member. The reduction shall not exceed 40 percent for horizontal members, 60 percent for vertical members, nor (R) as determined by the following formula:

$$R = 23.1(1 + D/L)$$

WHERE: R = *Reduction in percent*
          D = *Dead load per square foot of area supported by the member*
          L = *Unit live load per square foot of area supported by the member.*

For storage live loads exceeding 100 pounds per square foot, no reduction shall be made except that design live loads on columns may be reduced 20 percent.

The live load reduction shall not exceed 40 percent in garages for the storage of private pleasure cars having a capacity of not more than nine passengers per vehicle.

It should be apparent from the above that all the so-called specified live loads are not really very accurate insofar as their representation of the actual conditions at a given time. Furthermore, the reduction of these live loads by a formalized rule-of-thumb process, as specified above, is only a more-or-less reasonable substitute for the lack of precise knowledge and is made for convenience. Such values should not be taken blindly as absolutely correct, even though they are specified in building codes.

As stated in Section 1 of this chapter, the additional impact loads produced by movement of live loads is not often specified as such. It is generally assumed that the usual amount of impact produced by moving loads in buildings is already included in the specified equivalent static load for a particular occupancy type. For example, impact loads from automobiles in a garage are often included in the specified uniform load because that is the predominant occupancy condition. However, this is not true in the case of special-occupancy conditions such as those associated with a moving piece of heavy machinery or one that is fixed but may vibrate on top of a floor or roof. There may also be trucks moving on the floor of a warehouse.

Table 5-2
Uniform and Concentrated Loads

USE OR OCCUPANCY		UNIFORM LOAD[1] (psf)	CONCEN-TRATED LOAD
CATEGORY	DESCRIPTION		
1. Armories		150	0
2. Assembly areas[4] and auditoriums and balconies therewith	Fixed seating areas	50	0
	Movable seating and other areas	100	0
	Stage areas and enclosed platforms	125	0
3. Cornices, marquees and residential balconies		60	0
4. Exit facilities, public[5]		100	0
5. Garages	General storage and/or repair	100	3
	Private pleasure car storage	50	3
6. Hospitals	Wards and rooms	40	1000[2]
7. Libraries	Reading rooms	60	1000[2]
	Stack rooms	125	1500[2]
Manufacturing	Light	75	2000[2]
	Heavy	125	3000[2]
8. Offices		50	2000[2]
9. Printing plants	Press rooms	150	2500[2]
	Composing and linotype rooms	100	2000[2]
10. Residential[6]		40	0
11. Rest rooms[7]			

Table 5-2 (*Contd.*)

USE OR OCCUPANCY		UNIFORM LOAD[1] (psf)	CONCEN-TRATED LOAD
CATEGORY	DESCRIPTION		
12. Reviewing stands, grand stands and bleachers		100	0
13. Schools	Classrooms	40	1000[2]
14. Sidewalks and driveways	Public access	250	[3]
15. Storage	Light	125	
	Heavy	250	
16. Stores	Retail	75	2000[2]
	Wholesale	100	3000[2]

[1]See Section 2306 for live load reductions.
[2]See Section 2304(c), first paragraph, for area of load application.
[3]See Section 2304(c), second paragraph, for concentrated loads.
[4]Assembly areas include such occupancies as dance halls, drill rooms, gymnasiums, playgrounds, plazas, terraces and similar occupancies which are generally accessible to the public.
[5]Exit facilities include such uses as corridors and exterior exit balconies, stairways, fire escapes and similar uses.
[6]Residential occupancies include private dwellings, apartments, and hotel guest rooms.
[7]Rest room loads shall be not less than the load for the occupancy with which they are associated but need not exceed 50 pounds per square foot.
Source. Table 23-A. Reproduced from the 1976 edition of the Uniform Building Code, copyright © 1976, with permission of the publisher, the International Conference of Building Officials.

In the case of bridges or structures sustaining moving vehicles of various types, there are so-called impact formulas that would increase the static live load by a certain percentage, to take into account the dynamic effect.

In general, live loads on top of a roof are assumed to be much less than the live load on a floor. Except in special cases, where the roof is subjected to heavy equipment loading or car-parking, the live load is probably created by only a limited number of workmen or possibly some light construction equipment that may be used for maintenance on top of the roof. The *UBC* lists such requirements as shown in Table 5-3. Of course, these provisions do not include such items as heavy machinery, which require special considerations.

In addition to people and equipment, roofs must carry environmental

Table 5-3

Minimum Roof Live Loads

ROOF SLOPE	METHOD 1				METHOD 2	
	TRIBUTARY LOADED AREA IN SQUARE FEET FOR ANY STRUCTURAL MEMBER			UNIFORM LOAD[2] (psf)	RATE OF REDUC-TION r (Percent)	MAXIMUM REDUC-TION R (Percent)
	0 to 200	201 to 600	Over 600			
1. Flat or rise less than 4 inches per foot. Arch or dome with rise less than one-eight of span	20	16	12	20	08	40
2. Rise 4 inches per foot to less than 12 inches per foot. Arch or dome with rise one-eighth of span to less than three-eighths of span	16	14	12	16	06	25
3. Rise 12 inches per foot and greater. Arch or dome with rise three-eighths of span or greater	12	12	12	12		
4. Awnings except cloth covered[3]	5	5	5	5	No Reductions Permitted	
5. Greenhouses, lath houses and agricultural buildings	10	10	10	10		

[1]Where snow loads occur, the roof structure shall be designed for such loads as determined by the Building Official. See Section 2305 (d). For special purpose roofs, see Section 2305 (e).
[2]See Section 2306 for live load reductions. The rate of reduction r in Section 2306 Formula (6-1) shall be as indicated in the Table. The maximum reduction R shall not exceed the value indicated in the Table.
[3]As defined in Section 4506.
Source: Table 23-B. Reproduced from the 1976 edition of the Uniform Building Code, copyright © 1976, with permission of the publisher, the International Conference of Building Officials.

loads. For flat roofs, a certain amount of rainwater accumulation may have to be provided for. In other cases, snow load and especially compacted snow and ice buildup may be controlling factors, as shown in the following excerpt from the *UBC* ["Snow Loads," Section 2305(c), p. 122]:

> *Snow loads, full or unbalanced, shall be considered in place of loads set forth in Table No. 23-B where such loading will result in larger members or connections.*

> *Potential accumulation of snow at valleys, parapets, roof structures, and offsets in roofs of uneven configuration shall be considered. Where snow loads occur, the snow loads shall be determined by the Building Official.*

> *Snow loads in excess of 20 pounds per square foot may be reduced for each degree of pitch over 20 degrss by R_s as determined by the following formula:*

$$R_s = S/40 - 1/2$$

> *WHERE:* R_s = *Snow load reduction in pounds per square foot per degree of pitch over 20°.*
> S = *Total snow load in pounds per square foot.*

There is also the problem of wind load in the form of both pressure and suction, as will be discussed in the following section. However, in general, no live load will be considered to act simultaneously with these environmental loads such as wind and earthquake and normal allowable stresses can be increased by $\frac{1}{3}$.

SECTION 4:   Wind Loads

Wind pressure on a building surface depends primarily on its velocity, the slope of the surface, the shape of the surface, the protection from wind offered by other structures and, to a smaller degree, the density of the air, which decreases with altitude and temperature, and the surface texture. All other factors remaining unchanged, the pressure due to wind is proportionate to the square of the velocity and the density of the air:

$$P = C_D \times Q = C_D(\tfrac{1}{2}V^2 D).$$

Where P is the pressure on a surface in psf, V is the velocity of wind in feet per second, and D is the density of air in flux per cubic foot, C_D is a numerical shape coefficient (called drag coefficient) and Q is the dynamic pressure of moving air, equal to $(\tfrac{1}{2}V^2 D)$. At sea level, D is 0.002378 slugs per cubic foot, and for wind velocity V expressed in miles per hour, $Q = 0.00256\ V^2$ psf.

During storms, wind velocities may reach values up to or greater than 150 miles per hour, which corresponds to dynamic pressure Q of about 60 pounds psf. Pressure as high as this is exceptional, and in general, values of Q from 20 to 30 pounds psf are commonly used for wind loads on

FIGURE 5-2
RESULTANT WIND PRESSURE CRITERIA FOR THE UNITED STATES (FROM UNIFORM BUILDING CODE, 1976 SECTION 23, FIGURE 4).

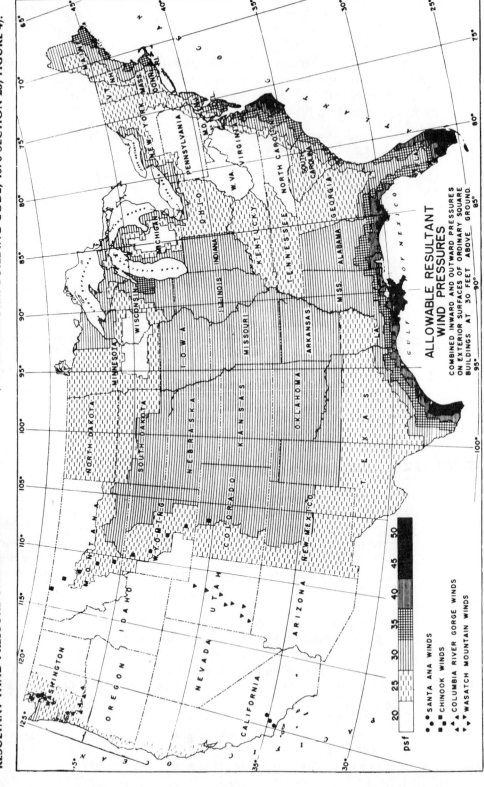

ALLOWABLE RESULTANT
WIND PRESSURES

COMBINED INWARD AND OUTWARD PRESSURES
ON EXTERIOR SURFACES OF ORDINARY SQUARE
BUILDINGS AT 30 FEET ABOVE GROUND.

● ● SANTA ANA WINDS
■ ■ CHINOOK WINDS
▲ ▲ COLUMBIA RIVER GORGE WINDS
▼ ▼ WASATCH MOUNTAIN WINDS

psf   20 25 30 35 40 45 50

buildings. The *UBC* gives a table of basic wind pressures on ordinary rectangular buildings for various areas of the United States, varying from 20 psf to 50 psf (see Figure 5-2 and Table 5-4). Table 5-4 also indicates that wind pressure should be increased for height zones above 50 ft and decreased for height zones less than 30 ft.

The drag coefficient for most rectangular buildings is unity (1.0); that for other structures is expressed in Tables 5-5 and 5-6. They give shape factors to be used for special solid and trussed structures of various configurations whereby the value of C_D varies between a minimum of 0.6 and a maximum of 2.4.

Table 5-4

Wind Pressures for Various Height Zones Above Ground[1]

HEIGHT ZONES (in feet)	WIND-PRESSURE-MAP AREAS						
	20	25	30	35	40	45	50
Less than 30	15	20	25	25	30	35	40
30 to 49	20	25	30	35	40	45	50
50 to 99	25	30	40	45	50	55	60
100 to 499	30	40	45	55	60	70	75
500 to 1199	35	45	55	60	70	80	90
1200 and over	40	50	60	70	80	90	100

[1]See Figure No. 4. Wind pressure column in the table should be selected which is headed by a value corresponding to the minimum permissible, resultant wind pressure indicated for the particular locality.

The figures given are recommended as minimum. These requirements do not provide for tornadoes.

Source. Table 23-F. Reproduced from the 1976 edition of the Uniform Building Code, copyright © 1976, with permission of the publisher, the International Conference of Building Officials.

Table 5-5

Multiplying Factors for Wind Pressures—Chimneys, Tanks, and Solid Towers

HORIZONTAL CROSS SECTION	FACTOR
Square or rectangular	1.00
Hexagonal or octagonal	0.80
Round or elliptical	0.60

Source. Table 23-G. Reproduced from the 1976 edition of the Uniform Building Code, copyright © 1976, with permission of the publisher, the International Conference of Building Officials.

Table 5-6

Shape Factors for Radio Towers and Trussed Towers
UBC, 1976—Table 23-H

TYPE OF EXPOSURE	FACTOR
1. Wind normal to one face of tower	
Four-cornered, flat or angular sections, steel or wood	2.20
Three-cornered, flat or angular sections, steel or wood	2.00
2. Wind on corner, four-cornered tower, flat or angular sections	2.40
3. Wind parallel to one face of three-cornered tower, flat or angular sections	1.50
4. Factors for towers with cylindrical elements are approximately two-thirds of those for similar towers with flat or angular sections	
5. Wind on individual members	
Cylindrical members	
Two inches or less in diameter	1.00
Over two inches in diameter	0.80
Flat or angular sections	1.30

Source. Table 23-H. Reproduced from the 1976 edition of the Uniform Building Code, copyright © 1976, with permission of the publisher, the International Conference of Building Officials.

Suction due to wind is seldom specified in building codes, but it is well known that a suction of *at least* 10 psf should be considered. For areas subjected to higher wind pressure, say from 30 to 50 psf, a suction effect of about one-half the pressure is usually considered. A more accurate recommendation on wind pressure and suction on sloping surfaces is given by the American Society of Civil Engineers ASCE, Figure 5-3.

An ASCE committee on wind forces made a comprehensive study and published a final report in the ASCE *Transactions.*[1] In the report it was recommended that, for tall buildings, the wind load should be 20 psf up to 300 ft height; for the portion above that limit, an increase of 2.5 psf should be made for every 100 ft increase in height. It further recommends that roofs and walls of buildings be designed for varying pressures, positive and negative, depending on their slope. If α is the slope of the roof to the horizontal, in degrees, the recommended wind forces on the windward

[1]"Wind Bracing in Steel Buildings." Final report of Subcommittee 31, Committee on Steel of Structural Division, *Transactions ASCE* 105, P. 1713 (1940).

FIGURE 5-3
WIND PRESSURES ON ROOFS & WALLS

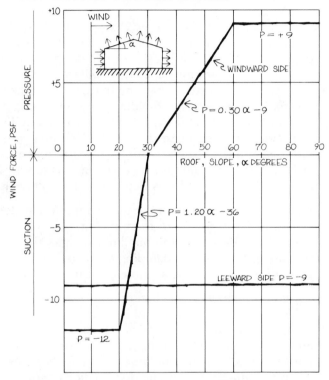

Source. (From *Design of Steel Structures*, Bresler, Lin, & Scalzi;
John Wiley & Sons, Inc., 2nd Ed, New York 1968)

slope are as shown in Figure 5-3. On the leeward slope, a suction of 9 psf is recommended for all values of α in excess of zero. These recommendations are based on an assumed wind velocity of about 78 mph, with due allowance for drag and suction effects.

SECTION 5: Earthquake Loads[2]

Earthquake loads are specified with two basic objectives in mind. One is to protect the public from loss of life and serious injury and to prevent buildings from collapse and dangerous damage under a maximum-intensity earthquake. The other is to insure buildings against any but very minor

[2]Text for this section is derived from *Design of Steel Structures*, Section 12-5, by Bresler, Lin, and Scalzi, John Wiley & Sons, Inc., Second Edition New York, 1968.

damages under moderate to heavy earthquakes. Equivalent static loads are specified so that these two objectives can be attained within reason and without excessive cost.

Earthquake resistance calls for energy absorption (or ductility) rather than strength only. If a building is able to deflect horizontally by several times the amount envisioned under the basic seismic design load and still maintain its vertical load-carrying capacity, it will be able to absorb earthquakes considerably heavier than the design earthquake. If such ductility is present, collapse of the building can be prevented even if the building is seriously damaged. Thus, in addition to seismic load design, the ductility and plasticity of a building should be given due consideration.

The seismic loads on the structure during an earthquake are due to internal inertia that results from ground accelerations to which the mass of the system is subjected. Actual loads depend on the following factors:

1. The intensity and character of the ground motion as determined at the source and its transmission to the building.
2. The dynamic properties of the building, such as its mode shapes and periods of vibration and its damping characteristics.
3. The mass of the building as a whole or of its components.

Great progress in earthquake engineering throws considerable light on the earthquake effects on buildings and is reflected in the seismic design codes. However, numerous uncertainties still exist. Among these are the probable intensity and character of the maximum design earthquake, the damping characteristics of actual buildings, the effects of the soil-structure interaction, and the effects of inelastic deformations. However, discussion of more than the fundamentals of earthquake engineering and of their relationship to practical design is beyond the scope of this book. Here, we will simply lay out the more basic design factors.

For convenience in design, an earthquake is translated into an equivalent static load acting horizontally on the building. Although it is not possible to predict the maximum earthquakes at a location, history and experience together with geological observations have shown that the maximum probable earthquakes do vary with different areas, and different seismic design loads can be specified. Thus the Continental United States has been divided into four zones of approximately equal seismic probability (Figure 5-4). Zone 3 is the heavy-earthquake zone, Zone 2 the moderate-earthquake zone, Zone 1 the light-earthquake zone, and Zone 0 has practically no earthquakes expected, see pages 146–147.

These concepts are incorporated in Section 23 of the 1976 *Uniform Building Code* to which the reader is referred for details. Briefly summarized, the *Code* specifies a minimum lateral seismic force for building design as follows:

$$V = ZKCW$$

where $Z = 1$ for Zone 3, $\frac{1}{2}$ for Zone 2, and $\frac{1}{4}$ for Zone 1. K has a basic value of unity, but may vary from 0.67 for ductile buildings to 3.00 for elevated tanks. $C = 0.10$ for one-story and two-story buildings and can be determined from the following formula for other buildings:

$$C = \frac{0.05}{\sqrt[3]{T}}$$

where T is the fundamental period of vibration of the structure in seconds in the direction considered and frequently may be computed from the following empirical formula:

$$T = \frac{0.05 h_n}{\sqrt{D}}$$

where D is the dimension of the building in feet in a direction parallel to the applied forces and h_n is the height in feet above the base to the uppermost level. In buildings with complete moment-resisting frames, $T = 0.10N$, where $N = $ total number of stories above exterior grade.

To account for the dynamic nature of earthquake forces, up to 15 percent of the total force V as computed from $V = ZKCW$ is considered as concentrated at the top story. The remainder is distributed throughout the height of the building in proportion to the product of the weight at each story times its height above the base (e.g., as in Figure 2-9). The force concentrated at the top (F_t) is given by:

$$F_t = 0.004 V (h_n/D_s)^2 \leqslant 0.15 V$$

where D_s is the plan dimension in feet of the lateral force resisting system.

For parts or portions of buildings, the seismic force is given by a slightly different formula:

$$F_p = Z C_p W_p$$

where W_p is the weight of the part or portion of a structure and C_p varies from 0.10 for roofs and floors acting as diaphragms to 1.00 for cantilever parapets, ornamentations, and appendages, and up to 2.00 for attached exterior elements.

Although these provisions are largely empirical, they do represent an acceptable method for computing seismic loads. Corresponding to these loads, the normal allowable stresses can be increased by one-third (as for wind load), and under the action of these seismic loads the building structure still may not reach yielding. Should the seismic force greatly exceed these design loads, the plastic behavior, or ductility, of the structure will be called into action, thus providing reserve energy absorption in preventing collapse or catastrophic damage. The implications of plastic behavior will be discussed further in Section 7.

FIGURE 5-4
THE CONTINENTAL UNITED STATES, ALASKA, AND HAWAII AS ZONED FOR SEISMIC RISK BY THE *UNIFORM BUILDING CODE*

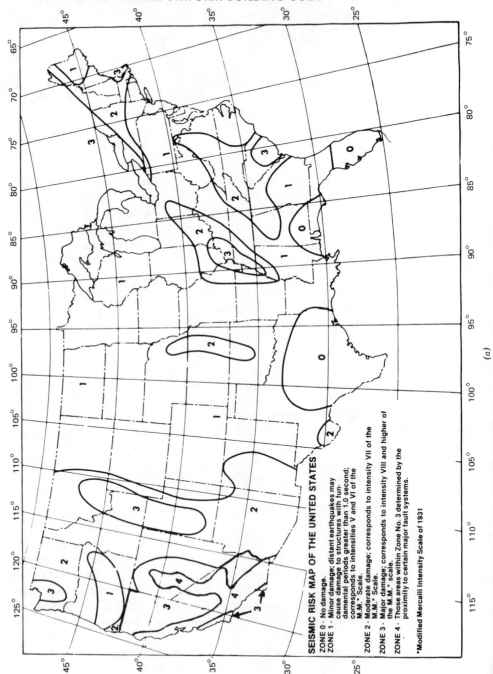

SEISMIC RISK MAP OF THE UNITED STATES

ZONE 0 - No damage.
ZONE 1 - Minor damage; distant earthquakes may cause damage to structures with fundamental periods greater than 1.0 second; corresponds to intensities V and VI of the M.M.* Scale.
ZONE 2 - Moderate damage; corresponds to intensity VII of the M.M.* Scale.
ZONE 3 - Major damage; corresponds to intensity VIII and higher of the M.M.* scale.
ZONE 4 - Those areas within Zone No. 3 determined by the proximity to certain major fault systems.

*Modified Mercalli Intensity Scale of 1931

(a)

Figure 5-4 (*Contd.*)

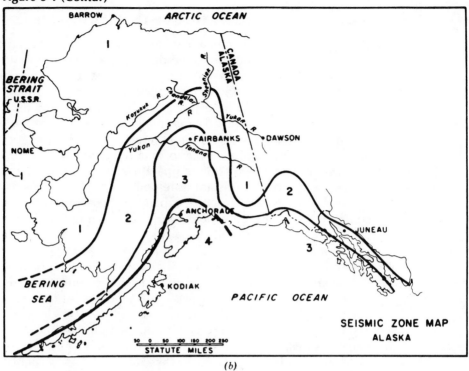

(b)

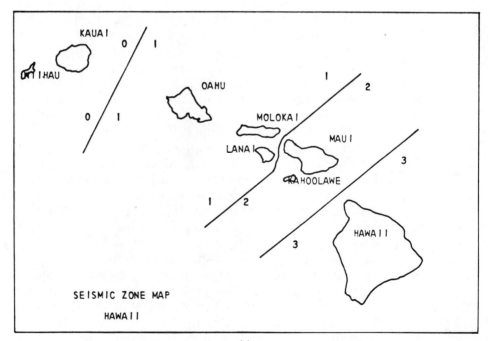

(c)

SECTION 6:   Internal and External Movements in Structures

Internal movements (strains) in a structure can be produced as a result of differential movement due to temperature variation across the structure. If a building is entirely free to expand and contract under temperature changes, then there may be no internal stresses produced. Unfortunately, different parts of a building will be exposed to, and will respond differently to, environmental conditions. For example, the portions of the building above ground will change with the temperature, while that portion beneath the ground will not. Also, the exterior of a building will tend to change its dimensions with the outside temperature, while the inside of a building may remain more-or-less constant in temperature. This means that when the exterior surfaces of a structure expand and contract with temperature, strains and stresses are produced in that building. Similarly, the sunny side will expand relative to the shady side.

To minimize such stresses and strains, provisions should be made to either facilitate differential movement of the building or to strengthen and tie together its various parts to resist such movement and resulting stresses. A common approach is to provide so-called expansion and construction joints, particularly along the roof lines and the outside walls of a building. Such provisions are often unsightly and expensive, but they can be a necessary evil. Empirical rules for the provision of these joints have been provided in certain publications; see, for example, "Expansion Joints in Buildings," National Research Council, National Academy of Sciences, Technical Report No. 65, 1974. Generally they must be provided for buildings such that each segment does not exceed 200 ft in length, but sometimes 300 ft or more is permissible, whereas at other times expansion joints are provided every 150 ft or less.

For tall buildings in cold climates, say, exceeding 30 or 40 stories, it may be necessary to stiffen the outside walls against the inside walls and shafts in order to avoid overshortening of the facade, which may result in a noticeably sloping floor on the higher stories. This can be achieved by designing heavy frames across the top of a building, or by using special cladding for skin of the building, which minimizes the temperature differential between the outside and inside structural components.

Apart from temperature changes, certain materials such as concrete, and particularly prestressed concrete, tend to shrink and/or creep under load as time goes on, thereby producing differential strains of one floor versus the other. This is not significant in the upper levels of a multistory building, since the floors will tend to shorten more-or-less equally. But often there is a problem between the second floor and the ground floor. A shortening of the floors above ground will tend to pull the vertical elements such as walls and columns inward toward the center of the building, while the ground floor may resist such movement. This can create heavy stresses in various parts of the structure. To minimize such movements and their effects, one may try to control the construction sequence so as to reduce the amount of shortening and shrinkage. Also, one may provide slender elements above the ground floor that are more free to bend laterally. Viewed from this perspective, stiff walls and columns should be located nearer the center of the building rather than at its corners. Along the outside of the building, these vertical elements should be slender in the direction of the movement.

Forces may also be created by unequal settling of the foundations. Generally speaking, a uniform settling will not create serious forces within a structure. On the other hand, if one part of the foundation settles more than the other part, a relative vertical movement will be produced among the floor systems of the building, which will thereby be subjected to undesirable stresses and strains. Chapter 12 will treat the issues of foundation design as a separate subject.

SECTION 7: **Response of Structures**

Under the action of the various forces and loadings described in the previous sections, the structure must be able to respond with proper behavior and prescribed stability. This can perhaps be best described by the load-behavior history of a structure (Figure 5-5).

As various loads are applied to a structure, it deflects both vertically and horizontally. In Figure 5-5, the vertical axis represents the increase in loading during various stages, and the horizontal axis measures the deflection, which is one measure of the response of the structure to the loadings.

Under the application of dead load only, the structure usually has very little deflection, if any, in the lateral direction; but various portions of it will have a certain amount of vertical deflection. For example, the floors will deflect; the walls, columns, and shafts will shorten somewhat. Generally under the dead load all portions of the building will be stressed only to a limited amount, and deflect relatively little.

When live load is added to the building, more deflection and higher stresses are produced locally. Although live load is normally only a fraction of the dead load and should not produce any serious additional movement, it may cause undesirable deflections and vibrations.

As far as the total structure is concerned, the horizontal effect of wind or earthquake can be quite serious in contrast to the dead and live load situations. When either wind or earthquake load is applied to a building, there will be appreciable lateral deflection of the structure as a whole. Consequently, higher forces and stresses will be produced in various components of the structure. Under such conditions, the deflection and vibration, as well as stresses, should be within given limits, although these

FIGURE 5-5
LIFE HISTORY OF A STRUCTURE.

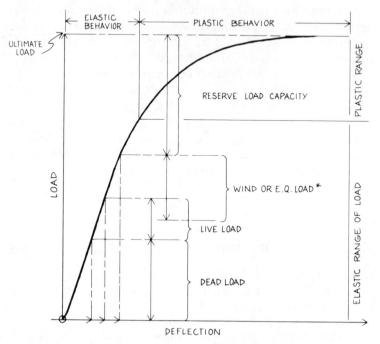

* ONLY PARTIAL OR ZERO LIVE LOAD CONSIDERED TOGETHER WITH WIND OR E.Q. LOAD

limits may be higher than those used under gravity load alone. Almost all building codes specify that a one-third increase in allowable stresses is permitted when considering wind or earthquake loads, since these loads occur rather infrequently.

It is not necessary to consider a situation where wind load and earthquake load act at the same time, since the probability of this occurring is quite low. There is no record in history of an extreme wind and a catastrophic earthquake affecting a building at the same time. Furthermore, full live load is generally not considered when either the design wind or the earthquake forces act.

It is also noted that there is a reserve load capacity above and beyond the specified combination of dead load, live load, and wind or earthquake load. This reserve load capacity is necessary in order to resist unexpected high wind, hurricane, or catastrophic earthquake loads. This may be termed the "margin of safety" provided in buildings.

The reserve load capacity not only supplies the additional margin of safety to take care of catastrophic forces, but it also keeps the behavior of the structure within tolerable limits of movement and strain under the normally expected high wind or earthquake condition. These limits are usually proscribed by the so-called elastic behavior range of a material. Therefore, it is expected that under the action of wind or ordinary earthquake load, of course in combination with dead load and some live load, the building structure will still behave within the elastic range. Hence the vertical and horizontal deflections of the building may not be excessive and can be predicted by the usual elastic or linear behavior.

In the case of catastrophic earthquake resistance, a different situation exists. As explained previously, the code-specified earthquake forces represent only the action of a moderately heavy earthquake under which we would expect to maintain elastic behavior of the structure. But a catastrophic earthquake can produce forces or movements several times, say three or four, that of the prescribed code earthquake. Although it may be advocated that we should design for this catastrophic earthquake, with the entire structure acting within the elastic limit, it is known that this will require undue expenses and cost increase of a building structure. Most of the time the owner of the building will not be willing to pay such a high premium to insure that a building will behave within the elastic limit under catastrophic earthquakes. Therefore, it is generally agreed that under catastrophic earthquakes the building will be permitted to extend into the plastic range so that certain portions of the building will suffer minor damages, provided the stability of the structure as a whole is still assured. This practice brings the behavior of the structure occasionally into the plastic range (Figure 5-5). But, fortunately, the peak force produced by earthquakes is of short duration, and therefore can be more easily absorbed by the movement of the building than a sustained static load.

You can see from Figure 5-5 that the design of a structure must take into account the various stages of loading conditions, so that under each stage a different behavior of the building is permissible. In addition to this regular life history of a structure, special conditions should be considered. Certain parts of a building may be subjected to repeated loads, for example under the action of a moving truck, wind agitation, or earthquake vibration. Such repeated loads may produce fatigue failures not obtained in a single cycle of loading. Certain portions of the building may be under sustained loads such as heavy dead loads and other storage loads, which may produce creep strains in portions of the structure, thus resulting in excessive or undesirable movements.

Another environmental effect is temperature change, as mentioned in Section 4 above. Repeated temperature changes of large magnitude may also result in fatigue failures of portions of the building, and they should also be taken into account.

It is desirable that the life history of a building structure approximate the curve shown on Figure 5-5. It should start off with a linear elastic response up to a point beyond the normally expected load combinations. At the same time, it should possess sufficient ductility to absorb energy under catastrophic earthquakes and still insure no collapse of the structure.

FIGURE 5-6
THE LIFE HISTORY CURVE IS OFFSET BY PRESTRESSING TO BALANCE DEAD LOAD FLEXURAL DEFLECTION.

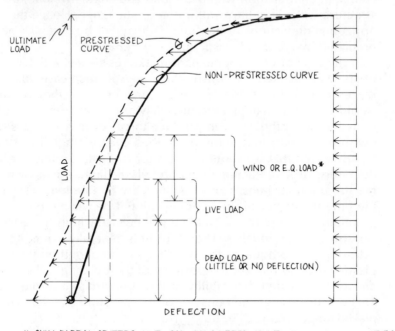

* ONLY PARTIAL OR ZERO LIVE LOAD CONSIDERED TOGETHER WITH WIND OR E.Q. LOAD

Figure 5-6 illustrates the effect of prestressing on the life history curve of a flexural member. Note that there is little or no deflection under dead-load conditions. But if dead load did not exist, or were greatly reduced, there could be undesirable camber (upward deflection).

SECTION 8: Building Codes, Structural Behavior and Strength

It is normal practice to design buildings according to building code requirements. Codes set up minimum requirements and serve as a rough guide for design. However, a word of caution must be mentioned with respect to the structural design regulations usually provided in building codes. Most codes specify a set of minimum loadings to be considered, and a set of maximum stresses not to be exceeded. As is evidenced from previous sections, the specified loadings are at best a set of empirical approximations. What is worse is the fact that there are no absolute rules, formulas, or methods for the computation of the so-called stresses. Building codes usually state that "accepted" methods of calculation must be used. However, there is no definition of what makes a method acceptable. Methods of calculation differ from time to time, from place to place, and from person to person. One can open up a book or paper, derive a new method of calculation, or make different assumptions, and consequently reach a different set of computed stresses, even though the loading conditions remain the same. This makes the usual code-specified allowable stresses rather crude as a measure of overall design effectiveness.

It is possible for any engineer to take any building, designed by some other engineer, and to "prove" by some calculation method that many parts of that structure are "overstressed," and therefore not in accordance with code requirements. Frank Lloyd Wright ran into the problem with his mushroom column design for the Johnson Wax Building (see Figure 1-11). He had to build and prove by testing that his design was better than a "code" computation would indicate. There are other examples of designs for bridges and buildings in which the proposed structures were considered more likely to collapse by comparison to "accepted" or code method designs, but which, in fact, out-performed "acceptable" designs. Conversely, one can find code-designed buildings that do not perform well. Nonetheless, it must be acknowledged that the great majority of code-designed buildings generally behave quite well.

Modern building codes have changed the "allowable stress" approach to a "load factor" method, known as "ultimate strength" design. Thus a code-specified load has to be multiplied by a factor to be equated to the ultimate strength of the structure. For example, a load factor of 1.7 would indicate an (ultimate load) margin of safety of 70 percent against failure. To provide an additional safety margin it is further stated that the computed ultimate

Table 5-7

Maximum Allowable Deflection for Structural Members[1]

TYPE OF MEMBER	MEMBER LOADED WITH LIVE LOAD ONLY (L.L.)	MEMBER LOADED WITH LIVE LOAD PLUS DEAD LOAD (L.L. + K D.L.)
Roof Member Supporting Plaster or Floor Member	$L/360$	$L/240$

[1]Sufficient slope or camber shall be provided for flat roofs in accordance with Section 2305 (f).
L.L. = Live load
D.L. = Dead load
K   = Factor as determined by Table No. 23-E
L   = Length of member in same units as deflection
Source. Table 23-D. Reproduced from the 1976 edition of the Uniform Building Code, copyright © 1976, with permission of the publisher, the International Conference of Building Officials.

FIGURE 5-7
VIBRATION AS A SOURCE OF COMFORT AND DAMAGE

(a) HUMAN SENSITIVITY TO VIBRATION

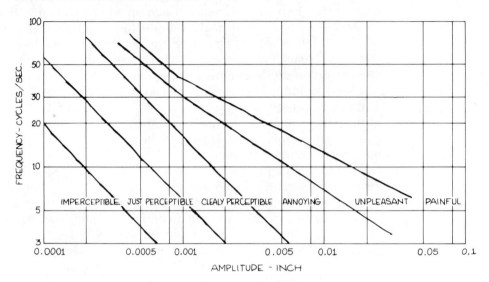

(b) VIBRATION INTENSITY AND POSSIBLE DAMAGE

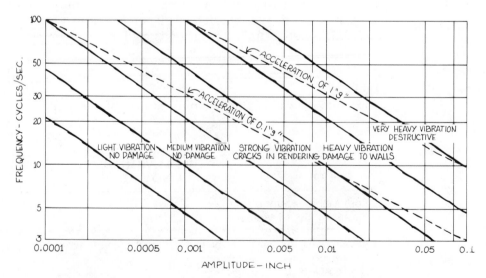

Source. Reproduced from "Vibrations In Building," *Building Research Station Digest*, No. 78, London, England, June 1955. See Also F. K. Chang, "Human Response to motions in Tall Buildings," *Journal of the Structural Division*, ASCE, Vol. 99, No. ST6, June 1973, pp. 1259–1272.

collapse strength of a structure should be reduced by some 10 or 20 percent to take care of other unknown elements, such as variation in material properties, dimensions, workmanship, and computation discrepancies. For component design this is a more realistic approach than "allowable stress." However, the problem still remains that it is quite difficult to predict the actual load capacity of a building that actually acts as a total space-structure. That is, all the various components will work together in a rather complicated manner and not as independent elements designed out of the total system context. Nevertheless, this ultimate strength method is now "accepted" as a more rational approach and will be discussed in Chapters 8 and 9.

Even when using the ultimate strength design method, one must check to be certain that under normal loadings, such as dead load and/or live load (and in combination with normal wind and earthquake load), the structure should behave properly. There should be no excessive deflections, vertical or lateral, and no excessive vibrations, cracks, and other undesirable movements. Such checking is usually accomplished by stress and strain calculations, although not limited by specified stresses (Table 5-7).

Although all buildings vibrate to a certain extent under moving live, wind, or earthquake load, it is a rather difficult problem to determine how much vibration is undesirable. Studies have been made of human sensitivity to vibrations as well as the possibility of structural damage produced by vibrations. These are illustrated in Figure 5-7. Note that the higher the frequency, or the larger the amplitude, the more uncomfortable human beings would feel and the higher are the chances for possible damage. In the actual design of a building structure, one cannot generally afford the time and does not often have the knowledge to decide whether the vibration is objectionable. Hence, the usual approach is to limit the depth-span ratio of structural components in a floor (Appendix B, Table I) and the aspect ratio (height-to-width ratio) of a building or its support structure (Figure 5-1). For ordinary structures, comparison to existing ones that have behaved well would give a fair guidance. For unusual structures, special studies should be made, and these are beyond the scope of this book.

In conclusion, a building designer should use building codes only as a reference and as a guide. They should not be taken as a bible to be followed blindly. Some code requirements are excessively conservative; others are not sufficient (particularly when applied to unusual designs). It is only through a thorough understanding of the strength and behavior of structures as total systems and of the requirement for subsystem and component interaction that one can design a safe and economical structure to fit the various functional requirements and environmental conditions found in modern architecture.

6
Overall Design of Horizontal Subsystems

SECTION 1: Introduction

In preceding chapters we suggested that there is a basic conceptual simplification to be gained in structural analysis by considering a building form as a whole, a total structural system. Here we show how this same approach can be applied to the design of major structural subsystems. This is important because it means that one does not have to learn one set of concepts for design of an overall structural scheme and then a different set for design of its subsystems. Thus, a change in the level of space-form thinking will not change the basic hierarchy of structural design thinking.

For example, the functional needs of habitation generally require that floor and wall surfaces be relatively flat. In order to supply such surfaces, buildings are usually made up of major horizontal or vertical structural subsystems that are also flat. One can visualize the horizontal subsystems as 2-D wholes that act vertically to carry the floor or roof loads in bending, and act horizontally as diaphragms and/or column connectors. Similarly, the vertical subsystems can be visualized as wholes that act to pick up loads from the horizontal subsystems and to resist horizontal forces. Roofs and ceiling subsystems may be either totally flat, as will be discussed here, or curved, as will be discussed separately in Chapter 11.

Horizontal surfaces can be designed as slab, beam, grid, or truss subsystems and can be realized in various materials. But note that the design and construction of horizontal subsystems is related to the arrangement of the supporting vertical subsystems. And these may consist of a fairly regular organization of columns, frames, bearing walls, and/or shafts. Hence, when *actually designing*, both types of subsystems must be considered more-or-less simultaneously. However, for convenience and clarity, we treat horizontal subsystems in this chapter, with only general reference to support conditions, and focus on vertical subsystems in the next chapter.

In terms of the structure alone, it would generally be more economical to space the vertical supports rather close together, say 10 or 15 ft apart, to minimize the horizontal subsystem span. However, in the more comprehensive context of architectural performance needs, longer spans are called for to increase the openness of enclosed space and the flexibility of its use. Therefore, it is often desirable to space the vertical supports much farther apart.

Obviously, the longer the span between vertical supports, the deeper horizontal subsystems must be. As a result, more structural material will be required for longer spans than for shorter spans, even though there is some saving in the number of vertical supports. Thus, a skillful designer will seek to provide the maximum usable space with the least obstruction, and yet to minimize the extra amount of structural material or construction energy needed to achieve openness. In other words, he will have to attempt to optimize the overall design by considering both spatial and engineering performance objectives.

With this goal in mind, the following sections will discuss the basic requirements for efficient design of a variety of floor subsystem types. Flat-roof subsystems can be treated similarly, the difference being that they are usually designed to carry smaller dead and live loads but extra environmental loads, such as wind, rain, and snow. In addition, roofs must sometimes be of much longer span than typical floors, because very large open areas for special activities often take place in a one-story building or are desired on the top floor of a multi-story building. Such long-span roof systems, along with curved subsystems, will be discussed separately in Chapter 11.

Keep in mind that this chapter focuses on only the more basic structural issues of subsystem design, and other physical requirements must be considered to complete the design of horizontal subsystems. Some of them are the following:

1. Support for nonstructural building components such as collection and distribution of services, mechanical equipment, and finish materials (floor and ceiling finish, piping, ducts, wiring, lighting, etc.).
2. Vibration resistance, acoustic transmission and absorption properties.
3. Protection from or resistance to damage caused by fire, exposure to sun, heat, freeze/thaw, and aggressive chemical environment causing corrosion.
4. Ease and accessibility to maintenance and repair.

SECTION 2: Overall Structural Behavior of Horizontal Subsystems

The structural behavior of horizontal subsystems will now be analyzed in terms of overall bending and in terms of the concentration of shearing forces around the supports. Such an overall approach is necessary at schematic and preliminary stages because the actual distribution of local shear and bending conditions across some subsystems can be quite complicated.

Basically, the overall approach begins by assuming even distribution of bending resistance across the subsystem acting as a whole. This assumption is in close correspondence with actual conditions for one-way flat slab or slab-and-beam action. But for two-way action, the local picture for bending and shear resistance, especially around supports, is more complicated.

FIGURE 6-1
A ROUGH APPROXIMATION OF FLAT SLAB BEHAVIOR.

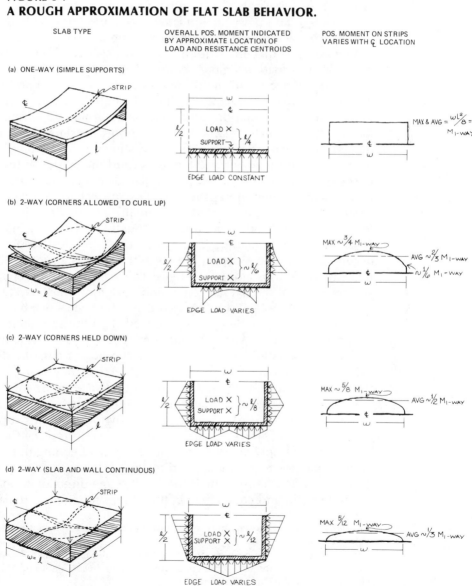

SLAB TYPE

OVERALL POS. MOMENT INDICATED BY APPROXIMATE LOCATION OF LOAD AND RESISTANCE CENTROIDS

POS. MOMENT ON STRIPS VARIES WITH ₵ LOCATION

(a) ONE-WAY (SIMPLE SUPPORTS)

(b) 2-WAY (CORNERS ALLOWED TO CURL UP)

(c) 2-WAY (CORNERS HELD DOWN)

(d) 2-WAY (SLAB AND WALL CONTINUOUS)

However, in practical terms, most two-way subsystems are constructed of constant depth. Thus it will be adequate for schematic and preliminary purposes to estimate the *average* moment for both one- and two-way subsystems. The overall design can then be refined to deal with local variations by varying the amount of material used rather than the overall dimensions.

Figure 6-1 shows a simple *square* plan for one-way and two-way flat slab subsystems, simply supported on walls. In Case *a* the action is one-way and any strip, or the whole slab, can be designed to carry its full tributary load. In Cases *b, c* and *d,* the action is two-way, and the total load will be carried *one-half* in each of two orthogonal directions.[1] The overall design approach simply assumes that any strip of a two-way slab can be designed to carry one-half of its load in simple bending. This will be a close approximation for moments in Case *c* and slightly low or high for Cases *b* and *d* respectively. The accompanying moment diagrams illustrate how the overall (or average) positive moment along the center line is determined by the centroid location of the total of slab load and support resistance. The midspan moment for strips in either direction is shown to vary from the average, being lower at the edges and higher toward the center of the subsystem for Cases *b, c* and *d.* The circles on the slab diagrams indicate that, in actuality, there will be a zone of negative bending, especially in the corner areas.

After the load is transmitted horizontally by shear and bending resistance to the walls, the load goes directly through the walls and into the foundations. Note that the kind of slab behavior illustrated in Figure 6-1 occurs *because* the walls act as very stiff supports against the downward movement of strips along the edges. The walls could be replaced with *very* stiff beams, and the overall action and design of the subsystem would be similar (Figures 6-2*a* and 6-2*b*). However, the beams will have to be designed to pick up the appropriate loads and transmit them into the supporting columns with very little deflection in order to approximate the wall support conditions. In Figure 6-2*a,* the beams must be designed to take at least half the overall moment in the direction they span.[2] In Figure 6-2*b,* the beams must carry the full moment of $wL^2/8$ at midspan.

Now suppose we take away the very stiff beams from the one-way subsystem (Figure 6-3*a*). Then we have only four columns supporting the slab (Figure 6-3*b*). Here the slab itself will have to replace the action of the beams that are now deleted; therefore, overall bending in the slabs will occur equally in two orthogonal directions. Hence the slabs will have to be designed to first transmit their *full* load in one direction as in Figure 6-3*a,* and then again and completely in the other direction, to replace the two

---

[1]As will be discussed subsequently, equal distribution of loads may not be assumed for rectangular plans.

[2]This is true because, for each elevation, the beams and slab will share the moment carrying roll: roughly $\frac{1}{2}$ when the slab is not continuous with the beam, and roughly $\frac{2}{3}$ on the beams when the slab and beams are continuous.

FIGURE 6-2
VERY STIFF BEAMS CAN APPROXIMATE IDEALIZED (WALL) SUPPORT CONDITIONS.

(a)
2–WAY SLAB

(b)
ONE–WAY SLAB

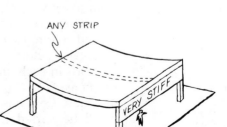

FIGURE 6-3
A COMPARISON OF SLAB AND BEAM WITH SIMPLY SUPPORTED SLAB ACTION.

(a) ONE–WAY SLAB ACTION WHEN BEAMS CARRY
ALL MOMENT IN ONE DIRECTION

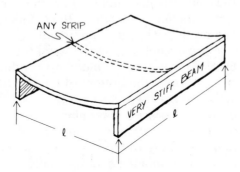

1. SLAB STRIPS CARRY TOTAL LOAD TO BEAMS

2. STIFF BEAMS CARRY TOTAL LOAD TO SUPPORTS

3. OVERALL M_{1-WAY} IS $\frac{w\ell^2}{8}$
   (ASSUMING EVENLY DIST. LOAD)

4. OVERALL SLAB MOMENT IS DISTRIBUTED
   EVENLY FROM EDGE STRIP TO EDGE STRIP:

(b) 2–WAY, SIMPLY SUPPORTED, SLAB ACTION WHEN
BEAMS REMOVED

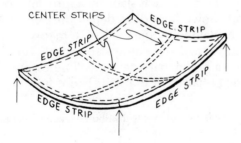

1. SLAB STRIPS IN EACH DIRECTION MUST
   CARRY LOAD

2. OVERALL M_{2-WAY} IS $\frac{w\ell^2}{8}$
   (ASSUMING EVENLY DIST. LOAD)

3. OVERALL MOMENT DISTRIBUTATION IS
   SLIGHTLY LARGER FOR EDGE STRIP
   THAN FOR THE CENTER STRIP

beams, as shown in Figure 6-3*b*. This is a point not often realized by designers, and yet it is always true when taking an overall approach to the preliminary design of simple column-supported flat slabs. The *average* moment on a strip will be equal to that of a one-way slab, $M_{one-way}$. But actually (for a slab of uniform thickness) the moment on each strip varies from roughly $1.1M_{one-way}$ at the slab edge to roughly $0.9M_{one-way}$ at the center.

Now let us consider a large flat slab supported on a *grid* of equally spaced columns (Figure 6-4*a*). The situation is similar to Figure 6-3*b* in that the overall action is two-way, since the entire load is transmitted to the supports in each direction. First the load is transmitted in the direction of Columns *A*, *B*, *C* (Figure 6-4*b*). Then it is also transmitted in the other direction, Columns *E*, *B*, *D* (Figure 6-4*c*). In each direction, the entire load has to be carried in shear and in bending. Figure 6-4*d* shows the overall positive and negative bending behavior of the slab in the direction *D*, *B*, *E* (or *A*, *B*, *C*). Figure 6-4*e* illustrates that each slab bay can be divided (in each direction) into a column strip zone near its edge that carries approximately 75 percent of the total *DL* and a middle strip zone that carries only 25 percent. For approximate design this can be interpreted to mean that the average unit load on a column strip passing over the supports will be approximately $1.5w$, where (*w*) is the unit load for simple one-way action. Similarly, the average unit load in the middle strip will be approximately $0.5w$. Figure 6-4*f* shows a strip over (or near) Column *B*, indicating the moments and forces in the slab. The maximum ($-$) moments on a component strip will be $\frac{2}{3}$ that for simple bending, the ($+$) moment being $\frac{1}{3}$.[3] Thus, if (*w*) is the unit load on the slab, the ($-$) moment in the column strip (Figure 6-4*d*) will average approximately $\frac{2}{3}(1.5wL^2/8)$, and the ($+$) moment approximately $\frac{1}{3}(1.5wL^2/8)$. Note that negative moments produce tension on top of the slab and compression near the bottom of the slab. The middle strip can be treated similarly, using $0.5w$ in place of $1.5w$.

In summary, remember that the precise determination of localized (strip) behavior of the so-called two-way flat slabs is a complicated matter. We have only crudely indicated the nature of the variation, and will not deal with precise determination of local moments in this book.

Fortunately, for preliminary purposes, it is only necessary that a designer assure that the *total* moment resistance satisfies the conditions of statics. Hence, the overall method of design shown here will enable one to make reasonable preliminary designs for such slabs by approximation. Furthermore, the method shown here does depict the behavior of the slabs in the ultimate range as explained by the crack-line theory (i.e., when one or more crack-lines extend across the entire slab width). Therefore, the total strength supplied by such a preliminary design will be essentially correct. Thus, although the elastic behavior of the slabs is not fully described by the

---

[3]This is the convention for an assumption of evenly distributed loading on a fixed end span as will be treated in Chapter 8.

FIGURE 6-4
THE BEHAVIOR OF A FLAT SLAB SUPPORTED ON A SQUARE BAY COLUMN GRID.

(a) SLAB SUPPORTED ON SQUARE
GRID OF COLUMNS

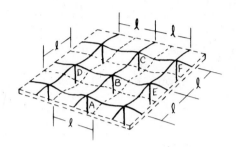

(b) OVERALL SLAB STRIP A B C
(TOTAL DL CARRIED)

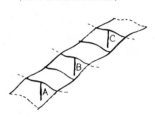

(c) OVERALL SLAB STRIP D B E
(TOTAL DL CARRIED)

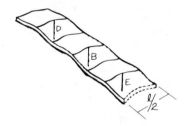

(d) OVERALL SLAB STRIP BENDING

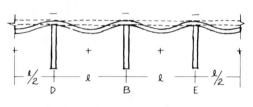

(e) COMPONENT STRIP LOAD DISTRIBUTION FOR A
SINGLE SLAB BAY (COLUMNS AT B, C, E, AND F)

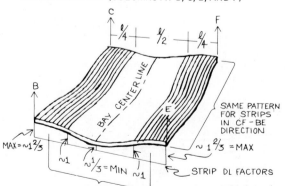

SAME PATTERN
FOR STRIPS
IN CF – BE
DIRECTION

MAX = ~1⅔   ~1   ⅓ = MIN   ~1   ~1⅔ = MAX

STRIP DL FACTORS

1. EDGE STRIPS CARRY ~75% OF TOTAL LOAD ON A BAY
(ie. 1.5 = AVG STRIP DL FACTOR)

2. CENTER STRIPS CARRY ~25% OF TOTAL LOAD
(ie. 0.5 = AVG DL FACTOR)

(f) STRIP MOMENTS IN
EDGE ZONE BC AND EF

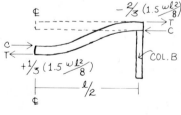

$$-\tfrac{2}{3}\left(1.5\,\tfrac{w l^2}{8}\right)$$

$$+\tfrac{1}{3}\left(1.5\,\tfrac{w l^2}{8}\right)$$

COL. B

method presented and a final design would require a more accurate analysis, the designer should be able to make a reasonable start on an initial layout and preliminary dimensioning of two-way schemes.

When long spans are required in only one direction, a combination slab-and-beam subsystem can be used (Figure 6-5a). Here the slab span is relatively small and will transmit its load to the longer-spanning beams by bending in one direction over and between the beams. For the design of this slab, one can again take any component strip width, say 1 ft, for computation. This will be applicable to the entire width of the slab. One can also take the entire slab, compute the total (−) and (+) moment over and

FIGURE 6-5
ONE-WAY SLAB AND BEAM BEHAVIOR.

(a) ONE-WAY SLAB AND BEAM ACTION

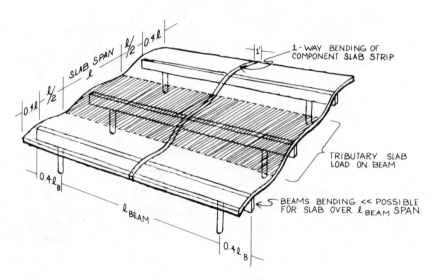

(b) BEAM MOMENTS ASSUMING CONTINUOUS BENDING

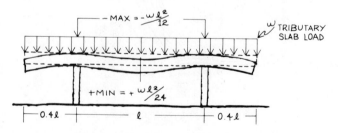

ELEVATION OF BEAM COMPONENT

between the beams, and distribute it evenly along the slab. Obviously, one gets the same answers.

The beams should be designed according to their tributary area loading as shown (Figure 6-5a and 6-5b). For an evenly distributed loading (w) on beams of uniform span length L, the maximum negative moment over a support is given by $wL^2/12$ for continuous spans. The maximum positive moment at midspan of these beams varies between $wL^2/16$ and $wL^2/24$, the total of (+) and (−) values being equal to or somewhat greater than the simple span value of $wL^2/8$. The reason for this will be seen in Chapter 8.[4]

If the beam and slab are connected for composite bending action, they will form a (T) beam that will span between Columns A and B, and should be designed accordingly (as will be discussed in Chapter 8). Here, it is sufficient to recognize that (T) beam action is stronger and stiffer than would be the case for the rectangular beam acting alone.

Figure 6-6 shows a simple *square grid* of beams supporting a flat slab. In this case, each slab bay is really supported equally on all four sides because the beams are assumed of equal length and stiffness. Therefore, it is a relatively stiff two-way slab with the total load and moment in each direction being carried by each beam and tributary slab, the proportion depending on the relative stiffness of the beams to that of the slab.

When a *rectangular* grid of very stiff beams supporting a flat-slab system exceeds the ideal span ratio of 1:1, by more than 1.5:1, the slab action will be basically one-way in the short beam direction. The potential for two-way action in a beam grid flat-slab design is approximated by the ratio L_1^4/L_2^4. Thus, when $L_1/L_2 = 1.5$, $L_1^4/L_2^4 = 5$, and the load carried along the L_1 direction is less than 20 percent.

In designing a more-or-less square system as a whole, the reader should consider bending of the system in two directions. First, as shown in Figure 6-6a and b, each beam will span in one direction, say over columns A, B, and C, essentially as a beam carrying the loads on its tributary area. Then the other beams will span in the other directions, say over columns E, B, and D, and carry the tributary load in that direction. Thus, if the beams are very stiff, and there is (T) beam action, they can be assumed to essentially carry the total moment in either direction (by either independent or T beam action). In this case, the slab is fully supported and can be designed as suggested in Figure 6-1d. However, if the beams are not very much stiffer than the slabs, the latter will carry more of the moment in both directions. In short, depending on the grid layout and the beam stiffness, the load and moment in each direction will be jointly carried by the beam and the slab. The actual behavior can be expected to fall between that of rigid edge support (Figure 6-6) and no edge support (Figure 6-4).

[4]Briefly the larger $+M$ assumes that the $(−M)$ will be less than $(wL^2/12)$.

FIGURE 6-6
TWO-WAY SLAB AND BEAM ACTION.

(a) SIMPLE 2-WAY BEAM GRID SUPPORTING FLAT SLAB

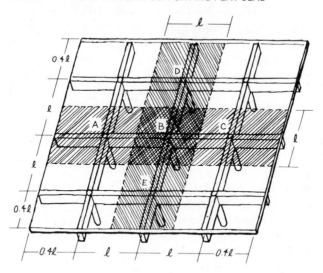

(b) IF BEAMS ARE ASSUMED TO CARRY 2/3 OF TOTAL M
IN EACH DIRECTION, THE REMAINING 1/3 M WILL BE
DISTRIBUTED OVER THE SLAB ACCORDING TO
SOMETHING LIKE THE FOLLOWING PATTERN:

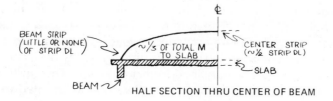

The actual proportion of the total moment carried by the slab as opposed
to the beam in each direction is a complicated matter to determine, and
beyond the scope of this book. But as a rule of thumb, it is often assumed
in preliminary design that the central two thirds of the slab will carry about
one-third of the total moment in each direction. The beams carry about
two-thirds of the total moment also in each direction (Figure 6-6b). The
beams can be designed with or without composite action, but T-beams will
carry even more of the total load and M, and the central slab strip is critical
for *slab* design.

A similar simplified visualization of structural action can be used for
approximate design of a *waffle grid* and slab system (Figure 6-7a). The use

FIGURE 6-7
BEAM GRID OR WAFFLE SLAB ACTION

(a) 2-WAY WAFFLE GRID AND SLAB SCHEME

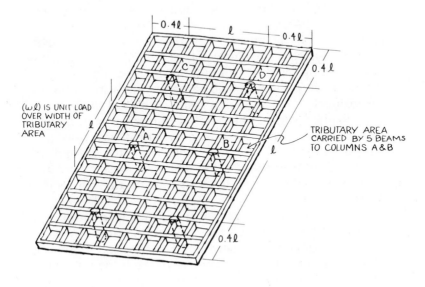

$(\omega\ell)$ IS UNIT LOAD OVER WIDTH OF TRIBUTARY AREA

TRIBUTARY AREA CARRIED BY 5 BEAMS TO COLUMNS A & B

(b) BEAM BENDING IN EACH DIRECTION:

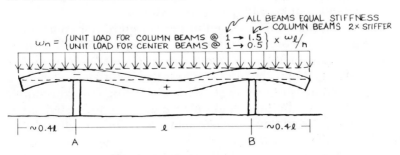

ALL BEAMS EQUAL STIFFNESS
COLUMN BEAMS 2× STIFFER

$\omega_n = \begin{Bmatrix} \text{UNIT LOAD FOR COLUMN BEAMS @ } 1 \rightarrow 1.5 \\ \text{UNIT LOAD FOR CENTER BEAMS @ } 1 \rightarrow 0.5 \end{Bmatrix} \times \omega\ell/n$

$\sim 0.4\ell$ ℓ $\sim 0.4\ell$
A   B

(c) MOMENTS AT CENTER OF BEAMS AND AT COLUMNS:

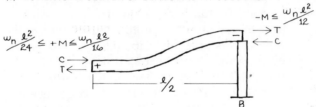

$-M \leq \dfrac{\omega_n \ell^2}{12}$

$\dfrac{\omega_n \ell^2}{24} \leq +M \leq \dfrac{\omega_n \ell^2}{16}$

$\ell/2$

B

of more than two beams in two directions creates a waffle grid system for carrying the slab load between Columns A, B, C, and D. And the waffle system is analogous to a flat slab in behavior. The entire width of a bay (e.g., the tributary area to Columns A and B) actually participates in carrying the load. Further, the beams nearer or over Columns A and B are usually made stiffer because, being closer to the supporting columns, they carry more load than the beams away from the columns (i.e., as in Figure 6-4e).

For final design, an elastic analysis of a grid subsystem (which can be performed by standard computer programs) should be made to determine the actual proportion of load carried by the various beams. But as an approximation, it can be first assumed that all the beams in a grid carry the same load if they are of equivalent stiffness. Or it can be arbitrarily assumed that the beams nearer the columns carry at least twice as much as those away from the columns if the beams along the column strip are only one to two times stiffer than the other beams. If these beams are very much stiffer, they will carry more (say, three quarters of the load as in Figures 6-4e and 6-7b).

In any case, it is important that the total load producing bending along the line AB be resisted by the summation of the strength in all the tributary beams. As long as the total strength is supplied, there will be sufficient ultimate strength to prevent failure of the horizontal system. Thus an approximate design can be made at the beginning. For example, Figures 6-7b and 6-7c isolate portions of the grid and indicate that, essentially, a tension and compression couple exists, both near midspan and around the columns. The grid (and flat slabs as well) must be designed to provide such resisting couples without overstressing.

Another essential point in the design of horizontal systems is the provision of sufficient material, or resisting strength, around the vertical supporting members to avoid failure in shear transfer of loads to the columns. When shearing forces are concentrated over a small area, they are referred to as punching shear. The shearing strength required around the columns can be quite high relative to other subsystem areas and must be supplied by adding sufficient material to increase the strength locally around the columns (Figure 6-8).

A flat slab with no additional material around the columns is subject to serious punching shear forces. This is particularly a problem if it is only a conventionally reinforced slab or slab-beam combination subsystem (Figure 6-8a). On the other hand, a prestressed concrete slab (page 174) would have cables that hang over the top of the columns. These cables will help to carry much of the shear directly to the columns. Furthermore, the compression produced by these cables onto the slabs helps to strengthen the slab against punching shear. The use of posttensioned flat-slabs for a 40-story building (Photo 6-1) reduced 1 ft. of floor-to-floor height for each story, thus lowering the building by 40 ft and resulting in large savings.

Figure 6-8b shows the addition of drop panels or capitals to increase the

FIGURE 6-8
DEALING WITH THE PROBLEM OF PUNCHING SHEAR RESISTANCE.

(a) FLAT SLAB ON COLUMN:

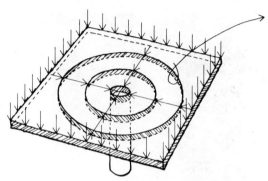

1. AS LOAD IS TRANSMITTED TO COLUMN, THE PUNCHING SHEAR RESISTANCE AREA DECREASES AND FAILURE CAN OCCUR

2. REENFORCING BARS OVER THE COLUMNS CAN HELP

3. PRESTRESSING CABLES DRAPED OVER THE COLUMNS WILL CARRY MUCH OF THE LOAD DIRECTLY TO COLUMNS

(b) LOCAL THICKENING OF SLAB MATERIAL HELPS

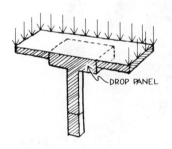

DROP PANEL

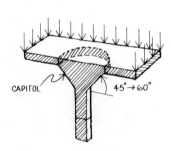

CAPITOL   45° → 60°

(c) BEAMS OVER COLUMNS

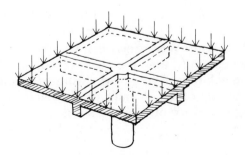

(d) WAFFLE PANEL FILLED IN

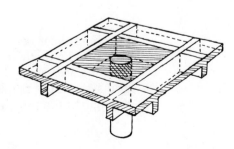

PHOTO 6-1
**UICD BUILDING NEAR COMPLETION, SINGAPORE. (URBAN
PLANNING GROUP, SINGAPORE, ARCHITECTS; T. Y. LIN
SOUTHEAST ASIA, SPECIAL STRUCTURAL CONSULTANTS)**

punching shear resistance. Figure 6-8c shows that when beams are
connected directly to the columns, then the beam section increases the
shearing area around the columns. Figure 6-8d illustrates that for a waffle
grid system, the panels on top of the columns can be filled solid to greatly
strengthen the punching resistance of these systems.

SECTION 3: **Flat-Plate Subsystems**

Now that we have discussed the overall behavior of various types of floor or
roof subsystems, we can look at each in terms of the relationship between
span, depth, and efficiency in the use of materials. For example, a flat-plate
subsystem is theoretically efficient, but a flat plate constructed of plywood
or steel cannot span a large distance between supports without getting too
thick and uneconomical. Plywood is light and strong, but an economical
thickness for plywood, say $\frac{1}{2}-\frac{3}{4}$ in. thick, cannot span more than 3 to 5 ft
because of the excessive deflections. A flat steel plate $\frac{1}{4}$ in. thick is stronger,
but it is more costly and it cannot span more than some 3 ft without getting
into vibration, deflection, and fire damage problems. Thus, steel, and
wood are both good structural materials but they are expensive in a
flat-plate form and designs tend to be of a slab-and-beam, grid,
space-frame, or truss type rather than solid flat slabs. For these reasons, the
relationship between theoretical span-to-depth ratios and practical spans for
various materials and subsystems are summarized in Appendix Table B-1
and will be discussed in this and the remaining sections of this chapter.

FIGURE 6-9
**TYPICAL REQUIREMENTS FOR SCHEMATIC DESIGN OF
CONCRETE AND STEEL FLAT SLABS SUPPORTED ON
COLUMNS.**

(a) CONCRETE AND STEEL PLACEMENT FOR FLAT SLAB ON COLUMNS

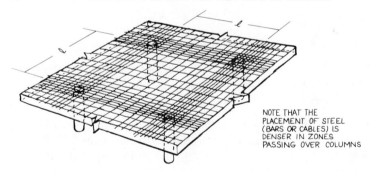

NOTE THAT THE
PLACEMENT OF STEEL
(BARS OR CABLES) IS
DENSER IN ZONES
PASSING OVER COLUMNS

(b) TYPICAL SPAN AND DEPTH FOR PC AND RC SLABS

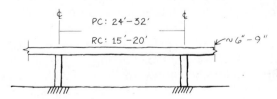

PC: 24′–32′
RC: 15′–20′
~6″–9″

A flat-plate of concrete is a weaker and cheaper material than steel. However, when reinforced or prestressed with a small amount of steel, it can economically span a reasonable distance, say from 15 to 30 ft or more, and supply desirable stiffness, sound insulation, and fire protection properties as well. Hence, efficient solid horizontal flat-plate subsystems can be made using concrete supported directly on columns (Figure 6-9a and 6-9b).

Reinforced concrete is an excellent combination of two materials. However, it represents a *passive* combination because the system must deflect before the reinforcing steel can begin to act (Figure 6-10a). If reinforced concrete construction is used for flat slabs, their span can be up to 15 to 24 ft, but this may still be too close spacing for the supports insofar as the degree of openness and flexibility of spatial use is concerned.

On the other hand, prestressed concrete *actively* combines two materials by prestressing high-strength steel against high-strength concrete. The two materials will now interact to control their stresses, to balance the load, and to reduce deflection (Figure 6-10b). By virtue of their curvature, the cables will carry much of the load directly to the supports. Therefore, much of the punching shear problem can be eliminated, and there will be only residual moment to be carried. As a result, prestressed concrete flat plates can be relatively thin and span up to 24 to 33 ft between columns with a thickness of not over 6 to 9 in.

A typical preliminary design for a post-tensioned prestressed concrete flat slab is shown in Example 6-1.

EXAMPLE 6-1:   Prestressed Concrete Flat Slab

A two-way post-tensioned concrete flat slab is supported on a grid of columns at 30 ft centers in each direction. It is required to make a preliminary design for this floor slab.

Solution: From previous experience, as well as calculations, it is known that the range of depth is 6 to 9 in., for two-way P.C. flat-slab spans 24 to 33 ft. Thus, Appendix Table B-1 indicates that a P.C. slab of 30 ft span should be about 8 in. thick, yielding a span-to-depth ratio of $360/8 = 45$. Since the cables in the slab are draped in two directions, as illustrated in the figure, a common design approach is to supply sufficient horizontal force into these draped cables in each direction so that their upward component will balance the weight of the slab. Assuming a concrete slab 8 in. thick weighs 100 lb psf, then there will be 30 ft width (per bay) $\times$ 100 psf = 3 klf (of span) overall load acting on the slab as a whole. For a parabolic curve to supply an upward force of 3 klf, we use the formula $F = wL^2/8h$, where $w = 3$ klf, $L = 30$ ft, and h is the sag of the cable. For our slab, (h) is 8 in. minus twice the top and bottom concrete protection of $1\frac{1}{2}$ in. each (8 in. $-$ 3 in.) = 5 in., and overall $F = 3 \times 30^2/(8 \times 5/12) = 810$ k. That is

FIGURE 6-10
**PASSIVE VERSUS ACTIVE DESIGN FOR BENDING IN
CONCRETE SUBSYSTEMS.**

(a) RC–(PASSIVE) BENDING REQUIRED TO BRING STEEL REINFORCEMENT INTO PLAY

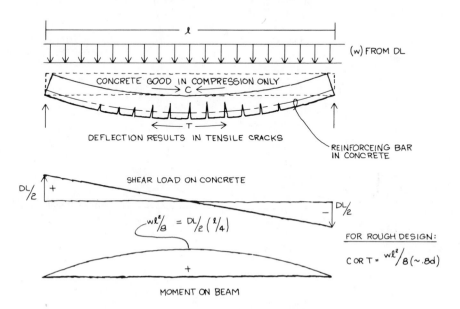

(b) PC–STEEL ACTIVATED BY PRESTRESSING

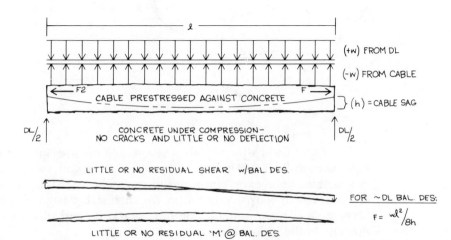

EXAMPLE 6-1

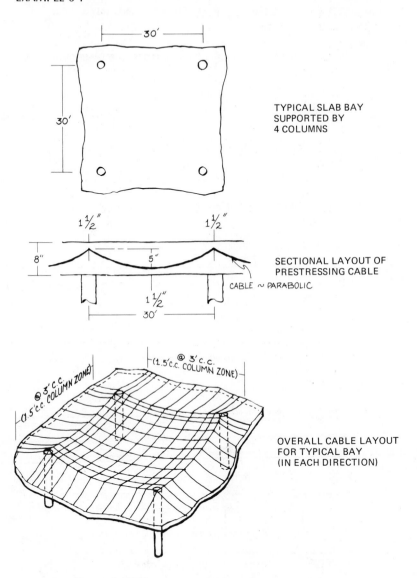

TYPICAL SLAB BAY
SUPPORTED BY
4 COLUMNS

SECTIONAL LAYOUT OF
PRESTRESSING CABLE

CABLE ~ PARABOLIC

OVERALL CABLE LAYOUT
FOR TYPICAL BAY
(IN EACH DIRECTION)

27 k per foot wide strip of slab. The average prestress in the concrete is computed as 27,000/(8 in. × 12 in.) = 281 lb per sq. in., which appears to be reasonable, since experience and theory indicate that an average concrete stress of 200 to 400 psi is optimum for such slab design (i.e. will result in an acceptable slab shortening, minimize cracking, and still yield sufficient ductility in the ultimate range).

As stated previously, 6 to 9 in. flat slabs made of ordinary reinforced concrete (without prestressing) can span only 15 to 24 ft economically and often (have to) employ an additional drop panel over the columns to increase resistance to punching shear (Figure 6-8*b*, p. 169). But drop panels are also useful, because, in effect, they shorten the slab span near supports and add to the stiffness and the strength of the slab as a whole. In fact, when the loadings on such slabs become extremely heavy, the additional use of a column capital may be required to further strengthen the shear as well as the bending resistance of a slab.

The provision of drop panels over the columns (Figure 6-11) can also allow post-tensioned concrete flat slabs to span even longer distances (35 to 45 ft)

FIGURE 6-11
PC SLAB WITH DROP OR HAUNCHED PANELS.

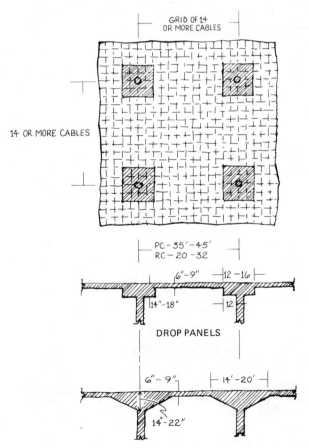

DROP PANELS

HAUNCHED PANELS

than shown in Figure 6-9. And, since the panels really act as haunches to increase the spacing between the columns, it is equally effective to provide tapered haunches rather than drops over the columns when longer spans are required (Figure 6-11).

Drop or haunched panels make longer spans of either R.C. or P.C. possible. But it should be recognized that it is often desirable for both architectural and construction reasons to design for shorter-span flat slabs that do not require drop panels. In any case, it will always be necessary to provide enough reinforcing steel or prestressing tendons to carry the load in both directions. Also note that the existence of cantilever or noncontinuous spans can make the problem of design more complicated, but that is beyond the scope of this book. For schematic and preliminary purposes, the reader need only recognize that these problems will require special treatment.

SECTION 4: Slab-and-Beam Subsystems

For more-or-less square bays, two-way slabs can span 20 to 30 ft or more between supports or stiff edge beams. When rectangular bays are involved, one-way slabs can span between edge beams that can efficiently span the

FIGURE 6-12
ONE-WAY SLAB AND BEAM SUBSYSTEMS IN P.C. OR R.C.

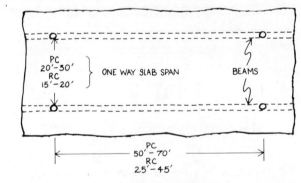

OVERALL BAY SPAN

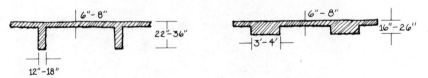

ALTERNATIVE SECTIONS

PHOTO 6-2
**PRECAST SLAB FOR A HIGH-RISE BUILDING. (COMPLIMENTS OF
CONCRETE ENGINEERING CO, LTD., HONOLULU, HAWAII)**

longer distance between supports. Generally, if L_1/L_2 is 1.5 or more, **only**
two beams are needed (Figure 6-12). Slabs can be cast in place or precast
concrete. Photo 6-2 shows a precast slab being lifted into position.

When the two beams are optimally spaced and made quite a bit **deeper**
than the slabs, the combination of edge beam and one-way slab **action can**
span much longer distances between supports with less material **and greater**
stiffness than would be the case for a flat slab alone.

Depending on the spacing between beams, a one-way slab can be **of**
plywood or timber planks (for slab spans under 4 ft), corrugated steel
decking (under 10 ft), or of reinforced or prestressed concrete for **longer**
slab spans (15 to 30 ft) as discussed above. Various combinations of **these**
slab-and-beam subsystems are possible, with the beams being of timber,
concrete, or steel, whichever is compatible for the system. Some of **the**
more practical ones are shown in Figure 6-13.

FIGURE 6-13
**SLAB AND BEAM DESIGNS MAY BE OF WOOD, STEEL, OR
ANY COMPATIBLE COMPOSITE CONSTRUCTION.**

(a) TIMBER-T AND G PLANKS OR PLYWOOD ON JOISTS

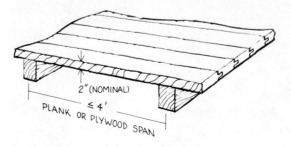

2" (NOMINAL)
≤ 4'
PLANK OR PLYWOOD SPAN

(b) SHEET STEEL (FOLDED) DECKING ON W STEEL BEAMS

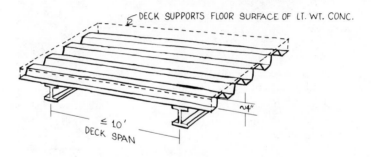

DECK SUPPORTS FLOOR SURFACE OF LT. WT. CONC.
~4"
≤ 10'
DECK SPAN

(c) PC OR RC SLAB ON W STEEL BEAMS (WITH SHEAR STUDS)

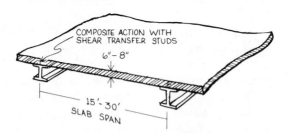

COMPOSITE ACTION WITH
SHEAR TRANSFER STUDS
6" - 8"
15' - 30'
SLAB SPAN

When the slab bays are of nearly square shape, beams may be added in
two directions at the slab edges (Figure 6-14). In this instance, the slab is
actually supported on all four sides and nearly optimum two-way action can
be achieved. As a result, a relatively thin slab can be used because the load
from the slab is distributed to the beams at approximately one-half in each
direction. A further benefit when the bay is nearly square is that the beams

FIGURE 6-14
**TWO-WAY SLAB AND BEAM ACTION WITH NEARLY
SQUARE BAYS.**

PLAN VIEW OF INTERIOR BAY OF TWO-WAY
SLAB AND BEAM SUB-SYSTEM

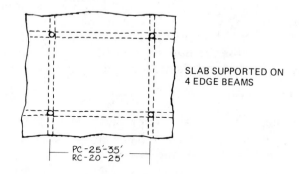

SLAB SUPPORTED ON
4 EDGE BEAMS

PC-25´-35´
RC-20-25´

SECTION

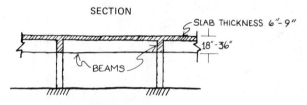

SLAB THICKNESS 6″-9″

18″-36″

BEAMS

BEAMS CAN CONNECT COLUMNS
TO GIVE FRAME ACTION

FIGURE 6-15
**IN RECTANGULAR BAYS WHERE $L_1/L_2 > 1.5$, THE SLAB ACTION IS
PRIMARILY (> 80%) ONE-WAY TO THE LONGER BEAM.
THEREFORE, SHORT BEAMS MAY BE OMITTED.**

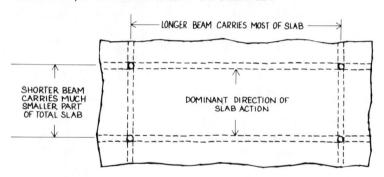

LONGER BEAM CARRIES MOST OF SLAB

SHORTER BEAM
CARRIES MUCH
SMALLER PART
OF TOTAL SLAB

DOMINANT DIRECTION OF
SLAB ACTION

can also be rigidly connected to the columns to form frames that resist lateral forces in both directions.

Such optimum two-way action can be achieved only if the typical floor bay is more-or-less square in proportion. But, as discussed in Section 2, rectangular bays with one side more than 50 percent longer than the other will result in the slab's carrying most of its load in its shorter direction to the longer beams, with only a nominal amount being transmitted in the longer direction to the shorter beams (Figure 6-15). For this reason, the shorter beams can either be utilized primarily as connectors for column-frame action or omitted and a simple one-way design adopted.

Example 6-2A

An office building has a typical interior bay of 20 ft × 40 ft with slabs spanning 20 ft supported on beams that span 40 ft between columns. Make a preliminary design for its floor subsystem.

A one-way R.C. slab can span 20 ft economically. To reduce the dead weight of floor, use lightweight concrete weighing 110 pcf., assuming it is available. From Appendix Table B-I, use span/depth ratio of 30; thickness of slab = 20 × 12/30 = 8 in. $DL = 110 × 8/12 = 73$ psf.

$$\text{Assuming: } LL = 50 \text{ psf}$$
$$\text{Partition} = 20$$
$$\text{Ceiling} = 5$$
$$DL = \underline{73}$$
$$TL = \overline{148} \text{ psf}$$

EXAMPLE 6-2A

(a) ONE-WAY RC SLAB AND BEAM LAYOUT

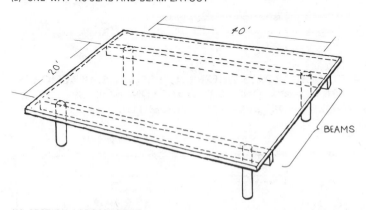

(b) SECTION ACROSS BEAMS

For a 1 ft wide slab strip

$$\text{Max} - M = wL^2/12 = 148 \times 20^2/12 = 4.9 \text{ k'/ft width}$$

Concrete protection to center line of steel $= 1\frac{1}{2}$ in.

$$\text{arm} = (8 - 1\frac{1}{2}) \times 7/8 = 5.4 \text{ in.}$$

$$C \text{ or } T = M/a = 4.9 \times 12/5.4 = 11 \text{ k/ft}$$

Therefore, $A_s = T/20 \text{ ksi} = 11/20 = 0.55 \text{ in.}^2$ per foot of slab and $p_s = A_s/bd = 0.55/(12 \times 6.5) = 0.7$ percent. Since .5 percent to 1 percent is a reasonable p_s for such slabs (as will be explained in Chapter 8), the design is considered acceptable.

Now for the beam we can use either concrete or steel. Assume steel, which can act by itself or in composite action with slab if shear studs are incorporated. Suppose we use steel W section by itself, with span/depth at 20, depth $= 40/20 \text{ ft} = 2 \text{ ft}$ (if composite design is desired, refer to Chapter 8), we have

LL tributary area for beam, 20 ft $\times$ 40 ft $= 800 \text{ ft}^2$; check for LL (large area) reduction

1. .08 percent $\times 800 = 64$ percent
2. $R = 23.1 (1 + D/L)$
    $= 23.1 (1 + 98/50)$
    $= 68$ percent
3. Max allowable is 40 percent reduction (see Chapter 5). Therefore, use 40 percent.

$$\begin{aligned} \text{Reduced } LL = 50 \times 0.6 = \quad &30 \text{ psf} \\ \text{Slab Load} = \quad &73 \\ \text{Partition} = \quad &20 \\ \text{Ceiling} = \quad &\underline{5} \\ \text{Total design load on beam} = \quad &128 \text{ psf} \end{aligned}$$

$$\begin{aligned} 20 \text{ ft spacing} \times 128 \text{ psf} = \quad &2.56 \text{ k/ft} \\ \text{Assume add weight of beam} = \quad &\underline{0.10 \text{ k/ft}} \\ \text{Total} = \quad &2.66 \text{ k/ft} \end{aligned}$$

If simple beam, $\text{Max} + M = wL^2/8 = 2.66 \times 40^2/8 = 532 \text{ kip-ft}$; use A-36 steel, $f_s = 24 \text{ ksi}$ (Appendix Table B-2).

Section modulus required $= M/f_s = 532 \times 12/24 = 266 \text{ in}^3$ (to be discussed in Chapter 8). And, from Appendix Table C-1, use 24 W 110, with S.M. $= 276 \text{ in.}^3$.

Example 6-2B

A concrete light storage floor has typical interior bays of 24 ft $\times$ 24 ft with two-way slabs supported by 24 ft-span beams along four sides. Proportion the slab-beam dimensions.

EXAMPLE 6-2B

(a) TWO-WAY SLAB AND BEAM SUB-SYSTEM

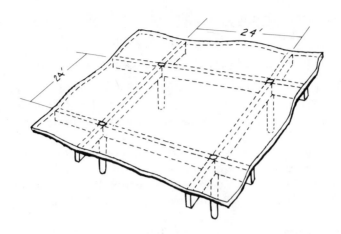

(b) SECTION THRU A BEAM (ASSUMING NO COMPOSITE ACTION)

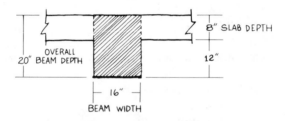

Use normal weight concrete of 150 plf

For two-way reinforced concrete slab, use span/depth ratio = 36 (see Appendix Table B-1).

Therefore slab is 24 ft × 12/36 = 8 in. thick

Assume span/depth for beam = 15 (Refer to Appendix Table B-1),

Assume LL = 100 psf.,

24 ft. × 12/15 = 19.2 in., use 20 in. overall depth.

For each bay of 24 ft, slab strip load is:

$$8 \text{ in. slab @ 100 psf} \times 24 \text{ ft. (width of strip)} = 2.4 \text{ k/ft}$$
$$\text{Assume weight of beam (12 in.} \times 16 \text{ in.)} = 0.2 \text{ k/ft}$$
$$LL \text{ (no reduction) } 100 \times 24 \text{ ft} = 2.4 \text{ k/ft}$$
$$\text{Total} = \overline{5.0 \text{ k/ft}}$$

Assume Max + M at $wL^2/16 = 5 \times 24^2/16 = 180$ kip-ft, $-M = wL^2/12 = -240$ kip-ft

Of total ± M, $\frac{2}{3}$ carried by beam = +120 kip-ft, −160 kip-ft,

$\frac{1}{3}$ carried by slab = + 60 kip-ft, − 80 kip-ft

Beam design is critical at $-M$ over column. Assume rectangular section, since slab not acting for $-M$, and use 2 in. concrete protection (to center line of steel bars):

$$a = (20 \text{ in.} - 2 \text{ in.} = 18 \text{ in.}) \times 7/8 = 16 \text{ in.}$$
$$T = M/a = 160 \text{ kip-ft} \times 12/16 = 120 \text{ k}$$

Allowing 20 ksi for steel bars,

$$A_s = 120/20 \text{ ksi} = 6 \text{ in.}^2$$

Over column, the beam has $p_s = A_s/bd = 6/(16 \text{ in.} \times 18 \text{ in.}) = 2.1$ percent, which is high, since optimum p_s for beams is closer to 1.5 percent. But we can add compressive A_s along the bottom of the beam if needed.

Slab design may be critical to carry $-M = 80$ kip-ft. Assume middle $\frac{2}{3}$ width of slab carries most of the slab load (i.e., $\frac{2}{3} \times 24$ ft = 16 ft effective width). Distribute slab $-M$ per ft over effective width, $80/16 = 5$ k'/ft.

$$\text{arm} = (8 \text{ in.} - 1\tfrac{1}{2} \text{ in.})7/8 = 5.6 \text{ in.}$$
$$T = M/a = 5 \times 12/5.6 = 11 \text{ k/ft width of slab}$$
$$A_s = 11/20 \text{ ksi} = 0.55 \text{ in.}^2/\text{ft width of slab}$$

Since effective depth of slab is 8 in. $-$ 1.5 in. protection = 6.5 in. $p_s = 0.55/5.6 \times 12 = 0.8$ percent, which is o.k. for slab.

SECTION 5: Joist and Girder Subsystems

In Section 4, we discussed flat slabs supported on beams. Now we will extend the discussion to include slabs (or decks) supported on joist and girder subsystems.

The terms joists, beams, and girders all refer to linear horizontal members that act in bending and transfer tributary floor or roof loading to vertical subsystems. Generally, *joists* refers to closely spaced beams (a few feet apart), which are thus relatively lightly loaded and small components. *Beams* usually refers to larger members, which are under heavier load than joists, spaced several feet to perhaps 30 ft apart. *Girders* are rather deep beams designed to pick up heavy loadings accumulated from many joists or beams. However, these terms are often used somewhat interchangeably.

When it is desired to use a very thin concrete slab for various reasons, such as economy or weight, it may be necessary to support these slabs with joists (or small beams) spaced at rather close intervals. When the joist spacing is from about 3 ft (with standard forms) to around 10 ft, the one-way slab or decking component can be 3 to 4 in. thick and the supporting joists can be of fairly long span (40–60 ft) and still not too deep (say less than 2 ft deep) (Figure 6-16 and 6-17). To avoid the need for closely spaced columns

FIGURE 6-16
R.C. JOIST AND GIRDER SUBSYSTEMS.

(a) PLAN OF JOIST AND GIRDER SYSTEM LAYOUT
    (PREFER RECTANGULAR PANELS, BUT CAN BE SQUARE)

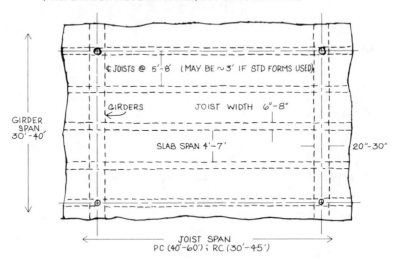

GIRDER
SPAN
30'-40'

⌀ JOISTS @ 5'-8' (MAY BE ∼3' IF STD FORMS USED)

GIRDERS   JOIST WIDTH 6"-8"

SLAB SPAN 4'-7'   20"-30"

JOIST SPAN
PC (40'-60'); RC (30'-45')

(b) SECTION THRU SLAB AND GIRDERS

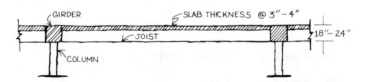

GIRDER   SLAB THICKNESS @ 3"-4"

JOIST   18"-24"

COLUMN

to support each joist, the joists can rest on heavy girders, which in turn can span a distance between columns of about 30 to 40 ft. Thus such a floor subsystem, comprised of one-way components running in perpendicular directions, is relatively lightweight, is capable of carrying considerable total load, and can be supported on columns rather far apart to provide necessary clear space for functional and other reasons. In comparison, the two-way flat-slab or slab-and-beam systems are more limited in terms of span and bay proportions, although they are simpler to construct. Of course, it is also possible to design slab-and-beam subsystems with the beams supported on long girders that span between columns.

For short spans of 2 to 4 ft between joists, T and G plank or plywood decking may be supported on timber joists that in turn rest on walls or girders, as shown in Figure 6-18.

Lightweight steel joists made of trussed members (or *H*-beams), supported either on walls or on girders, can be used as an alternative to wood (Figure

FIGURE 6-17
JOIST SPACING CAN BE QUITE SMALL WITH THE USE OF STANDARD TILE OR REMOVABLE FORMS.

(a) RC SLAB WITH STEEL JOISTS POSITIONED AS NEEDED

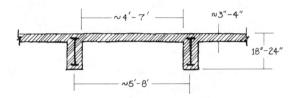

(b) STANDARD HOLLOW TILES USED TO FORM JOISTS AND SUPPORT SLAB

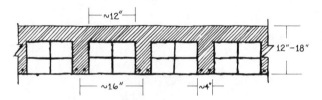

(c) RC SLAB AND JOISTS CAST OVER REMOVABLE FORMS

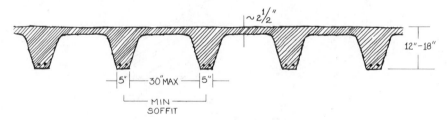

6-19). But since these joists are lightweight, they must not be spaced too far apart. These steel joists can be decked over with either wood, as discussed above, or with a thin concrete slab placed on lightweight corrugated steel slab forms.

For the longer slab spans or heavier design applications, a common type of steel-concrete subsystem involves the use of heavier corrugated metal decking simply laid on steel joists. If the slab is connected to the joists by shear-studs, composite action can be achieved. The metal deck serves as a form to allow concrete to be poured in place (Figures 6-20a and 6-20b). The corrugated decking supplies the total bending strength during construction while the poured-in-place concrete increases the slab stiffness after setting. Such deck construction is useful because small service ducts and conduits for wires may be easily buried into these slabs. In design, it is common practice to neglect the composite action between the decking and the

FIGURE 6-18
WOOD JOISTS WITH T AND G OR PLYWOOD DECKING.

(a) T AND G ROOF DECKING

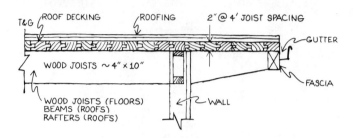

(b) PLYWOOD SUBFLOORING UNDER STRIP FLOORING

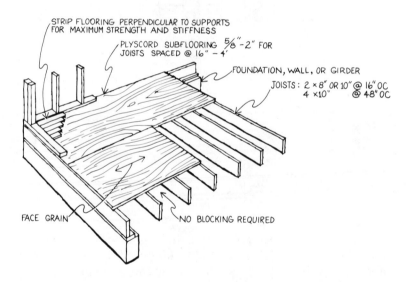

concrete (so as to permit holes and omissions of the concrete), although it certainly increases the overall bending strength considerably. Such decking commonly spans between beams or joists spaced at 6 to 8 ft apart, which in turn rest on girders. These girders can then span 30 to 50 ft between walls or columns, thus giving a wide-open activity space.

Different forms of the metal decking are available. Some of these, shown in Figure 6-21, will give strength and rigidity to the floor, enabling it to span 10 ft between the supporting beams and the girders.

FIGURE 6-19
LIGHTWEIGHT STEEL (TRUSS) JOISTS CAN BE USED AS AN ALTERNATIVE TO WOOD.

(a) LIGHTWEIGHT STEEL JOIST SUB-SYSTEM
WITH CONCRETE SLAB

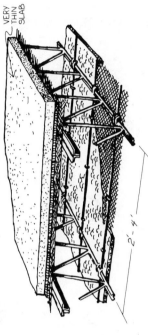

VERY THIN SLAB

2'-4'

(b) SECTION SHOWING TYPICAL FLOOR CONSTRUCTION

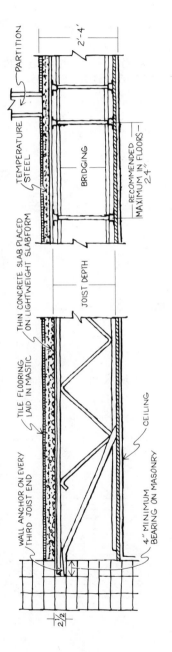

PARTITION

TEMPERATURE STEEL

2'-4'

BRIDGING

RECOMMENDED MAXIMUM IN FLOORS— 2.4"

THIN CONCRETE SLAB PLACED ON LIGHTWEIGHT SLABFORM

TILE FLOORING LAID IN MASTIC

WALL ANCHOR ON EVERY THIRD JOIST END

JOIST DEPTH

CEILING

4" MINIMUM BEARING ON MASONRY

(c) SECTION THRU JOISTS SHOWING FLANGE TYPES

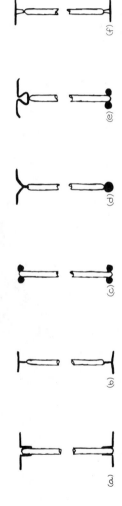

(a)

(b)

(c)

(d)

(e)

(f)

FIGURE 6-20
TWO TYPES OF CORRUGATED METAL DECKING FORMS FOR STEEL AND CONCRETE SUBSYSTEMS.

(a) SIMPLE CORRUGATION

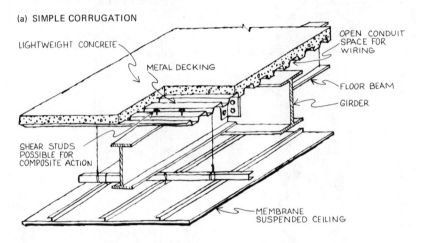

LIGHTWEIGHT CONCRETE

METAL DECKING

OPEN CONDUIT SPACE FOR WIRING

FLOOR BEAM

GIRDER

SHEAR STUDS POSSIBLE FOR COMPOSITE ACTION

MEMBRANE SUSPENDED CEILING

(b) TUBE FORM PROVIDES ELECTRICAL & AIR DISTRIBUTION AVAILABILITY

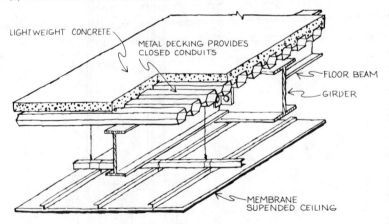

LIGHTWEIGHT CONCRETE

METAL DECKING PROVIDES CLOSED CONDUITS

FLOOR BEAM

GIRDER

MEMBRANE SUPENDED CEILING

FIGURE 6-21
METAL DECKING TYPES VARY.

(a) SHALLOW (1″ – 2″) FOR CLOSE (2′ – 4′) JOIST SPACING

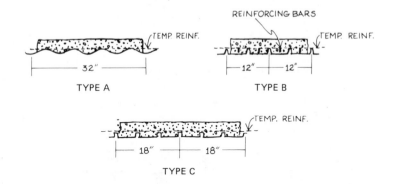

(b) DEEPER (3″ – 6″) FOR 5′ – 10′ (OR MORE) JOIST SPACING

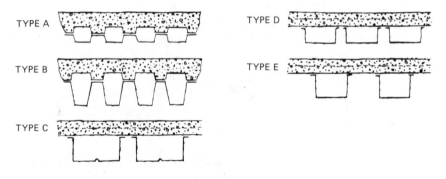

EXAMPLE 6-3

A typical interior panel of a building is shown, with a joist and girder bay of 30 ft × 40 ft. Design live load (*LL*) = 50 psf. Design for lightweight concrete placed on a 4-in. metal deck, the combined weight being 30 psf, and for W-joist and beam sizes use A36 steel. Assume the joists continuous over the girders, but the girder is simply supported at the columns. (As a start, assume girder to weigh 200 plf.)

1. Metal deck design: Span *L* = 7.5 ft
               DL of decking + concrete = 30 psf
                              LL = (no reduction) 50 psf
          Assume partition, ceiling, etc. = 20 psf

EXAMPLE 6-3
W JOIST AND GIRDER SUBSYSTEM.

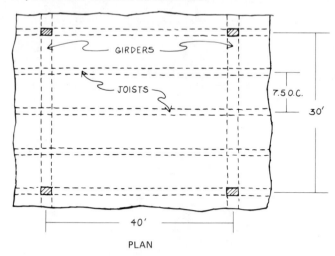

PLAN

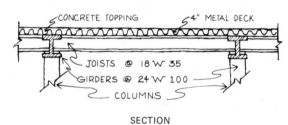

SECTION

Total load per ft width of deck = 100 psf. Assume simple spans:

$$M = wL^2/8 = 0.1 \times 7.5^2/8 = 0.7 \text{ k-ft}$$

Section Modulus required:

$$\frac{M}{f} = \frac{0.7 \times 12}{24} = 0.350 \text{ in.}^3$$

Refer to Appendix C-2*b*. Use 3″ deep Section 21-22:

$$+ \text{ Moment S.M.} = 0.386 \text{ in.}^3$$

$$\text{Span/depth ratio} = \frac{7.5 \times 12}{3 \text{ in.}} = 30$$

which is somewhat higher than the usual ratio of 25, but may be used if study of deflection and vibration permits.

2. Steel joist design: $L = 40$ ft, w from deck 0.1 k/ft × 7.5 ft = 0.75 k/ft

    Joist depth 40 × 12/25 ~ 19″, weight = 0.05 k/ft

    Total load = 0.80 k/ft

Make joist continuous over beam to reduce deflection.

$$-M = wL^2/12 = 0.8 \times 40^2/12 = 107 \text{ kip-ft}$$
Section modulus required $= M/f = 107 \times 12/24 = 53.5 \text{ in.}^3$, use 18 W 35.

3. Girder Design: $L = 30$, depth $= 30 \times 12/14 \sim 25''$

Dead load from slab 30 psf $\times$ 40 ft $= 1.20$ k/ft
Dead load from joists 40 ft $\times$ 35/7.5 $= 0.19$ k/ft
Dead load from girder, assumed $= \underline{0.20 \text{ k/ft}}$
Total dead load $= 1.64$ k/ft

Partition load 20 psf $\times$ 40 ft $= 0.80$ k/ft
Live load (no reduction) $= 50$ psf $\times$ 40 ft $= \underline{2.00 \text{ k/ft}}$
Total design load $= 4.39$ k/ft

$M_{max} = wL^2/8 = 4.39 \times 30^2/8 = 494$ kip-ft.
Section modulus required $= 494 \times 12/24 = 247 \text{ in.}^3$ Use 24 W 100
Check for girder deflection under LL; when $c = 12''$, $I = 250 \times 12 = 3000 \text{ in.}^4$, and from Appendix D-2
$\Delta = 5wL^4/384 \text{ EI} = 5 \times 2.00 \times 30^4 \times 12^3/(384 \times 29 \times 10^3 \times 3000) = 0.42$ in., which is less than 1/360 of span $= (30 \text{ ft} \times 12)/360 = 1$ in., and is considered to be O.K.

Note that the girder can be made continuous with columns by welding or bolting. Then $-M$ will be developed at the connections, and the $+M$ will be reduced. Furthermore, the girders and columns will then form a rigid frame that can help to resist lateral forces as a vertical subsystem (Chapter 7).

SECTION 6: Waffle Subsystems

The joist and girder system carries the floor load in one-way action along the joist and again in one-way action in the other direction along the girders. In contrast, a waffle system carries the load simultaneously in both directions because its beams are constructed to form a two-way grid. To be effective, this system requires that the column bays be approximately square rather than rectangular (Figure 6-22). When this is the case, the slab panels over the beam grid can be very thin (as little as $2\frac{1}{2}$ inches) because the span is short and the load is carried in two directions.

When there are *four* or more grid lines, the overall behavior and moment distribution of a waffle subsystem approximate that of a flat slab (see Figure 6-3 and Figure 6-4, pp. 161 and 163). But the waffle is more suitable for long spans because it attains greater overall structural depth than a flat-slab design while minimizing added material.

Concrete waffle systems were first introduced with small grids of approximately 3 ft square, using standardized forms. More recently it has been determined that such small waffles are not the most economical. It is possible to span 8 to 12 or 14 ft between grid lines with the thinnest *practical*

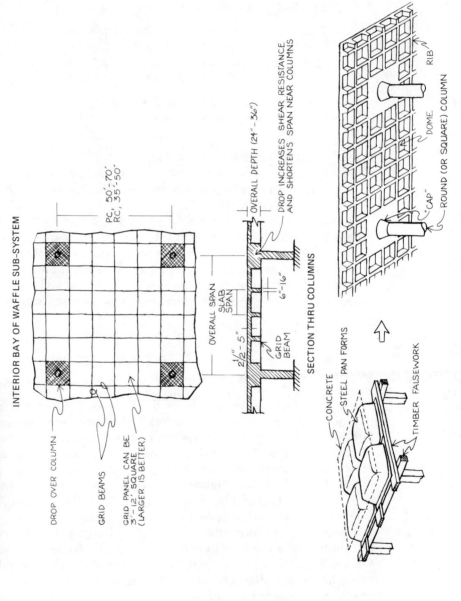

FIGURE 6-22
CONCRETE WAFFLE SUBSYSTEM.

INTERIOR BAY OF WAFFLE SUB-SYSTEM

PC, 50'-70'
RC, 35'-50'

DROP OVER COLUMN

GRID BEAMS

GRID PANEL CAN BE
3'-12' SQUARE
(LARGER IS BETTER)

OVERALL SPAN
SLAB SPAN
GRID BEAM
2½''- 5''
6''-16''

OVERALL DEPTH (24''-36'')

DROP INCREASES SHEAR RESISTANCE
AND SHORTENS SPAN NEAR COLUMNS

SECTION THRU COLUMNS

RIB

DOME

"CAP"

ROUND (OR SQUARE) COLUMN

RESULT WHEN FORMS ARE REMOVED

CONCRETE

STEEL PAN FORMS

TIMBER FALSEWORK

CONCRETE IS PLACED OVER PAN-LIKE FORMS ⟶

slab 4 to 5 in. as in Figure 6-23). The waffles can be square, somewhat
rectangular, triangular, or hexagonal in shape, and they can be formed of
materials suitable for reuse, such as plastic, plywood, or steel, so that
construction economy can be obtained. The concrete ribs can be reinforced
or prestressed. It is also possible to precast large concrete waffle forms that,
when topped with concrete, will be structurally integrated into the floor
construction.

A preliminary analysis of grid systems can be approximated by comparing
them to flat slabs, using the several T-beams in one way to provide the
equivalent of flat-slab bending resistance in each of two directions. This
means that for rough analysis, the total moment in one direction can be
distributed evenly to the beams across one bay. A more accurate analysis
would include consideration for the fact that beams along column lines
carry more load than those along middle strips. The solid panels around
columns are like drop panels in a flat slab. As will be discussed in Chapter 8,
the moment per beam can be resisted by reinforcement or prestressing for

FIGURE 6-23
LARGE WAFFLE SCHEME WITH BEAMS OVER COLUMNS.

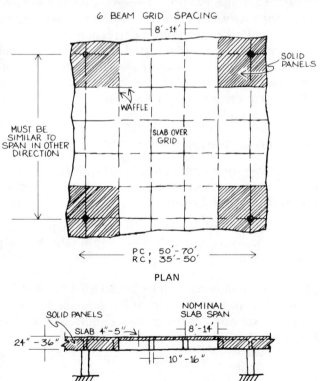

PLAN

SECTION THRU COLUMN

one of three loading conditions:

1. Ultimate load.
2. Working load.
3. Dead load (to be balanced by prestress).

Increased punching shear resistance can be provided by passing beams directly over the columns. But it is often more desirable to provide a solid panel as shown in Figure 6-22. This allows four beams immediately outside of the columns to carry most of the load and transmit it directly to the solid panel and thereby onto the columns. In addition, it has the effect of stiffening these beams.

Figure 6-23 illustrates an arrangement whereby one waffle beam runs directly through the center of the column. The single beams running over the columns could be made heavy enough to carry the greater part of the load and resist punching shear. But it may be more effective to use solid panels for all the four panels adjacent to the column. In this case, the three beams (one along the column and two adjacent) will carry most of the load. This can be considered by assigning at least twice as much moment to each of these beams as to the other beams.

Waffles in steel are also possible. However, waffles in steel are very expensive to construct when compared with concrete waffles because of the special joinery required at each grid intersection to achieve two-way action. Hence, H section grids are seldom used. But steel space frames used for very long spans are also analogous in overall behavior to the waffle grid (or flat-slab) layouts. Example 6-4 illustrates how to make a preliminary design for a prestressed concrete waffle subsystem.

EXAMPLE 6-4:   Waffle System

A typical interior bay for a large office building is to be built of 60 ft × 60 ft waffle layout using prestressed concrete construction. Obtain preliminary dimensions for the slab and beams.

$LL = 75$ psf, which includes partitions. Concrete 150 pcf; $f'_c = 4$ ksi
Grid at 12 ft. cc. For 2-way grid slab, span/depth ratio = 36 (Appendix Table B-1) 12 ft × 12/36 = 4 in.
Use 4 in. slab.
Similarly, for beam span/depth ratio = 24; 60 ft × 12/24 = 30 in. Use 30 in.

Use 12-in.-wide stem for beams in order to house tendons and take compression over supports. Assume beam equivalent weight of 4 in. average thickness over whole area, we have

Slab weight =   50 psf
Beam weight =   50 psf
LL =   75 psf
TL = 175 psf

EXAMPLE 6-4
P.C. WAFFLE SUBSYSTEM.

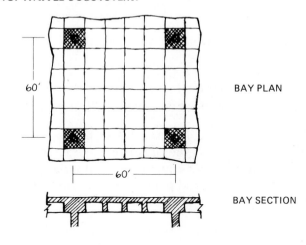

BAY PLAN

BAY SECTION

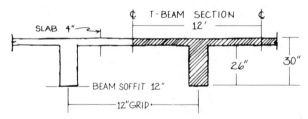

DESIGN GRID BEAMS @ T–SECTION

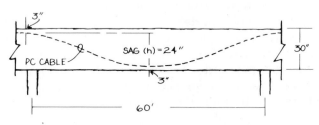

LAYOUT OF PC CABLE IN BEAM

As a preliminary design let us balance the dead load of 100 psf. For 60 ft width of bay, $DL = 0.1\,k \times 60 = 6\,k/ft$ in each direction, neglecting the effect of solid panels over columns. Cable sag $h = (30 - 6) = 24$ in. and total prestress required to balance total $DL(w)$ is

$$F = wL^2/8h = 6 \times 60^2/8 \times 2\ \text{ft} = 1350\ \text{k per 60 ft bay.}$$

If we assume the prestress to be equally divided among five beams, each takes $1350/5 = 270$ k.

A_c for each beam is   For slab, $4 \times (12 \times 12) = 576$ in.^{2}
For stem, $26 \times 12 = \underline{312 \text{ in.}^2}$
Total $= 888$ in.^{2}

$f_{avg} = 270$ k/888 $= 304$ psi, which is quite reasonable (200 to 400 psi being optimum).
Using $f_s = 150$ ksi for the high-tensile steel tendons,
$A_s = 270/150 = 1.8$ in.^{2} per beam.
At 3.4 plf/in.^{2}, steel weighs $1.8 \times 3.4 = 6.2$ lb/ft of beam; 6.2/12 ft $= 0.52$ lb/ft^2 of slab.

For two directions, we have 1.04 lb/ft^2 of slab. Somewhat more steel may be needed for exterior bays. But it is likely that no more than 1.0 lb of tendons will be required per square foot of floor area if we take into account the effect of the solid panel over the columns. This looks very economical.

SECTION 7: Space Truss Systems

To cover a very large clear-span area (on the order of 100-ft spans or more) with a *flat* floor or roof, one may use trusses running in one direction spanning between columns to serve as the main carrying members, and then span between these large trusses with smaller ones perpendicular to them (Figure 6-24). This is similar to the joist and girder system, but using trusses instead of joists and girders. If this is the case, then the basic design for a one-way system can be conceived like a joist-girder system (see Section 5). For truss design, the moment arms for the resisting coupled will be approximately the overall depth.

A space truss is a modification of the above truss-on-truss systems. By using *equal depth* trusses and *staggering*, in plan, all top chord members relative to the bottom chord members (in both directions), a two-way space truss layout is created (Figure 6-25). Since the top chord of the trusses will be connected to the bottom chord by web members, the result is a grid of

FIGURE 6-24
ONE-WAY TRUSS ON TRUSS SUBSYSTEM.

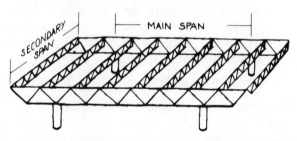

FIGURE 6-25
TWO-WAY SPACE TRUSS ACTION IS ACHIEVED BY STAGGERING TOP AND BOTTOM CHORDS IN BOTH DIRECTIONS TO GET INCLINED TRUSSES WITH COMMON CHORDS.

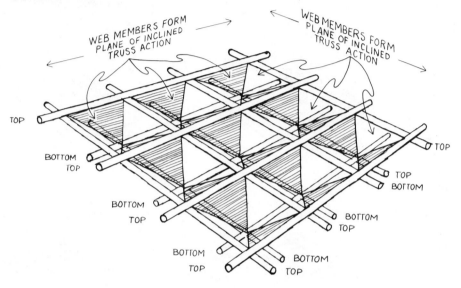

inclined trusses. In this manner, each web member serves as the diagonal for two chords in two directions, thus achieving some construction material economy, at the expense of added joinery costs. Because of the two-directional nature of the web system, this is often termed a space-structure, a space truss system, a space-frame, and so on. But two-way action could also be achieved if the trusses in two directions were vertically intersected. In either case, the overall behavior (and thus design) of a two-way truss scheme is very similar to that discussed for a waffle (or flat-slab) subsystem.

Figure 6-26 shows several arrangements of a two-way space truss system. Figure 6-26a illustrates a frame supported on four corner columns only. Figure 6-26b shows a cantilever layout, with only one large column at the center. Figure 6-26c is an economical arrangement, using cantilevers to decrease the bending moment in the main center panel. Figure 6-26d illustrates a rectangular layout where the action is primarily one-way. In Figures 6-26c, 6-26e, and 6-26f, the existence of deep edge supports suggests design by analogy to an edge-supported flat-slab subsystem.

When the bay is oblong (as in Figure 6-26d), more columns may be added along the long sides so as to minimize the main truss bending moment. When the bay is nearly square and large openings are not needed on the periphery, columns can be placed along all sides, thus cutting down considerably the bending moments in both directions (Figure 6-26e). In

FIGURE 6-26
ALTERNATIVE LAYOUTS FOR TWO-WAY SPACE TRUSS SUBSYSTEMS (PRIMARY TRUSSES SHOWN).

(d) RECTANGULAR (ONE-WAY)

(a) SIMPLE BAY

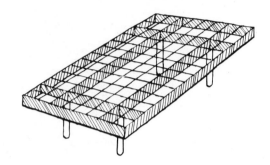

(b) CANTILEVER

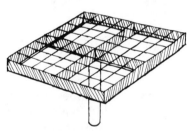

(e) SIMPLE EDGE SUPPORTED BAY

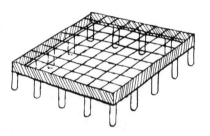

(c) BAY WITH CANTILEVER

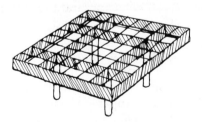

(f) EDGE SUPPORTED BAY WITH CANTILEVER

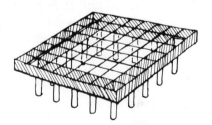

addition, cantilevers can be constructed on all sides to further minimize the maximum center moments (Figure 6-26f).

A rigorous analysis and design of these frames is best accomplished by computer programs. However, one can make a reasonably economical layout by using the flat-slab analogy to obtain preliminary dimensions for comparative studies and for supplying input to the computer. A sample calculation is shown in Example 6-5.

EXAMPLE 6-5: Steel Space Truss System

Solution: Since span/depth ratio for such trusses is usually between 12 and 20, we will assume truss depth of 160/16 = 10 ft.

Assume weight of steel = 10 psf

Roofing = 5

Mechanical equipments = 5

LL = 30

Total = 50 psf

Total load on roof (W) = 50×160^2 = 1280 k.

Moment at midspan, referring to freebody diagram (c below) with resultant of overall load and reaction as shown,

Max moment = $W/2 \times 20$ ft = $1280/2 \times 20$ = 12,800 kip-ft

For an overall truss depth of 10 ft, assume a lever arm of 9 ft, then the

EXAMPLE 6-5

(a) SECTION

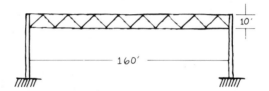

(b) QUARTER PLAN

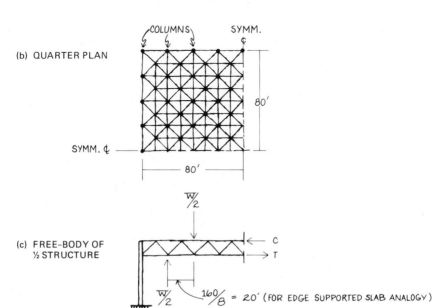

(c) FREE-BODY OF ½ STRUCTURE

total compressive or tensile force to be supplied in all top chords or all bottom chords is:

$$12,800/9 = 1420 \text{ k}$$

Assuming average allowable stress of 20 ksi for A36 steel,

$$A_s = 1420/20 = 71 \text{ in.}^2; \times 3.4 \text{ lb/in.}^2 \text{ ft} = 242 \text{ lb/ft}$$

For 160-ft width, steel per sq ft $= 242/160 = 1.5$ lb/ft^2 at points of maximum moment.

Assuming 80 percent $\times$ 1.5 lb/ft^2 $= 1.2$ lb/ft^2 as the average steel weight, we have:

One direction total for both top and bottom chords $2 \times 1.2 = 2.4$ psf

Add web, say 60 percent of chords $= \underline{1.4}$ psf

Total A_s in each direction $= \underline{\underline{3.8}}$ psf

For two directions, $2 \times 3.8 = \overline{7.6}$ psf

Add connection, details, and allowance for bending

on chords, at 20 percent $= \underline{1.5}$ psf

Total $= \underline{\underline{9.1}}$ psf

which is close enough to the assumed weight of 10 lb/ft^2.

7
Vertical Subsystems

SECTION 1: Introduction

We have divided a building's total structural system into two groups of subsystems, vertical and horizontal. As discussed in the previous chapter, horizontal subsystems must be supported on vertical subsystems. At the same time, vertical subsystems are generally slender in one or both section dimensions (relative to the overall building height) and cannot be very stable by themselves. They must be held in position by the horizontal subsystems.

As introduced in Chapter 3, horizontal subsystems pick up and carry floor and roof loads through bending, and horizontal loads through diaphragm action, to the vertical subsystems. Horizontal subsystems may also serve to connect the various vertical subsystems or their components and make them work together as frames. One must also remember that there are broader architectural needs to be considered along with those of structural design. As was discussed in Section 1 of Chapter 6, service requirements and other spatial and symbolic considerations must be included when actually designing vertical and horizontal subsystems. Thus, it is only for the convenience of focusing on the more basic issues of designing vertical subsystems that we discuss the two basic subsystems separately. In actual design, horizontal and vertical thinking must be synthesized to produce an efficient interaction of the subsystems.

In terms of an overall capability to carry both vertical and horizontal loads to the foundation, there are *three primary types* of 2-D or 3-D vertical subsystems in buildings: (1) *wall subsystems*; (2) *vertical shafts*; and (3) *rigid beam-column frames* (Figure 7-1). Walls are very rigid subsystems in their plane and may be made up of solid masonry, paneled or braced timber, steel trusses, and so on. Shafts are usually made up of four solid or trussed walls forming a tubular space structure that houses elevators, staircases, and/or other vertical trunk conduits for ventilation and services. As 3-D tube-structures, shafts can constitute rather stable and stiff vertical elements on their own. They can carry the vertical loads tributary to it, and also serve as excellent horizontal-force resisting elements (refer to Chapter 3, Section 3). Normally rigid frame subsystems consist of linear vertical components (columns) rigidly connected by stiff horizontal components (beams or girders). The rigid connection thus causes the columns to interact in bending to form a relatively stiff plane of overall resistance to vertical and

FIGURE 7-1
THREE BASIC TYPES OF VERTICAL SUBSYSTEMS.

(a) WALLS: SOLID   TRUSS

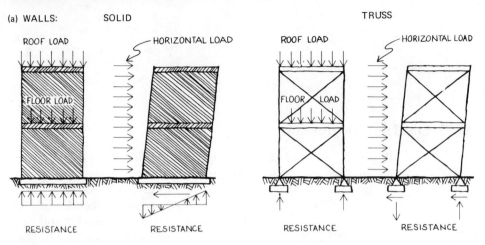

(b) SHAFTS: (WALLS CAN BE SOLID OR TRUSSED)

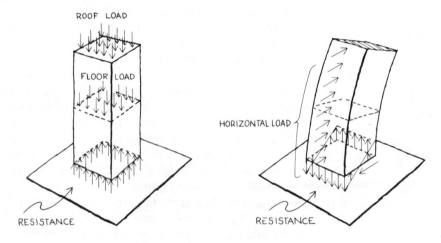

(c) RIGID FRAMES: (REQUIRE STIFF COLUMN CONNECTORS)

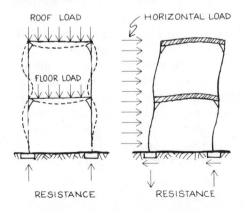

horizontal forces, as compared with the case for pin connections (refer to Chapter 3, Section 4). It is also possible to create megaframe structural schemes by using very large (mega) girders to rigidly connect large shafts at several floor intervals.

A set of pin-connected slender columns can carry only vertical loads. However, the set can be looked at as a *linear* acting vertical subsystem. And there is often good reason to use pin-connected columns as a *fourth* basic vertical subsystem type because stability under horizontal loading can be provided by composite design using one of the other three primary types. In fact, in most good-sized buildings, the overall capabilities for vertical and horizontal load resistance are usually made up of a combination of the above four subsystem types, as illustrated in Figure 7-2. In each example it should be noted that the relatively slender vertical subsystems can be directly tied to one another by horizontal subsystems available at each floor level and at the roof.

Figure 7-3 illustrates the important role that can be played by a shear- and moment-resisting shaft in providing a means of using pin-connected columns to achieve an open-facade design within the context of a sound overall structural scheme. Figures 7-4a and 7-4b illustrate the advantage of strategically using heavy trusses at the top and possibly the middle of a building to brace the exterior columns against a core shaft. This will achieve a frame-like action in the shaft, equalize the temperature shortening of vertical components, and reduce deflection under horizontal loads.

The three primary types of subsystems will now be discussed separately in the following sections. Linear column action will be treated in Chapter 9.

SECTION 2: Wall Subsystems

Exterior walls serve to enclose a building form, and interior walls serve to partition the building space. Both can also serve as major structural subsystems for carrying both vertical and horizontal loads. Photo 7-1 shows a group of precast concrete walls three to four stories high erected for a dormitory building.

As mentioned previously, walls are generally constructed of masonry, timber, concrete, or steel. In all cases, when the walls are braced by floors or roofs, they can provide excellent resistance to horizontal loads in the plane of the walls. But, being thin, they are relatively weak against horizontal forces applied in the direction of the wall thickness.

For example, most walls are only a few inches thick, but many feet wide, and in each plan dimension the rigidity of these walls is proportional to the moment of inertia (I) of the section. Briefly, the stiffness of rectangular wall sections (as measured by I) varies with the area times the square of the depth (d) in the direction of action (refer to Chapter 3, Section 3). For walls, having a rectangular section, $I = Ad^2/12$.

206

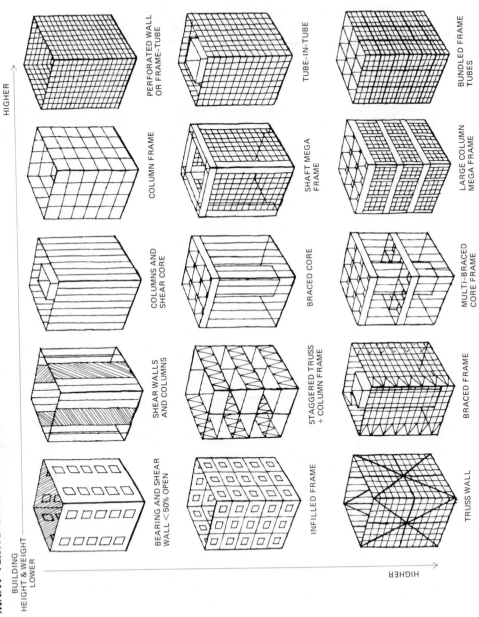

FIGURE 7-2
MANY VERTICAL SCHEMES ARE POSSIBLE BY COMBINING THE FOUR BASIC SUBSYSTEM TYPES.

BUILDING
HEIGHT & WEIGHT
LOWER

HIGHER

HIGHER

PERFORATED WALL
OR FRAME-TUBE

COLUMN FRAME

COLUMNS AND
SHEAR CORE

SHEAR WALLS
AND COLUMNS

BEARING AND SHEAR
WALL < 50% OPEN

TUBE-IN-TUBE

SHAFT MEGA
FRAME

BRACED CORE

STAGGERED TRUSS
+ COLUMN FRAME

INFILLED FRAME

BUNDLED FRAME
TUBES

LARGE COLUMN
MEGA FRAME

MULTI-BRACED
CORE FRAME

BRACED FRAME

TRUSS WALL

FIGURE 7-3
VARIOUS SCHEMES WITH INTERNAL SHEAR RESISTING CORE-SHAFTS.

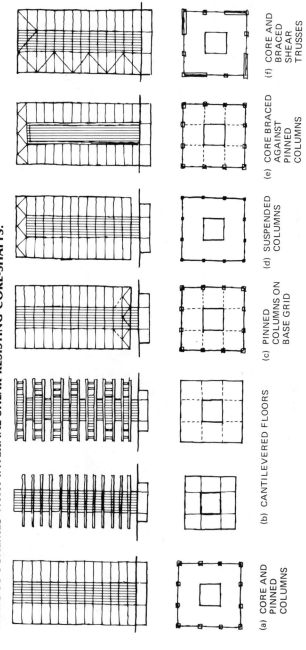

(a) CORE AND PINNED COLUMNS

(b) CANTILEVERED FLOORS

(c) PINNED COLUMNS ON BASE GRID

(d) SUSPENDED COLUMNS

(e) CORE BRACED AGAINST PINNED COLUMNS

(f) CORE AND BRACED SHEAR TRUSSES

FIGURE 7-4
BRACING EXTERIOR COLUMN REDUCES DEFLECTION.

(a) CORE BRACED AGAINST COLUMNS BY TRUSS AT TOP:
ROTATION OF THE TOP TRUSS IS RESTRAINED BY
LINES OF EXTERIOR COLUMNS

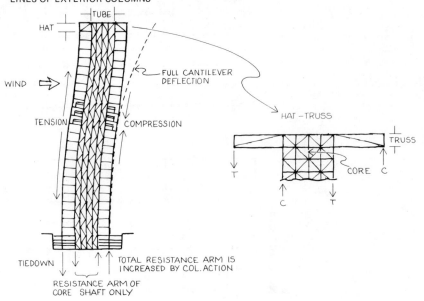

THE WINDWARD COLUMNS ARE STRETCHED BY HAT'S ROTATION AND ARE SUBJECTED
TO TENSION. THE LEEWARD COLUMNS ARE COMPRESSED BY THIS ROTATION
(SUBJECTED TO COMPRESSION). THESE FORCES—TENSION AND COMPRESSION—
PRODUCE PARTIAL REVERSAL OF ROTATION IN THE CENTRAL BRACED FRAME AND
MODIFY FORCES IN THE TRUSS.

(b) OVERALL DEFLECTION OF BUILDING AS REDUCED BY BRACING COLUMNS AGAINST
CORE AT BOTH TOP AND MID-HEIGHT.

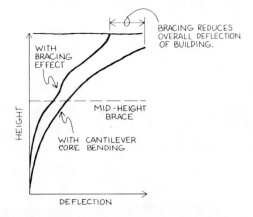

PHOTO 7-1
PRECAST CONCRETE WALLS FOR DORMITORY BUILDING, UNIVERSITY OF CALIFORNIA, DAVIS. (T. Y. LIN INTERNATIONAL, STRUCTURAL ENGINEERS).

Since the depth, for in-plane horizontal loads, is the length of a wall (say, 30 ft), and for transverse loads, it is only the thickness (say, 1 ft), the stiffness ratio is 900 : 1. Thus, the potential for resistance to lateral forces is high along the length of a wall but quite weak across its thickness. For this reason, the transverse resistance of walls to horizontal forces is usually neglected and two or more walls must be aligned more-or-less orthogonally (at right angles) to provide resistance to all lateral loads. When a lateral force acts in an oblique direction, it can be resolved into two orthogonal vector components, each of which will act in plane to some of the walls and be resisted (Figure 7-5).

When shear-wall subsystems are used, it is best when the center of orthogonal shear resistance is close to the centroid of lateral loads as applied due to surface or mass properties of a building form. If this is not true, there will be a horizontal moment (torsion) design problem. Note that Figure 7-5a and b represent unstable arrangements of walls to resist horizontal forces. In part a the walls supply no stiffness in the X direction; and in part b the centroid of resistance does not coincide with the center of load application, and there is almost no stiffness against torsional rotation. Arrangements in Figure 7-5c to f are quite satisfactory. In Figure 7-5d there is horizontal torsion produced by load in the X direction, but the two walls in the Y direction form a couple that can provide torsional or rotational resistance. In Figure 7-5e, the tube form offers excellent resistance for horizontal loads in any direction. The arrangement in Figure 7-5f is not only satisfactory with respect to horizontal and rotation resistance, but it has the additional advantage of permitting the corners of the building to move quite

FIGURE 7-5
PLAN VIEWS OF SHEAR WALL LAYOUTS.

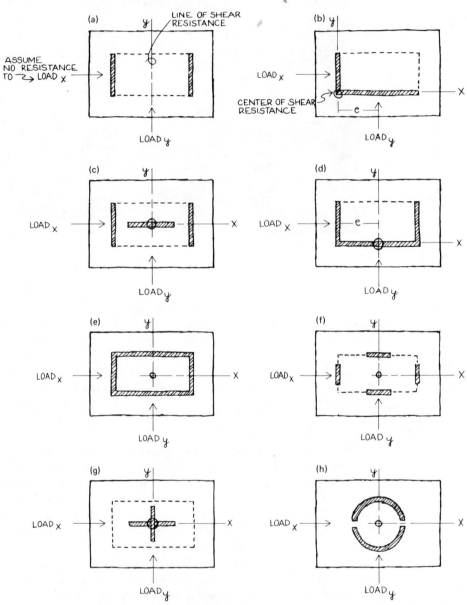

NOTE : AXIAL CENTROIDS, NOT SAME AS LATERAL CENTROIDS. FOR LATERAL CENTROIDS,
DETERMINE IN TERMS OF <u>EFFECTIVE</u> SHEAR RESISTING AREAS :

A. $V_{RESIST.}$ $\propto$ STIFF. <u>IN DIRECTION OF</u> LOAD APPLICATION (ie X OR y ABOVE):

B. TO FIND SHEAR RESISTANCE CENTROIDS, INTERSECT LINES GIVEN BY:
ξ EFFECTIVE AREA $M_x / A_{x\,EFF.}$ = y LINE AND ξ EFFECTIVE AREA $M_y / A_{y\,EFF.}$ = X LINE

C. $M_{TORSION}$ = LOAD X ECCENTRICITY AND CAN BE RESISTED BY A WALL COUPLE ONLY.

a bit as influenced by temperature, creep, and shrinkage (as discussed in Chapter 5). The arrangement in Figure 7-5g represents an unusual case where sufficient shear, but not torsional resistance, may be afforded by the perpendicular walls. In fact, it is similar to Figure 7-5b in that the system as a whole provides little torsional resistance to an unsymmetrical horizontal force on the building such as due to wind turbulence or, in the case of earthquake, due to asymmetric mass distribution (as discussed in Chapters 2 and 3). Therefore, the arrangement in Figure 7-5g is not, in itself, a

FIGURE 7-6
OVERALL DESIGN OF SHEAR WALL SUBSYSTEM.

(a) LOADS AND OVERALL RESISTANCE FORCES

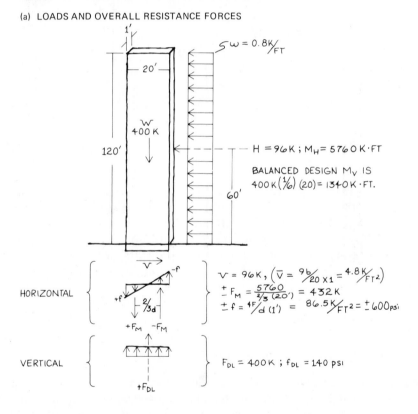

$$w = 0.8 \text{ K/FT}$$

$$H = 96 \text{ K}; M_H = 5760 \text{ K} \cdot \text{FT}$$

BALANCED DESIGN M_V IS
$$400 \text{ K} \left(\tfrac{1}{6}\right)(20) = 1340 \text{ K} \cdot \text{FT}.$$

HORIZONTAL

$$V = 96 \text{ K}, \left(\bar{V} = \tfrac{96}{20 \times 1} = 4.8 \text{ K/FT}^2\right)$$

$$\pm F_M = \tfrac{5760}{\tfrac{2}{3}(20')} = 432 \text{ K}$$

$$\pm f = \tfrac{4F}{d}(1') = 86.5 \text{ K/FT}^2 = \pm 600 \text{ psi}$$

VERTICAL

$$F_{DL} = 400 \text{ K}; f_{DL} = 140 \text{ psi}$$

(b) COMBINED STRESSES AT FOUNDATION

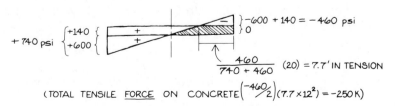

$$+ 740 \text{ psi} \begin{cases} +140 \\ +600 \end{cases}$$

$$\left.\begin{matrix} -600 + 140 = -460 \text{ psi} \\ 0 \end{matrix}\right\}$$

$$\frac{460}{740 + 460}(20) = 7.7' \text{ IN TENSION}$$

(TOTAL TENSILE <u>FORCE</u> ON CONCRETE $\left(\tfrac{-460}{2}\right)(7.7 \times 12^2) = -250 \text{ K})$

FIGURE 7-7
DESIGN OF A TRUSS WALL SUBSYSTEM.

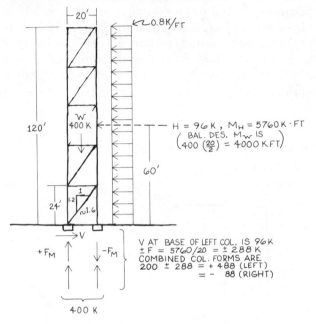

desirable layout. Curved walls (*h*) can offer good lateral resistance by virtue of shell action, especially if the floors serve as diaphragms stiffening the shell (to be discussed in Chapter 11).

It is possible to rigidly connect stiff floor or roof systems to walls and gain frame action in the transverse direction. In this case, the wall will act like a wide column in the transverse direction, and design will be similar to that discussed for rigid frames in Section 4 of this chapter. But in most cases a shear wall component will be designed to be stiff only in its longitudinal plan dimension (Figure 7-6) and may be of trussed construction (Figure 7-7).

EXAMPLE 7-1: Design of a Shear Wall

Referring to Figure 7-6: The reinforced concrete wall is 20 ft wide, 1 ft thick, and 120 ft high, with a vertical load of 400 k acting on it at the base. As a result of wind, we assume a uniform horizontal force of 0.8 kips per linear foot of vertical height acting on that wall. It is required to compute the bending stresses and the shearing stresses in the wall to resist the applied horizontal force. Computation follows:

$$V_{max} = wL = 0.8 \times 120 = 96 \text{ k}$$

$$M_{max} \text{ for cantilever} = \frac{wL^2}{2} = 0.8 \times \frac{120^2}{2} = 5760 \text{ k-ft}$$

This will produce an eccentricity of $5760/400 = 14.4$ ft which is way outside the balanced design eccentricity for a 20 ft wall ($20/6 = 3.3$ ft). Hence, there will be tension at the base (i.e. maximum balance $M = 400 \times 3.3 = 1330$ k-ft).

The wall has a moment of inertia:

$$I = \frac{bd^3}{12} = 1 \times \frac{20^3}{12} = 667 \text{ ft}^4$$

The maximum bending stress is

$$f_{max} = \frac{Mc}{I} = 5760 \times \frac{10}{667} = 86.5 \frac{k}{ft^2} = \pm 600 \text{ psi}$$

The average unit shearing stress is

$$\bar{v} = \frac{V}{A} = \frac{96}{(1 \times 20)} = 4.8 \frac{k}{ft^2} = 33.3 \text{ psi}$$

(or, using the rectangular section formula, $v_{max} = 3/2(V/A) = 50$ psi). Since a concrete wall with nominal reinforcement will carry at least 100 psi in shear (see Appendix D-1), these computed shearing stresses are certainly permissible.

At the bottom of this wall, the total vertical load of 400 k will produce a vertical compressive stress of $400 \div 20 = 20$ k/ft^2 or ~140 psi. This vertical stress will add to and subtract from the bending stresses previously computed at ± 600 psi. Thus the design will have to resist tension of -460 psi and compression of $+740$ psi at the foundation (Figure 7-6b). You can also get these values by using the balanced design M-ratio (as in Chapter 2)

$$\left(\frac{5760}{1330} \times 140 \simeq \pm 600 \text{ bending stress} \right).$$

Either way, the total compressive stress of 740 psi is quite low when compared with allowable values of around 1800 psi for concrete alone (Appendix D-1), and the tensile stress of -460 psi can be resisted with bars, an approximate design for which can be worked as follows.

If the stresses are -460 psi and 740 psi as computed, the length of the tension area is $[460/(740+460)]\,20 = 7.7$ ft, and the total tensile force in this triangular stress block is $(460/2) \times (7.7 \times 12) \times 12 = 250$ k. For allowable stress under wind (or seismic) load of 20 ksi $\times \frac{4}{3} = 26.7$ ksi in the reinforcing bars, $A_s = 250/26.7 = 9.4$ in.^{2} This amount of reinforcement should be provided at both ends of the wall section, since wind or earthquake can act in any direction.[1] In addition, the foundation will have to be designed to carry that tensile uplift force.

---

[1]You should rework this example for a typical *EQ* load of $w = 1.6$ k/ft at the top of the wall and assuming the building is of rectangular plan (refer to Figure 2-9).

EXAMPLE 7-2: Design of a Trussed Shear Wall

Referring to Figure 7-7: A 20 ft × 120 ft steel truss is erected in place of the concrete wall in the previous example and that it is desired to compute the major forces acting in the chord members of this truss at the foundation.

Solution: Using the maximum moment of 5760 kip-ft in Example 7-1 we can compute the C-T force in the chord (or column) for a lever arm of 20 ft,

$$\text{Chord Force} = \frac{M}{20\,\text{ft}} = \frac{5760}{20} = \pm 288\,k \text{ in the columns}$$

If we add the effect of 400 k vertical load on the two columns (400 k/2 = 200 k), we see again that the foundation design must be able to resist a

FIGURE 7-8
SHEAR WALLS MAY BE OF COLUMNS WITH CONCRETE OR MASONRY INFILL.

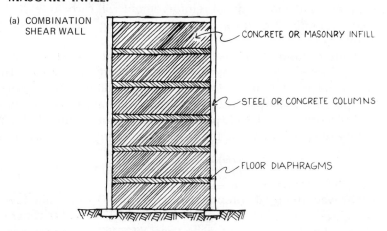

(a) COMBINATION SHEAR WALL

CONCRETE OR MASONRY INFILL

STEEL OR CONCRETE COLUMNS

FLOOR DIAPHRAGMS

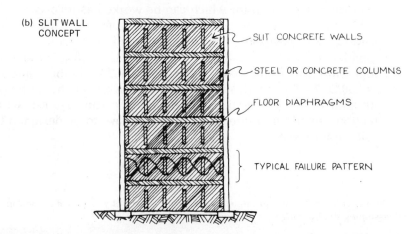

(b) SLIT WALL CONCEPT

SLIT CONCRETE WALLS

STEEL OR CONCRETE COLUMNS

FLOOR DIAPHRAGMS

TYPICAL FAILURE PATTERN

tensile force of $(-288 + 200 = -88 \text{ k})$ in the columns to prevent overturning. The force in the diagonal to resist the horizontal shear of 96 k is

$$\frac{\sqrt{20^2 + 24^2}}{20} \times 96 \text{ k} = 154 \text{ k}$$

One can determine for oneself the modifications required to produce a design that will have no tension in the foundation of the column.

The preceding two examples illustrate that some major preliminary calculations can be quickly made by working with overall form properties and dimensions of buildings and subsystems in order to get an idea whether or not a design concept is feasible or in what direction refinements must go to produce the greatest improvement.

In addition to solid or truss-wall constructions, we may also use in-filled frame walls. For example, we may have two steel columns having their space in-between filled with reinforced concrete (Figure 7-8a). Or we may have reinforced concrete columns filled in-between with brick masonry. Both schemes will provide a stiff wall, although their exact strength and behavior is not too well known. Both frame and shear wall behavior are involved. Figure 7-8b shows a special "slitwall" design where a concrete wall is slit vertically (similar to Figure 3-32) so that, in case of catastrophic earthquakes, predetermined breakage would occur between these slit lines. Such a design will allow the building to sway more than a solid design and thus sustain higher earthquake forces without brittle collapse. If the walls and columns are rigidly connected together without the slits, some abrupt or brittle shear cracking initiating in the walls may possibly extend into the columns, thus causing failure of the columns themselves and perhaps the total building.

SECTION 3: Shafts

Vertical shear-resisting shafts in a building act as tube structures and generally have a rectangular or square cross-section. But in the case of a round building, they may be circular (Photo 7-2). In any case, when there is only one shaft in a building, it is usually located central to a plan. When there is more than one shaft, they may be arranged in various locations, preferably symmetrically placed. Since structural shafts will often serve as vertical transportation and service cores, they will have a more-or-less limited number of service-system holes or door penetrations or, when located at the exterior, possibly windows. For the purpose of a preliminary design as structural tubes, the existence of these penetrations should be qualitatively considered, but their effect on the design may not be

PHOTO 7-2
**THIRTY-THREE STORY P. C. APARTMENT BUILDING UNDER
CONSTRUCTION AT LONG BEACH, CALIFORNIA. (T. Y. LIN
INTERNATIONAL, STRUCTURAL ENGINEERS).**

computed. For example, a tube with a moderate amount of holes (say, less
than 30 percent) will be somewhat reduced in strength and stiffness
compared to one without penetrations. But, for preliminary purposes, the
effect may be overlooked. If more than 60 percent of the shaft surface is
open, the action will be more like a frame tube and the strength and
stiffness will be reduced accordingly.

When a shaft is relatively short and wide say, with an aspect ratio under 1 or 2), the dominant structural action is that of a stiff shear-resisting tube. The moment or bending requirements of such a short shaft will usually not control the design. When the aspect ratio is higher (say, above 3 or 5), then shear forces may not be the controlling criterion, and the bending requirements may determine the design. For more slender shafts with

PHOTO 7-3(a)
THE KNIGHTS OF COLUMBUS BUILDING, DESIGNED BY KEVIN ROCHE, JOHN DINKELOO AND ASSOCIATES. FOUR CORNER SHAFTS CARRY BOTH VERTICAL AND HORIZONTAL LOADS.

PHOTO 7-3(b)
THE KNIGHTS OF COLUMBUS BUILDING.
DETAIL OF FLOOR FRAMING FROM CORE TO SHAFTS.

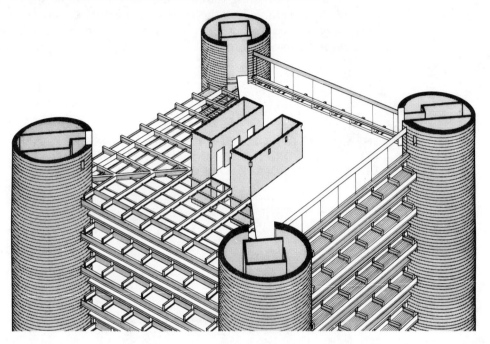

aspect ratios above 5, bending will definitely tend to dominate. At 7 or more, the design problem will be one of excessive flexibility and may require tying two or more shafts together with heavy connectors to get some overall mega-frame action as referred to previously (Photos 7-3a and 7-3b).

As wholes, shafts are space structures in that they are usually significantly stiff and strong in any direction. Accordingly, the computation of forces and the stresses produced in these shafts is somewhat more complicated than for walls. But, as shown in Chapter 3 and here in Figure 7-9, they can be approximated by using certain portions of the shafts as the effective shear-resisting area, and other portions of the shafts as the effective moment-resisting area. Greater accuracy can be obtained by using proper formulas as illustrated in Example 7-3.

FIGURE 7-9
ROUGH APPROXIMATION OF LOADS, MOMENTS, AND RESISTANCE STRESS IN A TUBE STRUCTURE.

(a) OVERALL MOMENT AND SHEAR (ASSUME WALLS ARE 1FT. THICK)

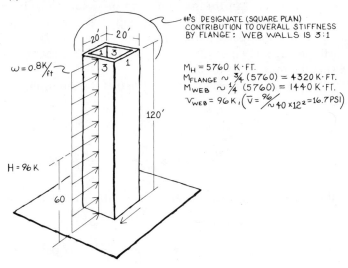

\#'S DESIGNATE (SQUARE PLAN) CONTRIBUTION TO OVERALL STIFFNESS BY FLANGE : WEB WALLS IS 3:1

$$M_H = 5760 \text{ K·FT.}$$
$$M_{FLANGE} \sim \tfrac{3}{4}(5760) = 4320 \text{ K·FT.}$$
$$M_{WEB} \sim \tfrac{1}{4}(5760) = 1440 \text{ K·FT.}$$
$$V_{WEB} = 96 \text{ K,} \left(\bar{v} = \tfrac{96}{\sim 40 \times 12^2} = 16.7 \text{ PSI} \right)$$

$\omega = 0.8 \tfrac{K}{ft}$

$120'$

$H = 96 \text{ K}$

60

$20' \quad 20'$

(b) ROUGH ESTIMATE OF OVERTURN STRESSES AT FOUNDATION

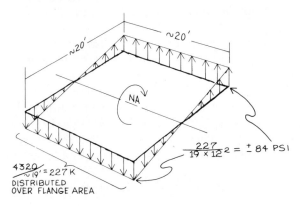

$\sim 20'$ $\sim 20'$

NA

$$\frac{227}{19 \times 12^2} = \pm 84 \text{ PSI}$$

$\frac{4320}{\sim 19'} = 227 \text{ K}$
DISTRIBUTED OVER FLANGE AREA

EXAMPLE 7-3: Design of a Tube Subsystem

Referring to Figure 7-9: The reinforced concrete shaft is 20 ft square, 120 ft high, and with walls 1 ft thick, is subjected to a lateral force of 0.8 kip per foot of height throughout its entire height. We will compare this shaft with the single wall in Example 7-1. Thus the total vertical load acting on the base of the 4-wall shaft is now increased to $4 \times 400 = 1600 \, k$, and it is required to determine the shear and bending resistance of this shaft. Referring to page 212, we have maximum moment $= 5760$ kip-ft and maximum shear at the base $= 96 \, k$.

By approximation, $\frac{3}{4}$ of the moment can be assumed to be resisted by flange action. Therefore, $\frac{3}{4} \times 5760 = 4320$ kip-ft. For a 19-ft lever arm between the flange centroids, each flange must withstand $4320/19 = \pm 227 \, k$ and the stress will be $227 \, k/(19 \, ft^2 \times 12^2) = \pm 84$ psi. The web stress due to moment will be ± 84 psi decreasing to (0) at the N–A.

More accuracy can be obtained by calculating the moment of inertia:

$$I = \Sigma \frac{bd^3}{12} = \frac{20^4}{12} - \frac{18^4}{12} = \frac{15.3^4}{12} = 4600 \, ft^4$$

$$f = \pm \frac{Mc}{I} = 5760 \times \frac{10}{4600} = \frac{12.5 \, k}{ft^2} = \pm 87 \, psi$$

In either case the average shear stress in the web is approximately given by:

$$\bar{v} = \frac{V}{A} = \frac{96}{(2 \times 20 \times 1)} = \frac{2.4 \, k}{ft^2} = 17 \, psi$$

The vertical load of 1600 k produces $1600/80 = 20 \, k/ft^2 = 140$ psi stress over the entire section. This stress can be combined with the above bending stress of ± 87 psi to obtain compressive stresses of 227 psi and 53 psi, and no tension will occur.

Comparing this example to Example 7-1, it is seen that much better stress conditions are produced in this shaft as compared to a single flat wall of 20 ft width.

Although there is no tensile stress in the concrete flanges and walls of the shaft, it would be desirable to reinforce the walls with steel so that greater strength and energy absorption is provided. As an extremely conservative approximation, one may neglect the effect of the axial load on the shaft and reinforce it to carry the entire $M = 5760$ kip-ft with a lever arm of about 19 ft.

$$T = \frac{M}{a} = \frac{5760}{19} = 302 \, k$$

$$f_s = 20 \, ksi \times \frac{4}{3} \, ksi \textit{ for resisting wind overturn.}$$

and
$$A_s = \frac{302}{26.7} = 11.3 \, in.^2 \text{ for each flange}$$

Compared to A_c of 20 ft^2 for each flange, this is obviously a very small percentage of reinforcement, indicating that the arrangement is quite strong to resist the lateral force imposed on it.

SECTION 4: Rigid-Frame Subsystems Under Vertical Loads

A rigid frame can carry both vertical and horizontal loads. Here we will discuss the transmission of loads and moments in a rigid frame as produced by vertical loads. Frame resistance to lateral loads will be discussed in Section 5. In either case, it will be necessary to understand the concept of moment distribution in a rigid frame. For a rigorous description of the theory, one should refer to a standard textbook on structural analysis, or to the original paper "A Modified Method of Moment Distribution," by T. Y. Lin, *Trans. ASCE*, 1937. Figures 7-10 and 7-11 illustrate the basic concept of moment distribution, which is based on three definitions. (For simplicity, we consider only members of uniform moment of inertia I):

1. A fully Fixed-End Moment $FEM = wL^2/12$ (for a uniform load w on a member with span L).
2. Relative stiffness (K) of a member given by the I/L ratio of the member and as modified by Far-End Fixity Factor (FFF) (Figure 7-10).
3. Carry-over moment factor (maximum equals one-half if member is fully fixed at far end). This value is modified to reflect a less than fully-fixed condition as indicated by Figure 7-10.

FIGURE 7-10
THE RELATIVE STIFFNESS OF MEMBERS AROUND POINT *A* AS DETERMINED BY FAR END FIXITY.

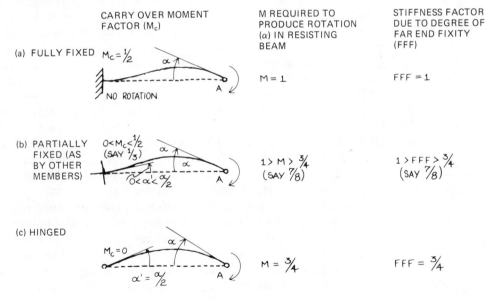

NOTE : THE PRODUCT OF FFF & $\frac{I}{L}$ YIELDS THE RELATIVE STIFFNESS (K) OF MEMBERS

FIGURE 7-11
THE BASIC CONCEPT OF MOMENT DISTRIBUTION.

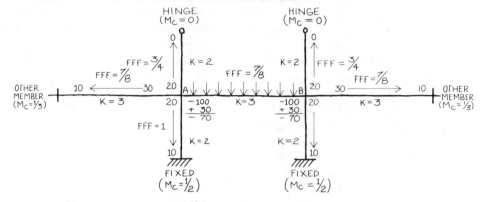

NOTE : 1) K = RELATIVE STIFFNESS OF EACH MEMBER AS
GIVEN BY A COMPARISON OF (FFF × I/L) VALUES.
(IN THIS EXAMPLE I/L VALUES ARE ASSUMED AS APPROPRIATE)

2) ΣK = TOTAL STIFFNESS FOR ALL MEMBERS AT A & B (i.e. 10 ABOVE)

3) RESTRAINT TO ROTATION IS PROPORTIONAL TO RATIO OF
$\dfrac{\text{ΣK FOR SUPPORTING MEMBERS AT JOINT}}{\text{ΣK OF ALL MEMBERS AT JOINT}}$) (TOTALS 70% ABOVE)

4) CARRY OVER FACTORS (M_C) DETERMINE PROPORTION OF RESISTING
MOMENT AT FAR END OF RESTRAINING MEMBERS

The sample case, Figure 7-11, represents a uniform load on member *AB* producing fixed-end moments of 100 units at each end. In relative terms, the four members at *A* and *B* represent a *total* (*K*) value of 10 units. If the total of relative *K* values for the *supporting* members at *A* and *B* is (3 + 2 + 2 = 7) as shown, the joint will rotate to the extent that the resulting moment at *A* or *B* is reduced from −100 (requiring full fixity) to a rotated value of −70. The restraining *M* of +70 at *A* or *B* is then distributed to the three supporting members in proportion to their relative stiffness factors (*K*) (i.e., in 3 : 2 : 2 ratio). The support member moments at *A* or *B* are then carried to their far ends by carry-over factors ranging from $\frac{1}{2}$ to 0, depending on the fixity of the far end. In this example, we have assumed "other members" at two extremities, which will allow some rotation (say $\frac{2}{3}$ fixity at the ends), and thus an *M* of 10 rather than 15 is obtained at each end.

It is recognized that the above offers an oversimplified explanation of moment distribution. It does not explain the effect of moment at *A* on that at *B* (or vice versa) and the distribution factors and carry-over factors have been more-or-less arbitrarily assumed. Uniform *I* values were assumed for each member (though allowing *I* to differ among members). The issue of side-sway due to asymmetry in stiffness or loading is neglected. And, finally, elastic behavior of all members and joints has been assumed.

But, from this simple illustration it can be inferred that in a regular rigid frame the vertical load on one beam will significantly affect only the members connected directly to it. Members farther away will be affected, but only to a rapidly decreasing extent. Thus, for the purposes of preliminary design, it is perhaps sufficient to make reasonable assumptions and examine a loaded member together with the stiffness of the immediately connected members only, as we have illustrated. This will be sufficient to give a basic idea of the overall requirements for design of a frame subsystem.

Some generalized statements can now be made. First, if the immediately connected members together are *much* more stiff than the beam in question (say, five or more times stiffer), then the beam can be assumed to approximate a fully-fixed-ended condition and the end moments can be easily estimated on that basis (e.g., $-M = wL^2/12$). Second, if the connected members together possess less than $\frac{1}{2}$ the total stiffness of the beam, they will offer only small rotational resistance and the loaded member will act more-or-less like a simply supported beam. When the connected members are of intermediate total stiffness (say, one to three times that of the beams), then somewhere in between will be the condition of end restraint and the end moments can be approximately estimated similar to Figure 7-11.

SECTION 5: Rigid-Frame Subsystems Under Horizontal Loads

As introduced in Chapter 3, when a rigid frame resists lateral forces, a method of approximate analysis is possible, although it is quite different from that for vertical loads. We now follow up on the overall frame action concepts in Chapter 3 and present the classical portal and cantilever methods, which have been traditionally used for the preliminary design and analysis of rigid-frame subsystems.

It should be emphasized that both the portal and the cantilever methods are approximate and are only sufficient for design of regular frames of simple and conventional proportions. Therefore, final design analysis should be made using available computer programs. However, the methods described here are needed to get the initial input into such programs, which requires approximate sizing of frame members. Furthermore, approximate methods are very useful at schematic and preliminary design stages in order to gain insight into the overall effectiveness and action of frame subsystems as various layouts are explored.

The portal method is more useful to depict columns and connecting girders bending at a particular building level as produced by lateral forces. The cantilever method is more useful in determining the overturning effect of lateral forces on tall frames and the axial loads on the columns produced by such overturning.

Portal Method

The portal method is based on the following major assumptions:

1. Each bay of a bent acts as a separate "portal" frame consisting of two adjacent columns and the connecting girder (Figure 7-12a).
2. The point of inflection for all columns is at midheight.
3. The point of inflection for all girders is at midspan.
4. For a multibay frame, the shears on the interior columns are equal and the shear in each exterior column is half the shear on an interior column.

Based on these assumptions, Figures 7-12, 7-13, 7-14, and 7-15 illustrate four simple cases of the portal method of analysis.

From these simple cases you should be able to deduce the basic portal method for approximate analysis of shear and moment in columns at any

FIGURE 7-12
CASE 1, A BASIC PORTAL FRAME.

(a) SINGLE STORY, SINGLE BAY FRAME

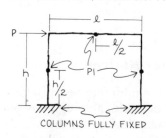

COLUMNS FULLY FIXED

(b) DEFORMED SHAPE

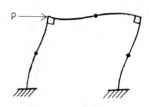

(c) COLUMN REACTIONS

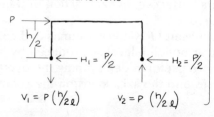

$$V_1 = P\left(\tfrac{h}{2\ell}\right) \qquad V_2 = P\left(\tfrac{h}{2\ell}\right)$$

APPROXIMATE ANALYSIS OF REACTIONS IN (c):

1. BY SYMMETRY; $H_1 = H_2 = H$
   $\Sigma H = 0$; $P - H_1 - H_2 = 0$, OR,
   $P - 2H = 0$; $\boxed{H = \tfrac{P}{2}}$

2. V_1 & V_2 (UPLIFT & THRUST) RESISTING OVERTURN MOMENT:
   $\Sigma V = 0$; $V_1 = V_2$
   $\Sigma M = 0 = P\left(\tfrac{h}{2}\right) - V_1\,\ell$; $\boxed{V_{1\,OR\,2} = P\left(\tfrac{h}{2\ell}\right)}$

3. (M) AT TOP (AND BASE) OF COLUMN:*
   $$\boxed{M_{COL} = H_1\left(\tfrac{h}{2}\right) = \tfrac{P}{2}\left(\tfrac{h}{2}\right) = \tfrac{Ph}{4}}$$
   * (BY SYMMETRY WHEN PI AT MIDPOINT)

NOTE: GIRDERS AND FOUNDATION WILL HAVE TO RESIST THE COLUMN MOMENTS

FIGURE 7-13
CASE 2, COLUMN BENDING MOMENTS IN A MULTIPLE-BAY FRAME.

(a) FOR A SUB-SYSTEM OF (n_b) BAYS, $(H_{EXT} = \dfrac{P}{2n_b}$ & $H_{INT} = \dfrac{P}{n_b}$;

MOMENTS ON EXTERIOR COLUMNS, $[\dfrac{P}{2n_b} (\dfrac{h}{2})]$, ARE ONE-HALF

AS LARGE AS ON INTERIOR COLUMNS $[\dfrac{P}{n_b} (\dfrac{h}{2})]$:

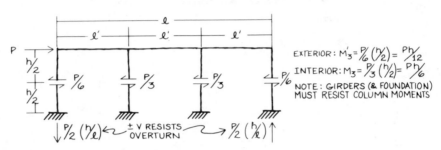

EXTERIOR: $M'_3 = \dfrac{P}{6} (\dfrac{h}{2}) = \dfrac{Ph}{12}$

INTERIOR: $M_3 = \dfrac{P}{3} (\dfrac{h}{2}) = \dfrac{Ph}{6}$

NOTE: GIRDERS (& FOUNDATION)
MUST RESIST COLUMN MOMENTS

(b) FOR THE BASIC CASE $(n_b = 1)$, $M_1 = \dfrac{Ph}{4}$. NOTE THAT FROM (a) ABOVE THE

MAXIMUM (INTERIOR) MOMENT $M_n = \dfrac{Ph}{2n_b}$ AND FOR

$(n_b \geqslant 2)$, $\boxed{\dfrac{M_n}{M_1} = \dfrac{2}{n_b}}$:

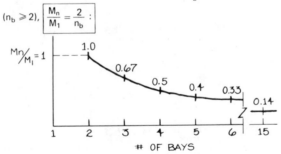

story of a building. But it must be remembered that the determination of column design requirements using this method assumes that the girders are designed to resist rotation of the columns. Thus, after the shears and moments are obtained for the columns, one should follow through with the determination of the moments produced in the girders. This can be done by referring to Figures 7-12 through 7-15 and to Figures 7-16 and 7-17. It should also be noted that moments on exterior columns are resisted by a girder on one side only while the moments on interior columns, which are twice as large, are resisted by girders on both sides of the column.

FIGURE 7-14
CASE 3, SINGLE BAY, MULTISTORY FRAMES.

(a) DISTRIBUTION OF TOTAL WIND LOAD (H) AND REACTIONS ($P = \frac{H}{2n_s}$)

ASSUME TRIBUTARY LOADS
CONCENTRATED AS SHOWN:

COLUMN MOMENTS:

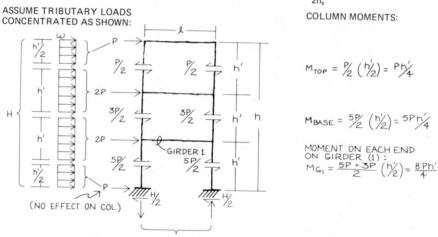

$$M_{TOP} = \frac{P}{2}\left(\frac{h'}{2}\right) = \frac{Ph'}{4}$$

$$M_{BASE} = \frac{5P}{2}\left(\frac{h'}{2}\right) = \frac{5Ph'}{4}$$

MOMENT ON EACH END
ON GIRDER (1):
$$M_{G_1} = \frac{5P + 3P}{2}\left(\frac{h'}{2}\right) = \frac{8Ph'}{4}$$

±V IS BEST DETERMINED FROM CANTILEVER METHOD

(b) WHEN THE LOADS AND MOMENTS FOR FRAMES y(n_s)
    STORIES ARE COMPUTED, IT BECOMES CLEAR THAT:

1) THE MAXIMUM HORIZONTAL FORCE H_n (AT THE 1ST STORY COLUMN MID POINTS)
   IS APPROXIMATED BY :
$$H_1 = \frac{(2n_s - 1)P}{2}$$

2) THE MAXIMUM COLUMN MOMENTS AT THE TOP & BOTTOM OF THE FIRST STORY IS
   GIVEN BY :
$$M_1 = \frac{(2n_s-1)P}{2}\frac{h'}{2} = \frac{(2n_s-1)Ph'}{4}$$

3) FOR ($n_s \geq 2$), THE FIRST STORY GIRDER MOMENT IS GIVEN BY :
$$M_{G_1} = (n_s - 1)Ph'$$

FIGURE 7-15
CASE 4, MULTIBAY, MULTISTORY FRAMES.

(a) DISTRIBUTION OF LOADS AND REACTIONS FOR
TWO STORY FRAME ($P = \dfrac{H}{2n_s}$ and $n_b = $ # BAYS):

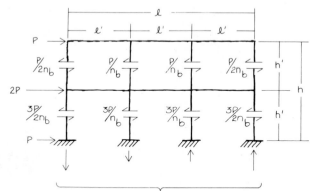

AT 1ST LEVEL
FOR $n_b = 3$:
$H_{EXTERIOR} = P/2$
$H_{INTERIOR} = P$
$M_{EXTERIOR} = P/2\left(h'/2\right)$
$M_{INTERIOR} = P\left(h'/2\right)$

AXIAL FORCES BEST DETERMINED BY CANTILEVER METHOD

(b) IF ($n_b = $ # OF BAYS) AND ($n_s = $ # OF STORIES):

1. H & M FOR FIRST STORY COLUMNS GIVEN BY

EXTERIOR —

$$H_{EXT.} = \frac{(2n_s - 1)P}{2n_b}$$

$$M_{EXT.} = \frac{(2n_s - 1)P}{2n_b}\left(h'/2\right)$$

INTERIOR —

$$H_{INT.} = 2\,H_{EXT.}$$

$$M_{INT.} = 2\,M_{EXT.}$$

2. FIRST STORY GIRDER MOMENT (FOR $n_s \geq 2$) :

$$\pm M_{G_1} = \pm \frac{(n_s - 1)Ph'}{n_b}$$

(c) A COMPARISON OF THE MAXIMUM (INTERIOR) FIRST STORY
COLUMN MOMENT (M_n) FOR MULTI-BAY, MULTI-STORY
FRAMES WITH (M_1) FOR A BASIC SINGLE FRAME CASE:

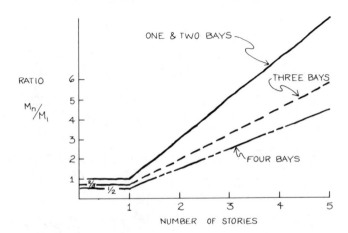

RATIO

M_n/M_1

ONE & TWO BAYS

THREE BAYS

FOUR BAYS

NUMBER OF STORIES

FIGURE 7-16

GIRDER MOMENT AS DETERMINED BY EXTERIOR COLUMN MOMENTS.

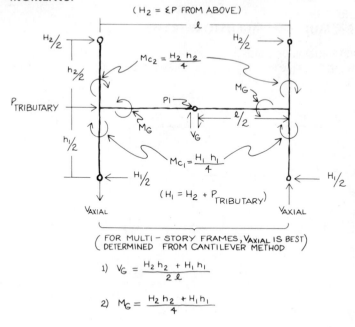

1) $V_G = \dfrac{H_2 h_2 + H_1 h_1}{2\ell}$

2) $M_G = \dfrac{H_2 h_2 + H_1 h_1}{4}$

FIGURE 7-17

GIRDER MOMENT AS DETERMINED BY INTERIOR COLUMN MOMENTS.

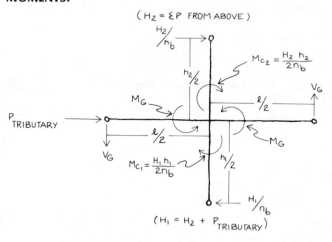

(a) FOR EQUAL GIRDER SPANS

1) $V_G = \left[\dfrac{H_1 h_1 + H_2 h_2}{2n_b \ell} \right]$

2) $M_G = \left[\dfrac{H_1 h_1 + H_2 h_2}{4n_b} \right]$

(b) FOR UNEQUAL GIRDER SPANS, M_G WILL BE UNEQUAL AS
    INDICATED BY MULTIPLYING (2) ABOVE BY THE GIRDER SPAN RATIO.

Cantilever Method

The cantilever method is particularly useful in determining axial forces in the columns due to overturning moment. It is assumed that as the building bends, for each story, the change in unit axial stress in each column is proportional to its distance from the center line of the building (Figure 7-18). Note that as the entire frame bends toward the left, Column 1 is

FIGURE 7-18
THE CANTILEVER METHOD OF APPROXIMATING AXIAL FORCES IN FRAME COLUMNS.

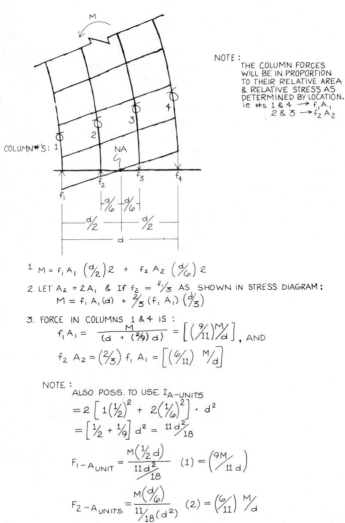

NOTE:
THE COLUMN FORCES WILL BE IN PROPORTION TO THEIR RELATIVE AREA & RELATIVE STRESS AS DETERMINED BY LOCATION.
ie #s 1 & 4 → $f_1 A_1$
2 & 3 → $f_2 A_2$

[1] $M = f_1 A_1 \left(\dfrac{d}{2}\right) 2 + f_2 A_2 \left(\dfrac{d}{6}\right) 2$

[2] LET $A_2 = 2 A_1$ & If $f_2 = \frac{f_1}{3}$ AS SHOWN IN STRESS DIAGRAM:
$$M = f_1 A_1 (d) + \frac{2}{3} (f_1 A_1) \left(\frac{d}{3}\right)$$

[3]. FORCE IN COLUMNS 1 & 4 IS:
$$f_1 A_1 = \frac{M}{(d + (2/9) d)} = \left[(9/11) \frac{M}{d} \right], \text{ AND}$$

$$f_2 A_2 = (2/3) f_1 A_1 = \left[(6/11) \frac{M}{d} \right]$$

NOTE:
ALSO POSS. TO USE $I_{A-UNITS}$
$$= 2 \left[1 \left(\frac{1}{2}\right)^2 + 2 \left(\frac{1}{6}\right)^2 \right] \cdot d^2$$
$$= \left[\frac{1}{2} + \frac{1}{9} \right] d^2 = \frac{11 d^2}{18}$$

$$F_{1-A_{UNIT}} = \frac{M \left(\frac{1}{2} d\right)}{\frac{11 d^2}{18}} \quad (1) = \left(\frac{9M}{11 d}\right)$$

$$F_{2-A_{UNITS}} = \frac{M \left(\frac{d}{6}\right)}{\frac{11}{18} (d^2)} \quad (2) = \left(\frac{6}{11}\right) \frac{M}{d}$$

highly compressed while Column 4 is highly tensioned. Interior Column 2 is compressed and Column 3 is tensioned, but to a smaller extent than the exterior columns. If the distances between these columns are the same, it can be assumed that the axial unit stress in Columns 2 and 3 is only one-third of that in Columns 1 and 4. Figure 7-18 also illustrates how the forces can be determined. For approximation purposes, this method is useful because it yields somewhat higher axial forces than the portal method, since bending moment in the columns is neglected.

The following example illustrates the use of both portal and cantilever methods.

EXAMPLE 7-4: Rigid Frame Analysis

An approximate rigid-frame analysis is to be made for a multistory office building, under the action of lateral forces. This example shows the tenth floor of a 40-story building. It is intended to determine the approximate moments in members on this floor as produced by the lateral forces. The loads on this particular floor are given:

The overturning moment = 75,000 kip-ft.
The total gravity load = 10,000 k.
The total story shear = 336 k above the floor and 348 k below the floor.

EXAMPLE 7-4
RIGID FRAME ANALYSIS.

$M_{OVERTURN} = 75,000\ K'$

GRAVITY

10,000 K

2000 K   3000 K   3000 K   2000 K

ΣP
(FROM ABOVE)

← 56 K   ← 112 K   ← 112 K   ← 56 K ⇐ 336 K

6.5'   25'   25'   25'

6.5'   ↙A   ↙½ A ₵   ↙½ A   ↙A

⇐ 12K
TRIBUTARY
(THIS LEVEL)

58 K →   116 K →   116 K →   58 K →

F_{EXT} F_{INT} F_{INT} F_{EXT}

|—12.5'—|—12.5'—|

|——— 37.5' ———|——— 37.5' ———|

AXIAL FORCES DUE TO $M_{OVERTURN}$

From previous analysis it has been ascertained that the gravity load of 10,000 k is shared by these columns in the following manner: interior columns carry 3000 k each, while the exterior ones carry 2000 k each.

Using the portal method, the shears of 336 k and 348 k are distributed among the columns with the interior ones carrying twice as much as the exterior ones.

Now let us compute the bending moments in the columns and the girders produced by the horizontal shear. The exterior columns will be subjected to an end moment of:

$$M_{\text{col above}} = 56 \text{ k} \times 6.5 = 360 \text{ kip-ft}$$
$$M_{\text{col below}} = 58 \text{ k} \times 6.5 = \underline{373 \text{ kip-ft}}$$
$$M_{\text{Gdr}} \qquad\qquad\quad = 733 \text{ kip-ft}$$

This shows that moment in the girder is 733 kip-ft at the exterior end. You should now try the interior columns and girders; you will find that the interior columns will be subjected to twice the moment compared to the exterior ones. However, the girders will have the same end moments.

Using the cantilever method, the total overturning moment of 75,000 kip-ft is resisted by change in axial unit stress (f) in the exterior column, which is three times that in the interior columns. Since each exterior column carries one-half the vertical load in the interior column, *plus the exterior walls*, we may assume that the steel area (A) of the exterior columns is about $\frac{2}{3}$ that of the interior column. Then fA is the force in the exterior columns, and $(f/3)[(3/2)A] = fA/2$ is the force in the interior columns. Thus we have:

$$fA(37.5)2 + (f/3)[(3/2)A](12.5)2 = 75{,}000 \text{ kip-ft}$$
$$75fA + 12.5fA = 75{,}000$$
$$87.5fA = 75{,}000$$
$$fA = 855 \text{ k}$$

This indicates that the *exterior* columns will be subjected to an axial force of 855 k compression on one side and tension on the other. The axial force on the *interior* columns will be half that value, or 427 k. Since either wind or earthquake force can reverse its direction, we must assume that each exterior column will be subjected to 855 k tension or compression due to overturning moment, and under combined effect of gravity load plus overturning moment, the columns take 2855 k and 1145 k compression.

Since each exterior column carries 2000 k under gravity loading, it can carry $\frac{4}{3} \times 2000 = 2670$ k axial compression under lateral loading. Hence the value of 2855 k compression would only require a slight additional amount of steel for axial resistance only. However, together with the bending moment at the column ends due to lateral loads, the high combined stresses will call for further strengthening of the column section.

SECTION 6: Approximate Lateral Deflections of Vertical Elements

Even at schematic or preliminary stages of design it is important that the designer appreciate the significance of lateral deflections of buildings (sometimes termed *horizontal drift*). Two important considerations make this necessary. First, the lateral deflection of buildings is often limited by code requirements. For example, some codes limit the deflection in each story (or building) to a drift index of *h*/500, where *h* is the height of the story (or building). This is important because occupants should not experience uncomfortable horizontal movement. Furthermore, a building that deflects severely under lateral forces may have damage problems associated with racking and vibration.

A third reason for determining lateral deflections is to get some idea of the relative horizontal load carried by the various vertical subsystems in a building. For the purpose of design, it is very useful to know how much load is carried by the shafts as compared to the frames and the walls in the building. This can be obtained by comparing their stiffnesses, as indicated by their deflections, since they will have to move together as a building unit. In general one can say that load goes to stiffness.

Admittedly this poses a difficult analytical problem, and must eventually be resolved by advanced computer programs. However, such programs require initiating input, and one will need to have preliminary dimensions for this purpose. Furthermore, such computations may yield insight that will affect the overall arrangements in a building scheme. Hence it can be helpful for one to know how to make approximate deflection analysis to arrive at some preliminary layouts and dimensions.

We will now briefly discuss and illustrate five cases in which overall deflections of wall, shaft, and frame subsystems can be roughly approximated. In each case, the lateral deflection in a building can be conceived as deriving from two sources. One is produced by the overturning moment (axial forces), and the other by horizontal and vertical shear (racking forces).

For example, in a squat wall, the shear deflection component predominates and may be approximated as shown in Case 1 below, neglecting moment deflection. For a slender wall, moment deflection predominates and the shear deflection component is usually neglected as in Case 2. "Short" walls are those with an aspect ratio less than 1 or 2.

When two slender walls are connected by lintels (if the lintels are heavy enough), the entire subsystem can be made to act as one wide cantilever supported at the foundation level. In that case deflections would be small. But if the lintels are lighter, their deformation may contribute a lot toward the lateral deflection. Thus it is possible to express the overall deflection of the entire system in terms of the lintel rotations, as illustrated in Case 3. In this case, the lintels resist the vertical shear between the two end walls and the cantilever deflection of the two wall components is neglected.

FOR PURPOSES OF ROUGH APPROXIMATION ONLY, FIVE BASIC CASES OF
OVERALL DEFLECTION MAY BE DEALT WITH AS FOLLOWS:

CASE (1)

A SHORT SOLID WALL IS DEFLECTED MORE BY SHEAR THAN BY BENDING DEFORMATION:

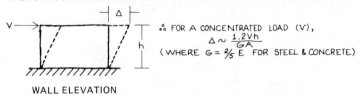

∴ FOR A CONCENTRATED LOAD (V),

$$\Delta \sim \frac{1.2Vh}{GA}$$

(WHERE $G = \frac{2}{5} E$ FOR STEEL & CONCRETE)

WALL ELEVATION

CASE (2)

A TALL SOLID WALL OR TUBE IS DEFLECTED MORE BY FLEXURE:

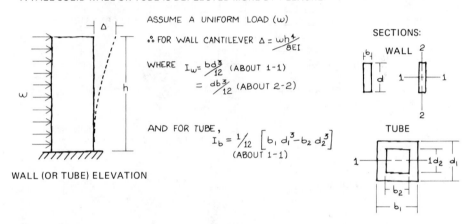

ASSUME A UNIFORM LOAD (ω)

∴ FOR WALL CANTILEVER $\Delta = \frac{\omega h^4}{8EI}$

WHERE $I_w = \frac{bd^3}{12}$ (ABOUT 1-1)

$= \frac{db^3}{12}$ (ABOUT 2-2)

AND FOR TUBE,

$$I_b = \frac{1}{12}\left[b_1 d_1^3 - b_2 d_2^3\right]$$
(ABOUT 1-1)

SECTIONS:

WALL

TUBE

WALL (OR TUBE) ELEVATION

CASE (3)

A TALL WALL PIERCED BY MANY DOOR OR WINDOW OPENINGS BECOMES TWO SLENDER WALLS
CONNECTED WITH LINTELS. OFTEN, THE DEFLECTION RESULTING FROM THE LINTEL DE-
FORMATION BECOMES MUCH MORE SERIOUS THAN THE WALL DEFORMATION ITSELF. AS AN
APPROXIMATION, WE CAN THEN COMPUTE ONLY THE DEFLECTION RESULTING FROM LINTEL
DEFORMATION.

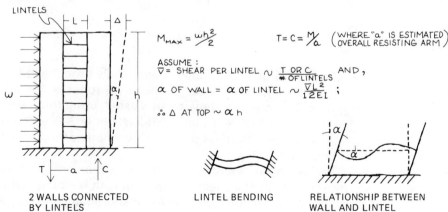

$$M_{MAX} = \frac{\omega h^2}{2}$$

$$T = C = \frac{M}{a}$$ (WHERE "a" IS ESTIMATED OVERALL RESISTING ARM)

ASSUME :

$\overline{V}$ = SHEAR PER LINTEL $\sim \frac{T \text{ OR } C}{\# \text{ OF LINTELS}}$ AND,

α OF WALL = α OF LINTEL $\sim \frac{\overline{V}L^2}{12EI}$;

∴ Δ AT TOP $\sim \alpha h$

2 WALLS CONNECTED
BY LINTELS

LINTEL BENDING

RELATIONSHIP BETWEEN
WALL AND LINTEL
DEFORMATION

CASE (4)

A REGULAR RIGID FRAME HAS ITS DEFLECTION PRODUCED BY (a) BENDING OF THE COLUMNS, AND (b) BENDING OF THE GIRDERS. FOR A TALL BUILDING, THE COLUMNS ARE HEAVY, HENCE BENDING OF THE GIRDERS CONTROLS THE DEFLECTION. FOR A SQUAT BUILDING, COLUMNS ARE LIGHT AND CONTRIBUTE MORE TOWARD THE DEFLECTION OF THE FRAME. FOR APPROXIMATION WE CAN REFER TO A TYPICAL BENT:

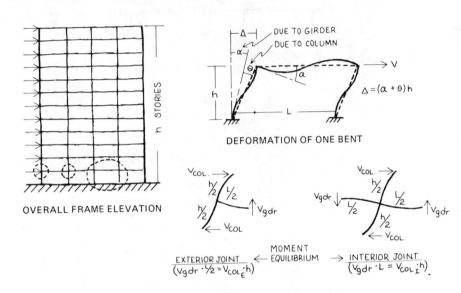

OVERALL FRAME ELEVATION

DEFORMATION OF ONE BENT

$\Delta = (\alpha + \theta) h$

MOMENT EQUILIBRIUM

EXTERIOR JOINT $(V_{gdr} \cdot \frac{l}{2} = V_{COL_E} \cdot h)$ INTERIOR JOINT $(V_{gdr} \cdot L = V_{COL_I} \cdot h)$

THEREFORE:

(a) Δ_n FOR FIRST STORY :

    1. EXTERIOR JOINT —

$$\Delta_{COL.} = \frac{V_{COL_E} \cdot h^3}{12 \, EI_{COL_E}} \quad ; \quad \Delta_{gdr} = \frac{V_{gdr} \, L^2 h}{12 \, EI_{gdr}} = \frac{2 V_{COL_E} \, Lh^2}{12 \, E \, I_{gdr}}$$

$$\Delta_{l_E} = \Delta_{col_E} + \Delta_{gdr} = \frac{V_{COL_E} \cdot h^2}{12 \, E} \left[\frac{h}{I_{COL_E}} + \frac{2l}{I_{gdr}} \right]$$

    2. INTERIOR JOINT —

$$\Delta_{COL.} = \frac{V_{COL_I} \cdot h^3}{12 \, E \, I_{col_I}} \quad ; \quad \Delta_{gdr} = \frac{V_{gdr} \cdot L^2 h}{12 \, E \, I_{gdr}} = \frac{V_{COL_I} \, Lh^2}{12 \, E \, I_{gdr}}$$

$$\Delta_{l_I} = \Delta_{col_I} + \Delta_{gdr} = \frac{V_{COL_I} \cdot h^2}{12 \, E} \left[\frac{h}{I_{COL_I}} + \frac{l}{I_{gdr}} \right]$$

(b) $\Delta_{TOTAL} = \frac{\Delta_l}{2} \times n_{STORIES}$ (FOR EITHER INTERIOR OR EXTERIOR JOINTS)

Deflection of a rigid frame is essentially produced by shear between stories, which, as discussed previously, produces vertical shear on the girders. This can be estimated by Case 4, using the portal method. Thus, a rigid frame will deflect somewhat like Figure 7-19 with the overall aspect ratio and girder stiffness determining whether the cantilever or portal mode is dominant.

The two major sources of lateral deflection are the bending of columns in resisting horizontal shear and of girders in resisting vertical shear. The cantilever deflection due to column shortening and lengthening (produced by overturning moment) is usually of secondary importance until the

FIGURE 7-19

DEFLECTION OF A RIGID FRAME SUBSYSTEM DERIVES FROM OVERALL SHEAR AND MOMENT-RESISTING ACTION IN COLUMNS.

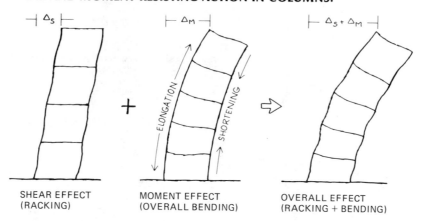

SHEAR EFFECT
(RACKING)

MOMENT EFFECT
(OVERALL BENDING)

OVERALL EFFECT
(RACKING + BENDING)

building is some 40 stories or higher. For a rigid frame, lateral deflection (also termed *side-sway*) may result from unsymmetrical vertical loading on the beam. For example, the frame in Figure 7-20 will deflect as shown. Such unsymmetrical loadings on various floors of a multistory frame will tend to balance themselves, and the side-sway so produced is usually neglected.

Trussed walls do not rely on rotational stiffness of frame joints and are more efficient than rigid frames in providing stiffness, because each member is uniformly and more fully stressed. Their deflections are also simpler to estimate, as shown in Case 5.

The Case 1-5 formulas can be useful, but the reader should understand that they are grossly simplified. They should only be used for the purpose of getting a very rough idea as to the relative stiffness of various subsystems and for obtaining an approximate measure of the lateral deformation of a building. They can also be very helpful when generating and comparing

FIGURE 7-20

SIDE-SWAY MAY RESULT FROM UNSYMMETRICAL LOADINGS.

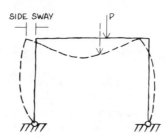

CASE (5)

DEFLECTION OF A TRUSS IS ESSENTIALLY DUE TO AXIAL ELONGATION
AND SHORTENING OF ITS MEMBERS:

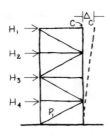

THE TOTAL Δ AT C IS GIVEN BY :

$$\Delta_C = \Sigma \; \frac{\overline{P} \; PL}{EA} \quad \text{WHERE,}$$

P = FORCE IN ANY MEMBER DUE TO LOADING ON THE WHOLE SYSTEM
L = LENGTH OF MEMBER
A = AREA OF MEMBER
E = MODULUS OF ELASTICITY OF MEMBER
$\overline{P}$ = FORCE IN SAME MEMBER DUE TO A UNIT
LOAD (1) APPLIED IN THE DIRECTION
OF THE DEFLECTION SOUGHT, AND
AT THE POINT IN QUESTION (ie. C IN THIS EXAMPLE). NOTE THAT
$\overline{P}$ IS A GEOMETRY FACTOR SINCE FORCE/UNIT FORCE IS DIMENSIONLESS.

OR THE ABOVE REPRESENTS:

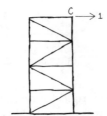

$$\Delta_C = \Sigma \overline{P} \; \delta \quad \text{WHERE,}$$

δ = CHANGE IN LENGTH OF ANY
MEMBER DUE TO LOADING ON SYSTEM.
$$= \frac{PL}{AE}$$

IF WE NEGLECT DEFLECTION DUE TO WEB MEMBERS, BUT CONSIDER
ONLY THE LENGTHENING AND SHORTENING OF THE COLUMNS
(THE CHORDS), WE MAY *APPROXIMATE* AS FOLLOWS:

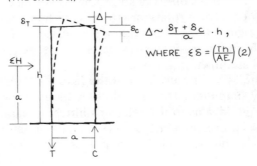

$$\Delta \sim \frac{\delta_T + \delta_C}{a} \cdot h \, ,$$

$$\text{WHERE} \quad \Sigma \delta = \left(\frac{Th}{AE}\right) (2)$$

schematic and preliminary design options, but eventually more accurate
methods must be applied to arrive at a final design evaluation. The above
will be illustrated in Example 7-5.

EXAMPLE 7-5: Transverse Deflection

This example is a continuation of Example 4-3 (Chapter 4, Section 5). It is
intended to determine the interaction between the central core and the
rigid frames, based on deflection calculations for these subsystems.

Typical plan, elevations, and floor section of the building are shown. The

EXAMPLE 7-5
TRANSVERSE DEFLECTION.

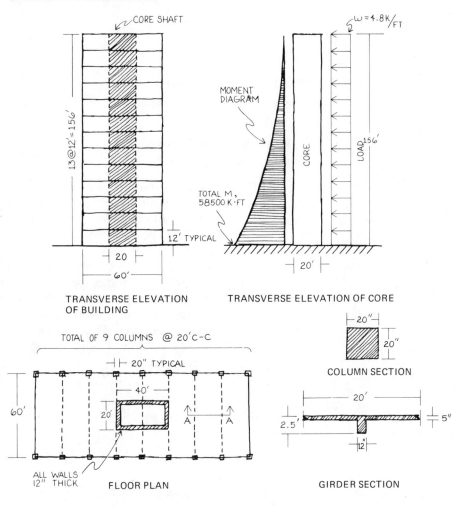

TRANSVERSE ELEVATION
OF BUILDING

TRANSVERSE ELEVATION OF CORE

COLUMN SECTION

FLOOR PLAN

GIRDER SECTION

lateral force-resisting elements are the center concrete shaft (20 ft × 40 ft in section and made up of four 12-in. walls) and the reinforced-prestressed concrete frames (made up of 12 in. × 30 in. T-beams and 20-in.-square reinforced concrete columns). Compute for $WL = 30$ psf in the transverse direction only.

The following simplifying assumptions are made:

1. Columns are of uniform sectional properties and height for all stories.
2. Shaft walls are of uniform thickness for all stories. Neglect wall openings.
3. (w) is uniform over the height of building.

(1) Compute bending deflection of top of shaft, as if it takes the entire lateral load *w*. Neglect shear deflection since shaft is slender.

$\Delta = wh^4/8EI$

$I = (1/12)(b_1 d_1^3 - b_2 d_2^3)$

$\quad = (1/12)(41 \times 21^3 - 39 \times 19^3)$

$\quad = (1/12)(380,000 - 267,000) = 9,400 \text{ ft}^4$

$E = 3 \times 10^6 \text{ psi} = 3 \times 10^3 \times 144 = 432,000 \text{ ksf}$

$w = 4.8 \text{ k/ft}, \ h = 156 \text{ ft}$

$\Delta = 4.8 \times 156^4/8 \times 432,000 \times 9,400 = 0.087 \text{ ft}$

$\Delta/h = 0.087/156 = 1/1800,$

which is much less than 1/500 as permitted by most building codes, and is within the usual index for concrete buildings, which ranges between 1/1000 and 1/2500. Also note that if the wall thickness is reduced, and if door openings are considered, the deflection will be correspondingly higher, maybe by some 10 or 20 percent.

Since the deflection due to moment increases rapidly at the top, the value of 1/1800 indicates only the average drift index for the entire building, whereas the *story* drift index may be higher, especially for the top floors.

(2) Compute deflection of top of frame, if each frame takes $\frac{1}{9}$ of total wind load moment and shear. Assume columns of lightweight concrete with $E = 3 \times 10^6$ psi (or 432,000 ksf) and neglect column shortening effect.

$$\Delta = \frac{\bar{V}_{col} h^2}{12 E_{col}} \left[\frac{h}{I_{col}} + \frac{2L}{I_{gdr}} \right]$$

$$I_{col} = \frac{(20/12)^4}{12} = 0.64 \text{ ft}^4; \text{ calculate } I_{gdr} \text{ as follows:}$$

N-A of girder section:

$$\begin{cases} 20 \times \dfrac{5}{12} = 8.33 \text{ ft}^2 \times \dfrac{-2.5 \text{ in.}}{12} = -1.74 \text{ ft}^3 \\[2mm] \underline{1 \times 2.08 = \ 2.08 \text{ ft}^2 \ \times \dfrac{12.5}{12} \quad = \ 2.17 \text{ ft}^3;} \\[1mm] \text{Totals:} \quad 10.41 \text{ ft}^2 \qquad\qquad\quad 0.43 \text{ ft}^3 \end{cases}$$

$$\frac{0.43}{10.41} = 0.041 \text{ ft below bottom of flange.}$$

$$\text{Therefore,} \quad 8.33 \left[\frac{(5/12)^2}{12} + (0.208 + 0.041)^2 \right] = 0.64 \text{ ft}^4$$

$$2.08 \left[\frac{2.08^2}{12} + (1.04 - 0.041)^2 \right] \quad = \underline{2.82 \text{ ft}^4}$$

$$I_{gdr} \text{ is } = 3.64 \text{ ft}^4$$

$$\text{For ground floor } \bar{V}_{col} = \frac{750}{2 \times 9} = 41.7 \text{ k/col.}$$

$$\Delta = \frac{41.7 \times 13^2}{12 \times 432,000} \left[\frac{12}{0.64} + \frac{2 \times 60}{3.46} \right]$$

$$= 0.00116[18.8 + 34.7] = 0.062 \text{ ft}$$

Since the story drift varies with the shear in the story, which decreases linearly to the top, the average drift will be 0.062/2 = 0.031 ft per story and the total deflection at top of building is approximately

$$13 \times 0.031 = 0.40 \text{ ft}$$

which indicates a drift ratio of

$$0.40 \text{ ft}/156 = 1/400$$

for the building as a whole, or

$$0.062/12 = 1/194$$

for the ground floor by itself.

   (3) Comparing the above 0.40 ft deflection with the shaft deflection of 0.087 ft, it is seen that this frame is about five times more flexible than the shaft. Furthermore, the frame would not be stiff enough to carry all the lateral load by itself. Proportioning the lateral load to the relative stiffnesses, the frame would carry about 1/6 of the load, 5/6 being carried by the shaft.

FIGURE 7-21
**COLUMNS RIGIDLY CONNECTED TO CORE WILL
RESULT IN A STIFFER SUBSYSTEM THAN FOR
CORE ACTING ALONE.**

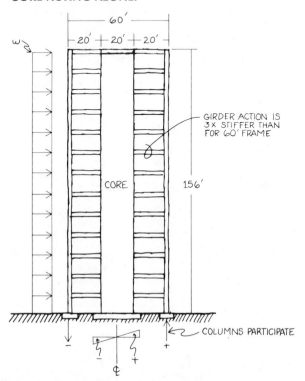

Increasing the column size will stiffen the frame, but in order to be really effective, the girder stiffness will also need to be increased, since the girders contribute about 2/3 of the deflection. Then the frames can be made to carry a larger portion of the load. Note that the deflected shapes of the shaft and the frames are quite different, so that the above simple comparison of top deflections is not an accurate assessment. But it does give some idea of their relative behavior.

It can also be observed that if the frame, as is, and the shaft act together, the overall stiffness of the building will not be much more than the shaft acting alone. However, if the shaft walls are made thinner and the frame members made larger, each would then contribute significantly toward the stiffness of the building.

We have not studied the effect of the shaft stiffened by the exterior columns, which are rigidly connected to the shaft walls and will act with the shaft as a unit (Figure 7-21). This would be quite effective because the girder length from column to core would be one-third that for the "standard" bent (60 ft) used in the example. A second line of defense against earthquakes is thus obtained by using both shaft and frames and connecting them together to resist lateral forces. The horizontal floor diaphragms will hold and force them to deflect together. The 20-ft short girders will also help. All told, it appears that the layout is an efficient one, and further analysis would refine and optimize the design.

8
Horizontal Linear Components

SECTION 1: Sectional Shapes and Proportions of Components

In Chapters 6 and 7 we discussed horizontal and vertical (2-D) subsystems using an overall design approach. In this chapter and the next we will focus on the selection and approximate design of individual (1-D) members that compose the components of the subsystems. As an example, Photo 8-1 shows a cored slab being lifted to serve as a floor component.

Horizontal components of structural subsystems are variously named as beams, joists, or girders. Essentially, these components carry vertical loads transversely (across their length). As a result, there are reactions concentrated at the supports that, together with the loads, produce shearing and bending action in these components. Sometimes a design can be such that loads are applied away from the vertical plane of the components and produce a twisting moment (torsion) in the member. But, such torsional bending is uncommon, and it will not be discussed. Therefore, in this chapter we will focus on typical design for bending and shear.

As discussed in Chapter 6, the design of individual components should start with a determination of their suitability for interaction in the context of an overall subsystem scheme. Having narrowed down to the issue of specific components, one may then proceed to design their size and shape. Before any calculation is made, the first step is to determine the approximate span/depth (L/d) ratios for these components. The L/d ratios will depend essentially upon the material and the loading. But they are also affected by the sectional shape, support conditions, and indeed the span length itself.

Appendix Table B-1 gives some L/d ratios for approximating component dimensions when designing floor subsystems. But, since the table is intended only as a gross approximation, its values should not be taken as an absolute guide. They are good to use as starting values in comparing options for choice of member types. Furthermore, they are intended for average loading conditions (whatever that may mean), and should be adjusted if loading is heavy (or lighter, as in the case of roofs).

It should also be noted that the ratios shown are intended to apply to spans that are simply supported. For continuous spans, that is, members having their ends continuous to adjoining members or supporting columns, it is possible to increase the ratios by about 10 percent. For roof construction, where loading is usually less than that for floors, it is again

PHOTO 8-1
**WELLS FARGO BANK BUILDING, OAKLAND,
CALIFORNIA (T. Y. LIN INTERNATIONAL,
STRUCTURAL ENGINEERS.)**

possible to add another 10 percent to these ratios. On the other hand, for
heavy or vibratory loadings, the L/d ratios must be reduced similarly. For
cantilevers, the L/d ratio must be much smaller—say, no more than 30
percent of those listed in Appendix Table B-1.

The table also assumes that the members are of uniform depth. In case of
haunched beams and slabs, it must be further modified to reflect the
stiffening effect of the haunches. Thus, given a depth limitation (d) around
midspan, L/d ratios greater than the limits cited may be obtained,
depending upon the length and depth of the haunch. For example, by
multiplying the area of the *haunch profile* by an effectiveness factor (say,

FIGURE 8-1
THE EFFECT OF HAUNCHES IS ROUGHLY INDICATED BY DETERMINING THE OVERALL *EFFECTIVE* DEPTH OF A BEAM.

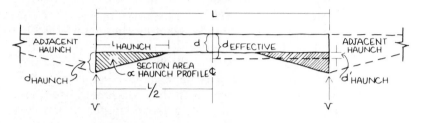

NOTES : 1. $d_{EFFECTIVE} = d + d'_{HAUNCH}$

2. $d'_{HAUNCH} = \dfrac{\varepsilon\, AREA_{HAUNCH} \times 1.5\, EFF.\, FACTOR}{L} = 1.5 d_h \times \dfrac{L_{HAUNCH}}{L}$

3. IF $d_{HAUNCH} > 2d$, SIGNIFICANT ARCH ACTION IS POSSIBLE

1.5) and averaging that value over the entire span, some idea of equivalent depth for haunch design is suggested (Figure 8-1). Thus, when using Appendix Table B-1 to get started, you should note that optimum L/d ratios are determined by taking the following considerations into account:

1. *Economics.* Generally, to satisfy economic requirements of structures alone, ratios greater than those listed in the table would be preferable. However, greater depth (or shorter spans) of these members would mean additional floor-to-floor depth (or more columns). This results in increased height of the building (or spatial obstruction), which would require more expenditure for other portions of the structure and spatial disadvantage. Therefore, the *average L/d* values shown are in common use today. To obtain the maximum values will require skillful design at all stages, including the provision for the interaction of components in a subsystem.

2. *Deflection limitations.* These will be discussed in Section 5 of this chapter.

3. *Vibration problems.* This is a complicated matter, but can be controlled approximately by controlling span/depth ratios to limit flexibility and avoid resonance.

4. *Aesthetic requirements.* These may require special span/depth ratios.

5. *Practical requirements*, such as the availability of sections, materials, or reliable workmanship.

6. *Possible movements of a structure*, such as uneven settlement of its supports that would tend to call for more slender and flexible members so that they will be less affected by any differential settling.

Furthermore, *L/d* ratios for each case will vary with the particular conditions, such as:

1. The nature, magnitude frequency and duration of live load.
2. The magnitude of the dead load—for example, sustained dead load may produce excessive deflections for certain members of certain materials due to creep and may call for a smaller L/d ratio.
3. The support conditions—whether the member is hinged, fixed, or continuous with others.
4. The shape of the section itself—a T-shape would require a lesser L/d ratio, while a box or I-shape can allow a larger ratio.
5. The L/d ratio will vary with the span length itself, and is not necessarily a constant for all span lengths. It is only a simplistic approach to a complicated problem, which involves periods of vibration, type of excitation, damping factors, creep and shrinkage effects, and so on.

In spite of the above complications, Appendix Table B-1 can be used as a preliminary guide to quickly establish a trial value. Then the trial section and proportion can be modified at later stages of overall design when more accurate calculations are required. The table also gives a range of practical spans applicable to each type of construction. These are again approximate values and should be adjusted under certain conditions.

Table B-1 lists different members, such as slab, beam, joist, and girder. As noted previously, these terms cannot be exactly defined. For example, a

FIGURE 8-2
COMMON ROLLED STEEL SECTIONS.

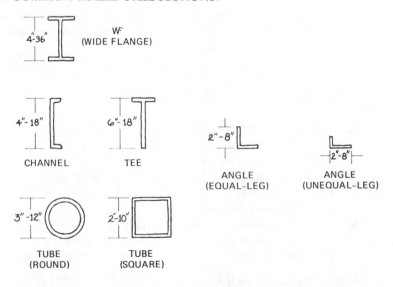

4"-36"   WF
(WIDE FLANGE)

4"-18"   6"-18"   2"-8"   2"-8"
CHANNEL   TEE   ANGLE (EQUAL-LEG)   ANGLE (UNEQUAL-LEG)

3"-12"   2"-10"
TUBE (ROUND)   TUBE (SQUARE)

FIGURE 8-3
P.C. COMMON SHAPES.

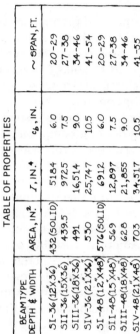

TYPICAL PRESTRESSED CONCRETE CORED SLAB SECTION.

TABLE OF PROPERTIES

BEAM TYPE DEPTH & WIDTH	AREA, IN²	I, IN.⁴	c_b, IN.	~ SPAN, FT.
SI-36(12×36)	432(SOLID)	5184	6.0	20-29
SII-36(15×36)	439.5	9725	7.5	27-38
SIII-36(18×36)	491	16,514	9.0	34-46
SIV-36(21×36)	550	25,747	10.5	41-54
SI-48(12×48)	576(SOLID)	6912	6.0	20-29
SII-48(15×48)	569	12,897	7.5	27-38
SIII-48(18×48)	628	21,855	9.0	34-46
SIV-48(21×48)	703	34,517	10.5	41-55

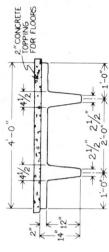

TYPICAL DOUBLE TEE SECTION.

TABLE OF PROPERTIES

TOPPING	AREA, IN²	I, IN.⁴	c_b, IN.
WITH 2" TOPPING	276	4456	11.74
WITHOUT TOPPING	180	2862	10.00

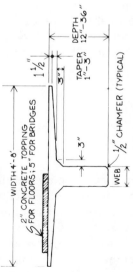

TYPICAL SINGLE TEE SECTION (LIN TEE).

TABLE OF PROPERTIES
(WIDTH = 6'; WEB = 8"; TAPER = 2"; NO TOPPING)

DEPTH, IN.	AREA, IN²	I, IN.⁴	c_b, IN.
12	265	2360	9.2
16	297	6170	11.8
20	329	11,730	14.5
24	361	19,700	17.0
28	393	30,400	19.5
32	425	44,060	21.9
36	457	61,000	24.2

TYPICAL PC I-BEAM SECTION FOR BRIDGES.

TABLE OF PROPERTIES
(WITHOUT IN-PLACE SLAB)

BEAM TYPE	AREA, IN²	I, IN.⁴	c_b, IN.	RECOMMENDED SPAN LIMITS, FT.
I	276	22,750	12.59	30-45
II	369	50,980	15.83	40-60
III	560	125,390	20.27	55-80
IV	789	260,730	24.75	70-100

joist is a lightly loaded small beam, a beam is a moderately loaded small girder, and a girder is a large and heavily loaded beam. How small is small, and how large is large? It's a matter that has not been closely defined.

For the practical purpose of preliminary sizing of members, it is useful to know what are the available shapes and sizes. For this purpose, Appendix C has been compiled to show commonly available (standard) sections for certain materials. Appendix Tables C-4 and C-5a list timber and plywood sections. Appendix Table C-5b shows glue laminated timber beams. Appendix Table C-3 for concrete and Appendix Tables C-1 and C-2 for steel are also included. All of these tables also include certain section properties often required for engineering computations.

Tables C-1 and C-2 are very partial lists of wide flange (WF or W) rolled steel shapes commonly available for building and other construction. Whereas the W shape is a most common section, rolled sections include channels, Ts, and angles of equal or unequal legs, square and round tubes, and so on, as shown in Figure 8-2. A complete set of tables for these sections as well as many others can be found in the *Manual of Steel Construction* by the American Institute of Steel Construction.

For poured-in-place reinforced concrete members, the size and proportions are not really limited. However, timber and plywood for forming, and steel bars for reinforcing (Appendix Table B-3), come only in standard sizes. In the choice of member dimensions for reinforced concrete, there is great liberty providing there is adequate space for proper placement of concrete and reinforcing bars within these members.

Precast concrete members are usually prestressed. They come in more-or-less standard shapes and dimensions, as indicated by available forms already proven effective in use. Some common ones are shown in Figure 8-3. However, provided there is sufficient repetitive use, it is not difficult to fabricate new forms, and almost any reasonable shape can be economically produced. Of course, the initial cost of the forms will vary with the outline of the section, and the per-member unit cost will depend upon the reuse of these forms.

SECTION 2: Moment Diagrams

Obviously, it is not possible to obtain an accurate dimensioning or even proportioning of a member based on the discussion in Section 1 and Appendix Table B-1. To arrive at more accurate dimensioning, it will be necessary to compute the maximum or controling bending moment in a component. To do this, one must be able to sketch, at least approximately, the moment diagram for a particular member. The plotting of this diagram will require the computation of the loading on the member, and the resulting reactions from the support. Sometimes shear or even torsion conditions will be controlling and may have to be computed to produce a preliminary design or dimensioning. However, for most members, the bending moment will control the preliminary design; hence, we discuss the moment diagrams more carefully.

Figure 8-4a shows a simple beam that is supported but allowed to rotate at the two ends. Let us assume that this beam will be loaded by a uniformly distributed load with an intensity of w pounds per linear foot of beam. An application of statics will indicate that the basic moment diagram for an evenly distributed load resisted by simple beam action is a *parabola* with a maximum ordinate given by $wL^2/8$ and occurring at midspan. Note that this term, $wL^2/8$, is a fundamental value for evenly distributed loading systems.

FIGURE 8-4
THE MOMENT CURVE FOR SIMPLE BEAM DESIGN IS TRANSLATED DOWNWARD WHEN END ROTATION IS RESISTED.

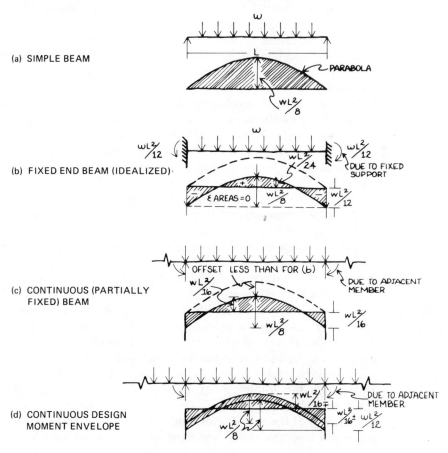

As will become evident, it is indeed basic to preliminary design of all horizontal components under beam bending, and will be used again and again in this and the following chapters.

For a simple beam, the overall maximum moment, $wL^2/8$, exists only at midspan, and decreases to zero as one approaches the supports. Nevertheless, it is common practice to design a member section that will be able to carry this maximum midspan moment and, for practical reasons, use the same section for the entire beam. In special cases, it may be economical to taper or curve the beam so that it will have a maximum depth or section at midspan only, but gradually decrease toward the ends.

Figure 8-4b shows a fully fixed-end beam with both ends fixed into some rigid body such as a wall shaft or some heavy columns to prevent rotation at the supports. In this case, two end moments will be produced to resist rotation of the ends of the beam and to keep them horizontal at these ends. Assuming a uniform section, these two end moments will produce a uniform counterbending, or negative moment $(-M)$, throughout the entire beam equal to $wL^2/12$. (If the section varies, the end moments will be different from $wL^2/12$.) Such a constant-section fixed-end beam can therefore be visualized as a simple beam with the addition of two end moments. Thus, by adding the parabolic moment diagram $(+M)$ in Figure 8-4a to the rectangular moment diagram $(-M)$ produced by the end moments, the net moment diagram is shown in Figure 8-4b. Note the center moment is now reduced to $wL^2/8 - wL^2/12 = wL^2/24$. It should therefore be noted that, other things being equal, a fully fixed-end beam of constant section has a center moment only one-third that of a simple beam, whereas its end moments are two-thirds of the midspan moment for the simple beam. Further, the *curve* of the moment diagram is the same as that for a simple beam (a parabola), but it has been *translated* vertically downward as shown in Figure 8-4b.

All told, the fixed-end beam is subjected to much less severe bending moments when compared to the simple beam. Also, note that the nature of bending is one of contraflexure with the point of inflection (zero moment) at approximately $0.17L$ from each end. As shown in the figure, the ends of a fixed-end beam are subjected to a convex curvature, designated as a negative moment. Its center portion is subjected to a concave curvature, designated as positive moment.

Figure 8-4c shows a beam that is continuous to other beams or fixed to less-than-rigid supports at its two ends. This beam is partially fixed. Since end rotation is only partially restrained, its end moments cannot be expected to be as high as for the fixed-end beam in Figure 8-4b. In a partially fixed case, the end moments will vary with the amount of fixity afforded by the adjoining beams or support system, and fall between 0 (Case a) and $wL^2/12$ (Case b). For simplicity in discussion, the end moments in this type of beam are often depicted as being on the order of $wL^2/16$. If this is so, then the center moment will also be $wL^2/16$ because the two moments must total $wL^2/8$. The net effect of partial fixity is therefore shown in Figure 8-4c with a relatively small amount of negative moment at the ends, and a somewhat larger positive moment at the middle as compared to Figure 8-4b. Of course, if the end conditions are very flexible, translation of the parabola would be even less and the point of inflection would move closer to the end. Thus, the idea of fixity is to resist rotation at the ends as much as possible in order to reduce the midspan moment.

Figure 8-4d shows a combination of Figure 8-4b and 8-4c, which is often required for a continuous beam analysis under different loading conditions. When the beam is loaded along its own span only, with no load on the adjacent spans (Figure 8-4c), it tends to bend more downward, and therefore a moment diagram with $(\pm)wL^2/16$ at end and midspan would exist. Should the adjoining spans be also loaded, then the $-M$ over the two supports will be higher, perhaps as much as $wL^2/12$. In this case, another moment diagram would occur, as shown by the lower curve in Figure 8-4d. This means that under one loading condition, the midspan $+M$ will control its design at $wL^2/16$, whereas under another loading condition the $-M$ over the support will control its design at $wL^2/12$. Hence a design condition *moment envelope* for a beam can be sketched as shown in Figure 8-4d for the purpose of design (particularly for the live-load effects). It will therefore be necessary to consider the extreme conditions that may exist under one loading for the midspan moment and under another loading for the moment over the support.

For *cantilever bending*, it will suffice here just to mention the case of uniform load (Figure 8-5). A simple application of statics will show that the moment diagram for this beam is also a parabola, with a maximum ordinate at the fixed support reaching a value of $-M = wL^2/2$. Note that in a cantilever beam, the section is almost always controlled by the moment requirement at the support, even if much smaller sections are used elsewhere along that cantilever. The negative moment of a cantilever can be profitably used to reduce the positive moment of an adjoining simple span. This is illustrated by Photo 8-2, whose roof beam has a total length of 150 ft.

In addition to the bending moment variation, sometimes the reactions

FIGURE 8-5
CANTILEVER BENDING MOMENTS.

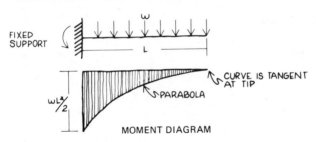

PHOTO 8-2
**PHOENIX AIRPORT BUILDING. (T. Y. LIN INTERNATIONAL, STRUCTURAL
ENGINEERS.)**

FIGURE 8-6
**SHEAR IN BEAMS AND
CANTILEVERS.**

(a) SHEAR (V) IN SIMPLE BEAMS

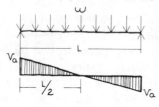

(b) SHEAR (V) IN CANTILEVERS

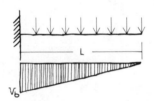

NOTE: FOR EQUAL L & ω, $V_b = 2 V_a$

and the shear may affect the design of the beam. Although this will not be discussed here, it may be well to show the variation of shear for two cases (Figure 8-6a and 8-6b). In Figure 8-6a a simple beam shear diagram indicates that the shear is maximum at the two ends and decreases to zero at the center. Similarly, in a cantilever beam (Case b), shear is zero at the free end, but increases to a maximum at the support. Note that in the case of a simple beam, shear is *lowest* where moment is greatest but the opposite is true for cantilevers.

Throughout the discussion of this book, only the simplest cases of uniformly distributed loading will be considered. With some understanding and practice, it is possible to reduce other forms of loading, such as a concentrated load or a uniformly varying load, to an equivalent uniform load as illustrated in Figure 8-7. Note that Case 4 produces two-thirds as much moment as an equivalent evenly distributed load (Case 3), Case 2 four-thirds, and Case 1 two times the moment of an evenly distributed load. When the maximum moment is known, the curve can be sketched as shown in Figure 8-7b to give an approximate, but in most cases sufficiently accurate, moment diagram for preliminary dimensioning and proportioning of members.

For more accurate analysis, refer to Appendix Table D; or, for special cases, refer to handbooks or compute on the basis of statics. Appendix Table D-1 shows a number of moment and shear design conditions for cantilevers and simple-span beams supported and loaded in various ways. Appendix Table D-2 also gives the controlling moment for these beams, the reaction from the supports, and the maximum deflection in these beams. It is self-explanatory, and the derivation of these values will not be shown herein.

To give some idea of moments in continuous beams, three examples are shown in Figure 8-8. Case a shows a two-span continuous beam carrying a uniform load. Here we obtain a maximum (−) moment over the center support, $wL^2/8$, whereas the maximum (+) moment at 3/8L from the end is $9wL^2/128$ (or about $wL^2/14$), the point of inflection occurring at $\frac{1}{4}L$ from the center support.[1]

Figure 8-8b shows a similar two-span continuous beam with cantilevers. Note that both the negative moment over center support and the positive moment at midspan are reduced when compared to Figure 8-8a. This shows that there is often an advantage in having a certain amount of cantilever at the end of a linear system. Of course, that cantilever cannot be excessive. For favorable moment conditions, a cantilever of about 0.4L would be optimum, but it will likely result in unfavorable deflections. Thus 0.25L or 0.3L is more practical. Beams can also be fixed to rigid columns and walls. Thus the rotation of beam ends can be restrained by a combination of arrangements.

---

[1]Also note that the tributary distributions of loads to the three supports is not symmetric as was assumed in earlier chapters.

FIGURE 8-7
SIMPLE BEAM MOMENTS VARY WITH LOAD DISTRIBUTION.

(a) SIMPLE BEAMS UNDER VARIOUS SYMMETRIC LOAD DISTRIBUTIONS

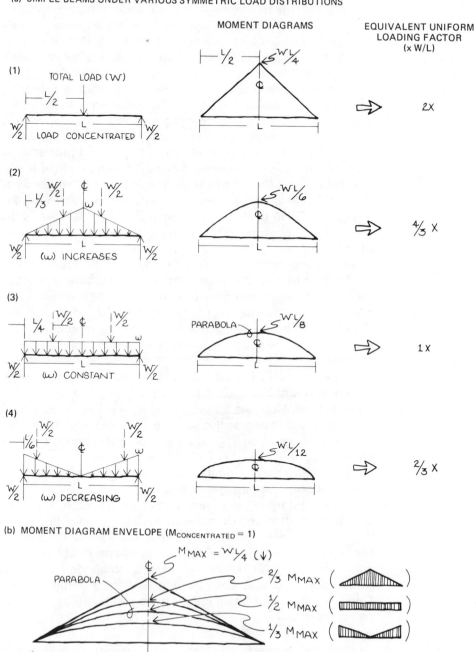

MOMENT DIAGRAMS

EQUIVALENT UNIFORM
LOADING FACTOR
(x W/L)

(1) ... 2X

(2) ... $\frac{4}{3}$ X

(3) ... 1 X

(4) ... $\frac{2}{3}$ X

(b) MOMENT DIAGRAM ENVELOPE ($M_{CONCENTRATED} = 1$)

FIGURE 8-7 (*Continued*)

(c) SIMPLE BEAMS UNDER VARIOUS ASSYMMETRIC LOAD

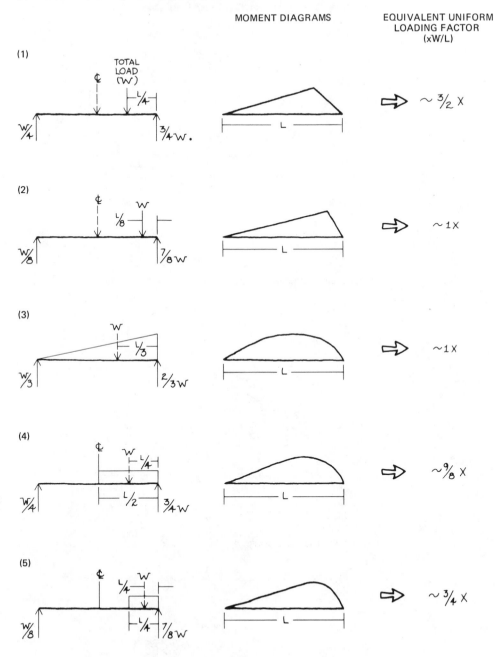

MOMENT DIAGRAMS

EQUIVALENT UNIFORM
LOADING FACTOR
(xW/L)

(1)

(2)

(3)

(4)

(5)

NOTE: $M_{MAX.}$ FOLLOWS LOAD ASSYMETRY—IN CASE OF A POINT LOAD $M_{MAX.}$ IS AT POINT OF LOAD.
IN CASE OF DISTRIBUTED ASSYMETRY, $M_{MAX.}$ IS SOMEWHAT LESS ASSYMETRIC THAN THE
LOAD CENTROID WOULD INDICATE.

FIGURE 8-8
MOMENTS ON CONTINUOUS SPAN BEAMS.

(a) 2-SPAN CONTINUOUS

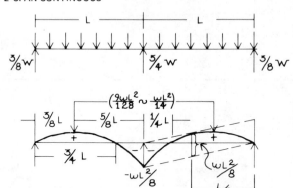

(b) 2-SPAN CONTINUOUS WITH CANTILEVER (IDEALIZED)

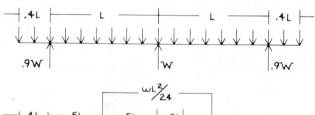

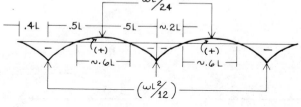

(c) 3-SPAN CONTINUOUS

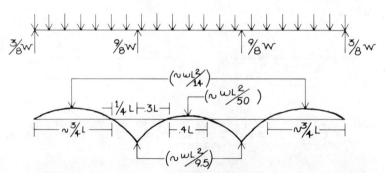

Figure 8-8*c* shows a three-span continuous beam. It is noted that the maximum positive moment in the end spans are higher than at center span. Therefore, it is often desirable to make the end spans shorter than the center span (by some 20 percent or more) if it is desired to use one uniform section for the entire beam.

The study of moment diagrams and perhaps also shear diagrams for beams of various span lengths, loading conditions, including such things as cantilevers and intermediate hinges, is an interesting problem, which is beyond the scope of this book. Refer to books on mechanics or theory of structures to obtain further data for study.

SECTION 3: **Internal Resisting Couple**

Bending moments produced by loading *on* a beam must be resisted by an internal resisting couple *in* that beam. A free body of one-half of a simple beam is shown in Figure 8-9*a*. By simple statics it is seen that the *internal*

FIGURE 8-9
MOMENTS *ON* BEAMS MUST BE RESISTED BY MOMENTS *IN* BEAMS.

(a) MOMENT RESISTANCE IN A SIMPLE BEAM

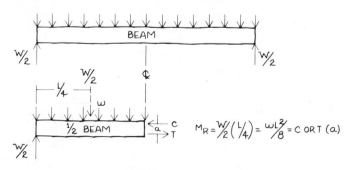

(b) MOMENT RESISTANCE IN A FIXED END BEAM

resisting couple for an *external* positive moment is made up of two equal and opposite forces (*C* and *T*), which are compressive in the top portion of the section and tensile in the bottom portion. They act over a lever arm (a) between them. The resisting couple, *Ta* or *Ca*, equals the externally applied moment, which is $wL^2/8$ in this case.

FIGURE 8-10
ELASTIC BENDING STRESS DISTRIBUTION FOR SYMMETRIC AND ASYMMETRIC SECTIONS.

(a) FOR ELASTIC BENDING, TRANSVERSE SECTIONS REMAIN PLANAR

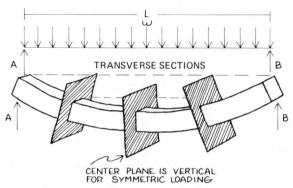

(b) STRESS DISTRIBUTION ACROSS CENTER SECTION (ie $+ M_{MAX}$)

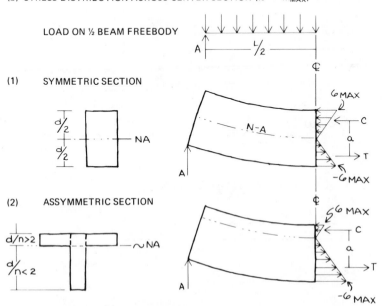

NOTE THAT C = T AND $^C/_{A_C}$ OR $^T/_{A_T}$ = AVERAGE STRESS.
∴ AS $- \sigma$ DECREASES (RELATIVE TO $+ \sigma$), A_C MUST INCREASE
& N-A MOVES UP IN CASE b-2.

FIGURE 8-11
THE ELASTIC THEORY FOR STRESS ANALYSIS.

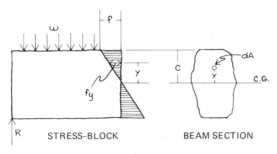

STRESS-BLOCK   BEAM SECTION

PORTION OF BEAM

In the case of a continuous beam or fixed-end beam, the $-M$ (over the support) is resisted by a T-C couple with a tension in the top of the beam and compression in the lower portion. If that $-M$ is $wL^2/12$, then that TC couple in Figure 8-9*b* must resist that moment.

One can easily recognize the necessity of the existence of T-C couple across any beam section to resist the external moment. But it takes a little more to visualize the distribution of either the tension or the compression force across a particular section. In order to deal with this question it will be well to discuss and visualize the stress distribution across the depth of a beam section. Note the similarity with discussions in Chapters 2 and 3 about stiffness as applied to whole building sections. The concepts are the same; only the scale changes.

Let us first assume that when a simple beam bends, a plane section across that beam remains plane so that for a $+M$, the top fiber is compressed most while the lower fiber is elongated most, and the variation remains a linear one (Figure 8-10). This behavior has been shown to be quite correct for the usual type of beams (except when the beams are very deep relative to their span with, say, a ratio of about 1:1). Using this approach, and further assuming that stress is proportional to strain (i.e., the amount of elongation or shortening would indicate the stress at that point in direct proportion as shown in Figure 8-11), we can derive the common flexure formula, $f = Mc/I$.

Flexural Formula—Elastic Theory

To prove that:

$$f = Mc/I,$$

Moment = M at a section of a beam under flexure,
Extreme fiber stress = f at top fiber, which is at distance $c = y_{max}$ from

centroid of section c.g.

$$I = \text{moment of inertia of section about c.g.} = \int y^2 \, dA$$

Portion of Beam

Proof: Internal moment = external moment M
Fiber stress f_y at any point y from c.g.
Force over area $dA = f_y \, dA$
Moment of force $f_y dA = y(f_y dA)$

Since, by elastic theory, $f_y = f\dfrac{y}{c}$

Moment of force over dA about c.g. is

$$y\left(f\frac{y}{c} \, dA\right) = \frac{fy^2}{c} \, dA$$

Total moment of force over whole section is

$$M = \int \frac{fy^2}{c} \, dA$$

$$= \frac{f}{c} \int y^2 \, dA$$

$$= \frac{fI}{c}; \quad f = \frac{Mc}{I}.$$

Also, at any point (y) from c.g., $f = My/I$.

The life history of structural subsystems discussed in Chapter 4 also applies to members. Thus, most members and materials behave elastically up to a certain point beyond which plastic behavior begins to appear. This is again illustrated in the stress-strain diagrams shown in Figure 8-12. This figure shows the relation between unit stress and unit strain of four different materials—timber, concrete, structural steel, and prestressing steel. In all these cases, an approximate stress-strain relationship shown by the dotted lines indicates that elastic behavior continues up to a certain point when it suddenly or gradually turns into plastic action. This means that when a material is stressed beyond a certain limit (often called the elastic limit or the yield point), it begins to act plastically instead of elastically. In the plastic range, the unit stress changes very little, if at all, even though the elongation continues. Thus the above formula, $F = Mc/I$, is no longer applicable because that formula assumes the stress-strain relation to be linear.

Plastic stress distribution can get rather complicated, although it can be approximated, as shown in Figure 8-12. Figure 8-13 diagramatically compares the bending stress distribution in a homogeneous rectangular section, in the elastic and in the plastic ranges. Figure 8-13a illustrates the elastic stress distribution, wherein the stress increases linearly from zero at the neutral axis to a maximum at the outer axis. In this case, the lever arm is between

FIGURE 8-12
THE ELASTIC AND PLASTIC RANGE OF BEHAVIOR.

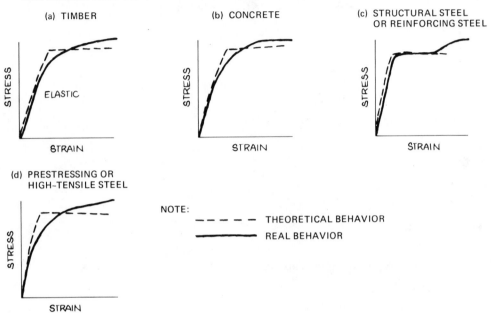

(a) TIMBER

(b) CONCRETE

(c) STRUCTURAL STEEL
OR REINFORCING STEEL

(d) PRESTRESSING OR
HIGH-TENSILE STEEL

NOTE:

$-----$ THEORETICAL BEHAVIOR

$———$ REAL BEHAVIOR

the centroid of the total tension or compressive *forces* (given by the stress
distribution). Thus the resisting forces C and T act with a lever arm of
($d = 2/3d$). Figure 8-13*b* shows the plastic stress distribution where the
stresses are uniform for the greater part of the section. As a result, the
average stresses, and thus the T-C values, are higher than in the elastic
range. However, the centroidal distance between the two couples is
reduced and *approaches* $d/2$ as a limit (when the entire section would be in
the plastic range). Of course, in spite of this reduction in the lever arm,
($\sim$25 percent) the total force C or T is much greater and the internal
resisting moment in the plastic range is higher by two or more times than in
the elastic range.

Figure 8-14 illustrates the bending stress distribution in an (H) section. In
this case, most of the forces are concentrated at the extreme fibers, or the
top and bottom flanges. In the elastic range, the lever arm approaches 0.9d
(of course the exact value will vary with the shape and the proportions of
the H section). In the plastic range, higher (average) stresses are obtained in
both the flanges and webs. Therefore the lever arm between the T-C
couple is somewhat reduced (to $\sim$0.8d). But again, the total resisting
moment in the plastic range is higher (by two or more times) than in the
elastic range because of the higher stresses developed in both the flanges
and in the web.

FIGURE 8-13
**ELASTIC VERSUS PLASTIC BENDING STRESS DISTRIBUTION IN
HOMOGENOUS RECTANGULAR SECTION.**

(a) ELASTIC STRESS DISTRIBUTION

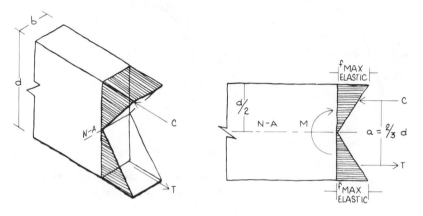

(b) PLASTIC STRESS DISTRIBUTION

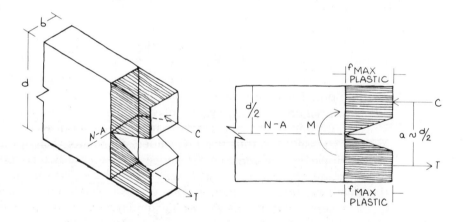

Figure 8-15 shows the bending stress distribution in a truss section. The essential moment resistance is provided by the compression in the top chord and the tension in the bottom chord. They act with a distance $a \simeq d$ in both the elastic and the plastic ranges. Therefore, a trussed arrangement gives an equally large lever arm, in both the elastic and the plastic ranges and the plastic resisting moment will be ~2 times the elastic moment.

In a reinforced concrete beam, the stress distribution is quite different (Figure 8-16). Since concrete cannot take any appreciable tension, its tensile force is usually neglected. It is assumed that the section cracks and reinforcing steel carries all the tension. Ideally, the designs will be balanced

FIGURE 8-14
ELASTIC AND PLASTIC BENDING STRESS DISTRIBUTION IN AN (*H*) SECTION.

(a) ELASTIC BENDING STRESSES

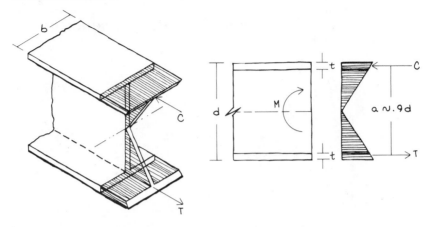

(b) PLASTIC BENDING STRESSES

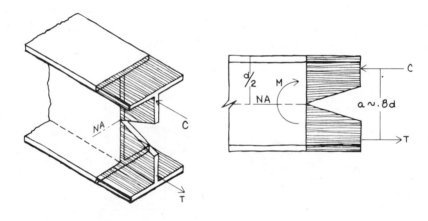

FIGURE 8-15
ELASTIC AND PLASTIC BENDING STRESS DISTRIBUTION IN A TRUSS ARE SIMILAR.

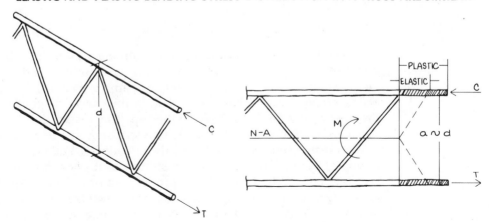

FIGURE 8-16
**ELASTIC AND PLASTIC BENDING STRESS DISTRIBUTION IN A
RECTANGULAR R.C. SECTION.**

(a) ELASTIC STRESS DISTRIBUTION

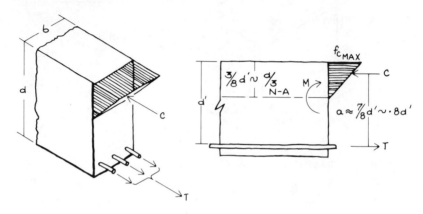

(b) PLASTIC STRESS DISTRIBUTION

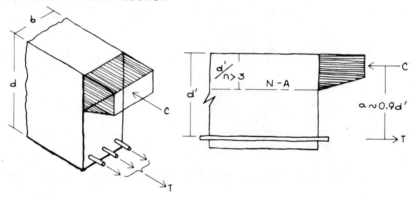

such that the concrete cracks up to a neutral axis, which is about $\frac{3}{8}$ of the *effective* depth (d') *from the compressive edge* of the section. Therefore, the lever arm between the *T-C* couple is approximately $\frac{7}{8}d'$ ($\sim 0.85d'$) in the elastic range (Figure 8-16a). Note that, in the plastic range (Figure 8-16b), the neutral axis tends to move further up. As a result, the lever arm between the *T-C* couple is somewhat *increased*, say, to approximately 0.9d'.

In a prestressed concrete beam, the situation is again different. For example, Figure 8-17 illustrates the stress distribution in a prestressed concrete beam. The beam, being prestressed or precompressed, will generally have no tensile stresses. Its stress distribution will begin from nearly zero in the bottom fiber up to a maximum in the top fiber.

Therefore, the center of compression is relatively low and the lever arm between the two *T-C* forces is on the order of 0.6*d'*. Note that although the lever arm is smaller, the stress block occupies the entire section of the concrete beam and therefore offers a greater compression resistance when compared to a conventionally reinforced concrete beam. In the plastic range, a prestressed beam will have stress distribution very similar to that shown in Figure 8-16*b*, previously indicated for a reinforced concrete beam. The reason is that in the plastic and near-ultimate range, the neutral axis will *move upward* to about .9*d'* in a prestressed beam; the lower portion of the beam will crack in a way similar to a reinforced concrete beam and therefore the stress distribution becomes quite similar.

If we consider a T-shaped beam, Figure 8-18, we will find that the compression is essentially and almost totally supplied by the top flange,

FIGURE 8-17
ELASTIC AND PLASTIC BENDING STRESS DISTRIBUTION IN P.C. BEAMS.

(a) ELASTIC STRESS DISTRIBUTION ('0' TENSION)

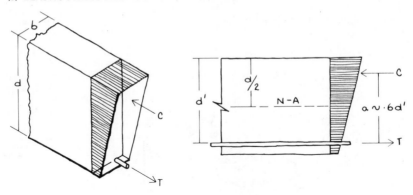

(b) PLASTIC STRESS DISTRIBUTION (SIMILAR TO R-C ACTION)

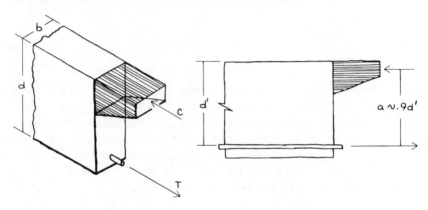

FIGURE 8-18

ELASTIC AND PLASTIC BENDING STRESS DISTRIBUTION IN R.C. OR P.C. 'T' BEAMS.

(a) ELASTIC STRESS DISTRIBUTION (----- INDICATES PC)

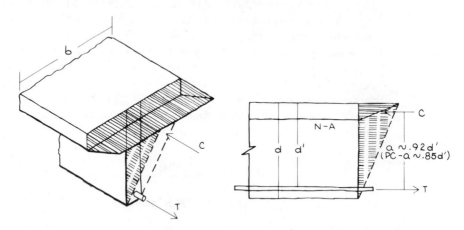

(b) PLASTIC STRESS DISTRIBUTION (R-C OR PC)

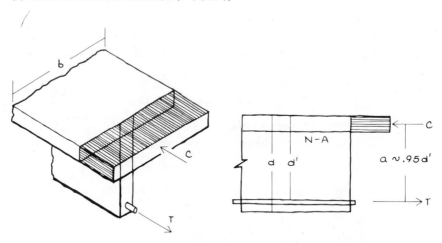

which is quite wide. In this case, the lever arm (a) will be quite high in the elastic range, and even higher in the plastic range for both *RC* and *PC*.

In each of the above examples, simply dividing the required resisting moment by the approximate lever arm (a) will yield the necessary total tension and compression forces to be supplied in the top and bottom portion of a beam. In the following section we will discuss how to use this information to size the beams accordingly.

SECTION 4: Designing with Allowable and Ultimate Stresses

In the previous section we discussed the approximation of sectional depth, moments, and the computation of total tensile and compressive forces required to serve as the internal resisting couple in a beam. We also discussed the approximate distribution of stresses across a beam section. Now we wish to determine what sectional properties are necessary to provide the required T and C forces. To do this we must determine either the allowable stresses or the ultimate stresses in different materials in order to arrive at sufficient sectional shape and area properties to provide these forces. Further, we must also distinguish between the so-called design loads and ultimate loads as well as the allowable stresses and ultimate stresses as provided in various code requirements.

Referring again to the life history of a structure (Chapter 5, Figures 5-5 and 5-6), note that the so-called working loads and the allowable stresses both refer to the elastic range, whereas the ultimate loads and the ultimate stresses refer to the plastic and near-ultimate range. The use of the working loads and the allowable stresses constitutes the so-called working stress methods and *have* been the common feature of practically all buildin{ codes. More recently, during the period 1960–1970, a change has taken place, and *now* most codes either permit or prefer the use of the so-called ultimate load design method.

As stated in Chapter 5, the ultimate loads are arrived at by multiplying the working loads by a load factor. Then the ultimate loads are compared with the ultimate resisting strength of the respective members. In contrast, the elastic design approach is to use actual or working loads and match them with allowable stresses. The basic idea in both cases is to design for proper structural behavior and also to provide an adequate safety factor.

Let us first discuss the working load method and the use of allowable or working stresses in design. Here a factor of safety is provided in the stresses allowed for computation in that they are significantly less than the ultimate stresses. Generally speaking, the allowable stresses are a fraction of the yield point stress or the ultimate stress (varying from 40 percent up to about 60 percent), the exact amount depending on the reliability of the material and other conditions. Assuming the allowable stress is 50 percent of the yield stress, it would appear that a safety factor of two is provided. But actually, because the ultimate stress distributions differ from the elastic stress distributions, and because of the changes in the lever arm of the internal resisting couple, the actual safety factor provided can be either greater or less than two. Nevertheless, it is a fairly satisfactory method. Also, by limiting the working stresses, the adequate behavior of the structure or the component is generally assured, since the stresses are quite a bit below the plastic range or yield point. However, this may not always be true.

For example, in slender beams, deflections can be quite high even though the stresses are low. In addition, the so-called stresses will have to be computed by some "accepted" or approximate method, which may or may not be correct. Furthermore, many critically stressed areas are not identified and thus not computed. Hence the so-called allowable stress method is in fact a gross approximation and does not give either the real stresses or the real factors of safety.

The ultimate strength method is supposed to yield the actual ultimate strength of the section or component, and therefore to provide sufficient strength to resist the so-called ultimate load. Since the ultimate load is a factored load based on the working load multiplied by a factor of 1.4, 1.8, or 2.0, it provides a factor of safety in terms of strength rather than elasticity. In order to allow for the actual as opposed to theoretical strength of a component or a section, such as due to the variation of material properties and dimensions, an additional reduction factor (ϕ) is applied to the computed ultimate strength (e.g., $\phi = 0.9$ for bending, and 0.7 for columns, etc.). This reduced strength has to be equal to or greater than the required ultimate strength so as to insure safety.

The ultimate strength method, while supplying the required factor of safety with respect to failure, would need an additional check on the behavior of the members (i.e., the deflection, camber, vibration, cracking, and fatigue failure characteristics). For the purpose of a preliminary design (generally using a load factor of 1.8 or 2.0), it may be sufficient to estimate ultimate stress values so as to approximate the ultimate strength of the section. For this purpose, tables are provided in Appendix B-2. These tables give the essential properties of major structural materials. They also give the allowable stresses for working strength design (*WSD*) as well as the ultimate stresses for ultimate strength design (*USD*). Since values in these tables are intended only for preliminary design, one who attempts to make a final design should refer to the respective building codes as well as other handbooks. In a final design there are also other stresses, such as shear, torsion, bearing, buckling, and so on that must be considered.

SECTION 5: Deflections

In addition to strength and stress requirements, deflections in structural members are important indicators of proper behavior and serviceability. Excessive deflections are undesirable because:

1. They may produce cracks in ceilings, floors, partitions—local damage, leakage.
2. They may cause discomfort to human occupants, for example, motion caused by wind, machinery, vehicles, and so on.
3. They may result in malfunction—for example, poor drainage ("ponding") of roof, elevators out of alignment, jamming of doors, malfunction of other mechanical and service systems.

4. They may cause secondary stresses—in connections, in columns, and so on.

5. They may cause undesirable aesthetic effects—visibly sagging roofs, cantilevers, beam lines.

Common limits in the form of maximum permissible deflections are often specified in building codes or related textbooks and specifications as follows:

1. With full dead load in place, the deflections (vertical) due to superimposed load (*LL*) should not exceed:

Floors   *L*/360
Roofs   *L*/240

2. Sum of superimposed load (*LL*) instantaneous deflections and long-term sustained (*DL*) deflections should not exceed:

Floors   *L*/240
Roofs   *L*/150

The deflection in a beam depends on the distribution of curvature along the entire length of that beam. It is also dependent upon the support conditions. The curvature in a beam is given by the formula:

$$\text{Angle } \phi_x = \epsilon \cdot \frac{dx}{y} = \frac{f_x}{E} \cdot \frac{dx}{y} = \frac{M_x y}{I} \cdot \frac{dx}{E \cdot y} = \frac{M_x}{EI} \cdot dx$$

Referring to Figure 8-19, the unit elongation (ϵ) is given by the unit stress (f_x) divided by Young's modulus E. Since the extreme fiber stress is ($f_x = M_x y / I_x$), where M_x = moment about the x-axis, y = distance to the extreme

FIGURE 8-19
THE GEOMETRY OF INCREMENTAL CURVATURE IN A BEAM.

(a) SIMPLE BEAM MOMENT DIAGRAM

(b) CURVATURE FOR ANY INCREMENTAL LENGTH OF BEAM (dx) IS GIVEN BY THE UNIT CHANGE OF LENGTH (εdx) FOR THE TOP OR BOTTOM FIBER DIVIDED BY 'y'

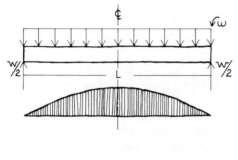

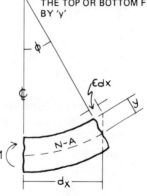

FIGURE 8-20
**APPROXIMATE GEOMETRY OF
OVERALL CURVATURE AND
DEFLECTION IN A BEAM UNDER
CONSTANT *M* (OVER *L*).**

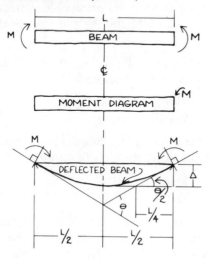

NOTE : FOR $\theta \leq 20°$,
$\Delta \approx \dfrac{\theta}{2}\left(\dfrac{L}{4}\right)$

fiber, and I_x = moment of inertia about the x-axis, the curvature at any point is defined as the change in angle per unit length:

$$\phi_x/dx = M_x/EI.$$

Figure 8-20 shows a beam under a constant moment M along its entire length. If the moment is constant, the curvature ϕ over the entire length is also constant (i.e., circular), and will result in the following angular change θ and deflection Δ:

$$\theta = \frac{M}{EI} \cdot L;$$

and for $\theta \leqslant 20°$;

$$\Delta = \frac{\theta}{2} \cdot \frac{L}{4} = \frac{M}{EI} \cdot \frac{L}{2} \cdot \frac{L}{4} = \frac{ML^2}{8EI}.$$

A general form of expression for deflection in beams is given by $\Delta = (k_3)WL^3/EI$; where k_3 for certain loading conditions is listed in Appendix Table D-2. Each coefficient will yield the maximum deflection in that particular beam. For example, a single-span beam simply supported at A and C will have $k_3 = 5/384$, and a maximum deflection at midspan B, equal to:

$$\Delta = \frac{5WL^3}{384EI}$$

It is often sufficient to note that the deflection on a uniformly loaded beam will vary directly with the *total* weight or loading W, with the third power of the span length L, and inversely with the E of the material and I of the section. Therefore, increasing the span length of a member will greatly affect its deflection, whereas increasing the weight or decreasing the E value or the I value will also affect the deflection, but only to the first degree. However, the moment of inertia of a section, I, varies with the area and second power of the depth and increasing the depth of the section will greatly reduce the deflection. Thus, for a rectangular section, I increases with the third power of an increase in (d), assuming the width remains constant.[2]

It should also be noted from Appendix C that, when a beam has both ends fixed, the deflection coefficient k_3 becomes 1/384 as compared with the coefficient for a simply supported beam (5/384). One may conclude that a fixed-end beam is five times stiffer than a simple beam, other things being equal.

For materials like steel, the E value remains almost constant with time. For concrete (or even timber, but to a smaller extent), the value E decreases with time. A decrease in E value, called creep effect, tends to produce more deflection in the beam as time goes on. This factor can be important under certain conditions. For example, in a concrete beam, the long-term deflection under sustained load may be two or three times greater than the initial deflection, and the DL design will have to reflect that fact.

It should be noted that in the above illustrations, the load was assumed symmetric and the maximum deflection occurred at midspan. When this is the case, k_3 values are computed by taking the moment area over $L/2$ times its centroidal arm from either support.

For example; k_3 for a simple beam with an evenly distributed load (w) would be:

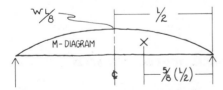

NOTE: DEFLECTION IS FROM BEAM ℄ TO END

$$k_3 = \left(\frac{1}{8}\right) \cdot \left(\frac{2}{3}\right)\left(\frac{1}{2}\right) \cdot \left(\frac{5}{8}\right)\left(\frac{1}{2}\right) = \left(\frac{1}{8}\right) \cdot \left(\frac{5}{48}\right) = \frac{5}{384}$$

Fortunately, any assymmetric load can be dealt with in the same manner because the maximum deflection will always be near the center line.

[2]In this case the area will increase in proportion to any increase in (d).

SECTION 6: Prestressing; Load Balancing

Prestressing means the intentional creation of permanent *internal* forces and stresses in a structure or assembly, for the purpose of improving its behavior and strength under various service conditions. For concrete structures, prestressing is commonly introduced by tensioning high-strength steel reinforcement. Since concrete is strong in compression and weak in tension, prestressing the steel against the concrete would put the concrete under compressive stress that could be utilized to counterbalance tensile stresses produced by external loads. The difference lies in the use of high-strength steel for prestressed concrete. If such steel is simply buried in the concrete without being prestressed, (*passive design*), concrete will have to crack seriously as with reinforced concrete before the strength of the steel is sufficiently developed (Figure 8-21). But by prestretching the steel and anchoring it against the concrete, (*active design*), we can achieve more efficient combination of the two materials (refer also to Figure 6-10 on p. 172).

There are basically two methods for prestressing concrete: pretensioning and posttensioning. In *pretensioning*, the tendons (steel wires or strands) are tensioned against abutments or molds, previous to the placement of concrete. After the concrete has set, the tendons are released from their temporary anchorages and their tensioning force transferred to the concrete, generally by bond action near the ends of the tendons. When pretensioning is done by the long-line process, the tendons are stretched between abutments that are several hundred feet apart (Photo 8-3). This enables the production of a number of elements in one operation. Pretensioning can also be accomplished by the *individual mold* process, wherein the tendons are first tensioned and anchored to the mold, then concrete is placed in it and cured, with the demolding done after the hardening of concrete.

In posttensioning, the tendons (steel wires, strands, or bars) are tensioned and anchored against the concrete after it has hardened. Such tensioning is usually accomplished by hydraulic jacks (Photo 8-4). The

FIGURE 8-21
PASSIVE VERSUS ACTIVE DESIGN OF CONCRETE BEAMS.

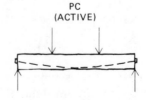

RC
(PASSIVE)

PC
(ACTIVE)

SIMPLY REINFORCED-CRACKS
AND EXCESSIVE DEFLECTIONS

PRESTRESSED-NO CRACKS
AND ONLY SMALL DEFLECTIONS

PHOTO 8-3
A PRETENSIONING BED IN OPERATION.

tendons are separated from the surrounding concrete before tensioning by conduits placed in the form. After tensioning, they may be bonded to it by grouting. But tendons can remain permanently unbonded to the concrete if they are greased and wrapped to achieve complete separation. In this case the tendons are shipped greased and wrapped (in plastic tubes, etc.) and may be placed directly in the forms.

With both pretensioning and posttensioning excessive creep must be avoided. Thus, precompression stresses are generally limited for beams to 600–800 psi and for slabs to 200–400 psi. Three different concepts of prestressed concrete behavior will now be explained:

First Concept: Prestressing to Transform Concrete into an Elastic Material. By this concept, one visualizes concrete being transformed from a brittle material into an elastic one as precompression is imparted to it. Concrete is considered as being subject to two systems of forces: internal prestress and external load. Since the tensile stress due to external load is counteracted by the compressive stress due to prestress, the cracking of concrete is prevented or delayed. As long as there are no cracks, concrete behaves like an elastic material.

Let us consider a simple rectangular beam eccentrically prestressed by a *straight* tendon (Figure 8-22) and loaded by external loads. The direct compression force due to prestress F produces a uniform stress across the section,

$$f_F = \frac{F}{A}$$

PHOTO 8-4
**HYDRAULIC JACK FOR PRESTRESSING TENDONS.
(POSTTENSIONED CONCRETE INSTITUTE.)**

where A = area of the concrete section. But there will also be a negative moment resulting from prestress $(-M_F = -Fe)$, where e = the eccentricity of C with respect to the centroidal axis of concrete (c.g.c.). The stress *at any point* due to this moment is

$$f_{M_F} = \pm \frac{Fey}{I}$$

where y is the distance from c.g.c., and I is the moment of inertia of the section. If M is the moment at a section due to the external load on and the weight of the beam, then the stress *at any point* across that section due to M is,

$$f_M = \pm \frac{My}{I}.$$

Hence, as shown in Figure 8-22, the actual stress at any point is given by,

$$f = \frac{F}{A} \pm \frac{Fey}{I} \pm \frac{My}{I}, \quad \text{or,} \quad f_{max} = \frac{F}{A} \pm \frac{Fec}{I} \pm \frac{Mc}{I}.$$

FIGURE 8-22
STRESS DISTRIBUTION ACROSS AN ECCENTRICALLY PRESTRESSED CONCRETE BEAM SECTION.

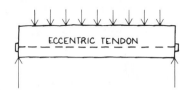

BEAM ECCENTRICALLY PRESTRESSED AND LOADED

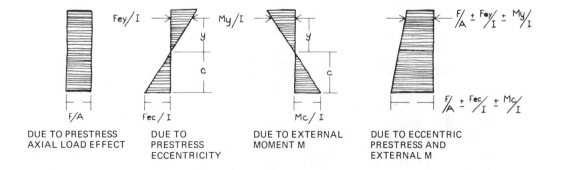

| DUE TO PRESTRESS AXIAL LOAD EFFECT | DUE TO PRESTRESS ECCENTRICITY | DUE TO EXTERNAL MOMENT M | DUE TO ECCENTRIC PRESTRESS AND EXTERNAL M |

The following example illustrates this.

Example 8-1

A rectangular prestressed concrete beam, 20 in. × 30 in., has a simple span of 24 ft and is loaded by a uniform load of 3 k/ft, which includes its own weight (Figure 8-23). A straight prestressing tendon is located as shown and produces an effective prestress force of 360 k. Compute fiber stresses in the concrete at the midspan section.

Solution: We have: $F = 360$ k, $A = 20 \times 30 = 600$ in.^{2} (neglecting any hole due to the tendon), $e = 6$ in., $I = bd^3/12 = 20 \times 30^3/12 = 45{,}000$ in.^{4}; $y = 15$ in.

FIGURE 8-23
ELASTIC DESIGN OF A P.C. BEAM.

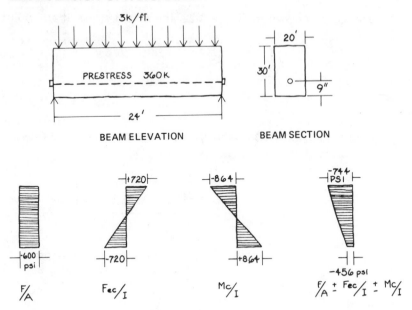

BEAM ELEVATION   BEAM SECTION

for extreme fibers. Since $M = 3 \times 24^2/8 = 216$ k-ft, we can write:

$$f = \frac{F}{A} \pm \frac{Fey}{I} \pm \frac{My}{I}$$

$$= \frac{-360,000}{600} \pm \frac{360,000 \times 6 \times 15}{45,000} \pm \frac{216 \times 12,000 \times 15}{45,000}$$

$$= -600 \pm 720 \pm 864$$

$$= -600 + 720 - 864 = -744 \text{ psi (compressive) for top fiber}$$

$$= -600 - 720 + 864 = -456 \text{ psi (compressive) for bottom fiber.}$$

Second Concept: Prestressing for Efficient Combination of High-Strength Steel with Concrete. Here the tendon is also *straight.* But the *behavior* of prestressed concrete is seen to be similar to that of reinforced concrete in that steel takes tension and concrete takes compression so that the two materials form an internal couple against the external moment.

However, in a prestressed concrete beam, the internal resisting couple (*C-T*) is active (Figure 8-24) and differs from that in a reinforced concrete beam. The *TC* couple has a lever arm (a), which *varies* with the magnitude of the external moment. In a prestressed concrete beam, the lever arm (a) is determined from the following formula, when the prestress force *F* and the external moment M are known; thus,

$$a = \frac{M}{F}$$

Note the similarity of this concept with the eccentricity concept discussed in Chapters 2 and 3 for whole buildings.

FIGURE 8-24
VARIABLE M-RESISTING ARM (a) IN P.C. BEAMS.

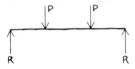

| (a) EXTERNAL MOMENT = 0; $a = 0$ | (b) SMALL EXTERNAL MOMENT; a IS SMALL | (c) LARGE EXTERNAL MOMENT; a IS LARGE |

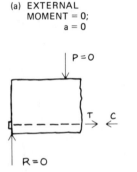 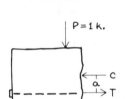 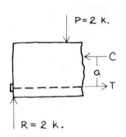

Having determined the lever arm (a) and knowing the center of the prestress force F (which is also the tension T), the center of compression C is located. Depending upon the location of C, various stress distributions in the concrete section can be obtained. Referring to Figure 8-25, it can be seen, for example, that when C is at the c.g.c., we have a uniform stress distribution. When the C is at the top or bottom *kern point* (center of elastic stress above and below N-A), triangular stress distribution is obtained. The value of the top and bottom kern distances k_t and k_b are computed from the formulas

$$k_t = \frac{r^2}{c_b}$$

$$k_b = \frac{r^2}{c_t}$$

where r = the radius of gyration, and $r^2 = I/A$ (e.g., $r^2 = d^2/12$ for a rectangular section). Since, respectively, c_t, c_b = distance from the c.g.c. to the top and bottom extreme fibers, $k_t = k_b = d/6$ for a rectangular section. Once C is located, eccentricity e is known and the stress at any point on

FIGURE 8-25
EFFICIENT DESIGN BY USE OF KERN POINT.

(a) C BELOW BOTTOM KERN POINT

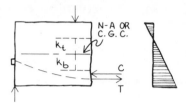

(d) C AT c.g.c.

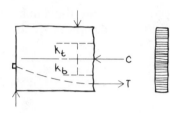

(b) C AT BOTTOM KERN POINT

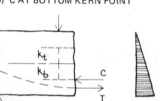

(e) C AT TOP KERN POINT

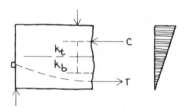

(c) C WITHIN BOTTOM KERN POINT

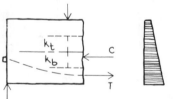

(f) C ABOVE TOP KERN POINT

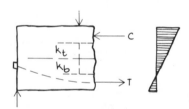

NOTE : KERN POINT FOR ANY SECTION IS THE CENTER OF ELASTIC STRESS ABOVE & BELOW N-A

FIGURE 8-26
DESIGN OF P.C. BEAM USING KERN POINT.

NOTES : SECTION IS RECTANGULAR ; $k_t = d/6 = 5"$
AXIAL (DIRECT) STRESS = -600 psi

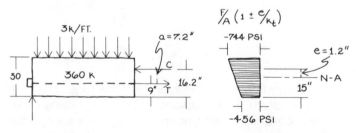

HALF ELEVATION OF BEAM   STRESS DISTRIBUTION AT MIDSPAN

the section is given by

$$f = \frac{F}{A} \pm \frac{Fey}{I}$$

(or by $e/k_{t\ or\ b}$ as introduced in Chapter 3 and illustrated in Figure 8-26)

Example 8-2

Solve the problem stated in Example 8-1 by applying the principle of the internal resisting couple.

Solution: Take one-half of the beam as a freebody, thus exposing the internal couple (Figure 8-26). The external moment at the section is

$$M = \frac{wL^2}{8}$$

$$= \frac{3 \times 24^2}{8}$$

$$= 216 \text{ kip-ft}$$

The internal couple is furnished by the forces $C = T = 360\,k$, which must act with a lever arm of

$$\frac{216}{360} \times 12 = 7.2 \text{ in.}$$

Since T acts at 9 in. from the bottom, C must be acting at 16.2 in. from it. Thus, the center of the compressive force C is located.

For $C = F = 360,000$ lb acting with an eccentricity $e = 16.2 - 15 = 1.2$ in., the extreme fiber stress f corresponding to $y = 15$ in. is

$$f = \frac{F}{A} \pm \frac{Fec}{I} \qquad \left(\text{or use} \pm \frac{e}{k_t} \cdot \frac{F}{A}\right)$$

$$= \frac{-360,000}{600} \pm \frac{360,000 \times 1.2 \times 15}{45,000} \quad \text{or,} \quad \left(\pm \frac{1.2}{5} \cdot 600\right)$$

$$= -600 \pm 144$$

$$= -744 \text{ psi (compression) for top fiber}$$

$$= -456 \text{ psi for bottom fiber}$$

Third Concept: *Prestressing to Balance External Loads.*   By this concept one visualizes the use of a *curved* tendon to carry a part of the load on a beam. That is, prestressing is primarily an attempt to balance a certain portion of the external loads on a member with an internal *counter load* induced by tensioning a curved tendon. In its simplest form, one can conceive a parabolic tendon in a simple beam prestressed so as to apply a uniform upward force on the beam. If this beam carries an external downward load of equal intensity, then the net transverse load on this beam is zero. Provided that the curve is such that C always coincides with the c.g.c. [(a) varies with M diagram], we are left with a constant compressive stress of $(f = F/A)$ across any section of the beam.

The net prestress F required to balance a uniform load of intensity w lb/ft is given by

$$F = \frac{wL^2}{8h},$$ where L = span length in feet and h = *cable sag* in feet (Fig. 8-27).

Due to friction losses, curved tendons require an *initial* prestress some 15% higher. When the external load differs from the balanced loading, only the moment M produced by the unbalanced load needs to be considered in computing bending stresses f and the usual formula is applied, $f = Mc/I$. For a *simple* beam the *extra* load can be (K_t/h) times the balanced load (i.e., with C at the upper kern point). This is illustrated in the following example.

Example 8-3

A concrete beam is to be prestressed with a parabolic cable located as shown in Figure 8-27. Compute the total amount of prestress required to balance an external load of 2.5 k/ft on the beam, which includes the beam's own weight.

Solution: For a sag $h = (15 - 9) = 6$ in. or 0.5 ft and $L = 24$ ft, we have

$$F = \frac{wL^2}{8h}$$
$$= \frac{2.5 \times 24^2}{8 \times 0.5}$$
$$= 360 \text{ kips}$$

Thus, if a prestress of 360 k is applied to the beam as shown, the beam will be under a uniform compressive stress of $360{,}000/600 = 600$ psi. The beam will be under no bending at all when subjected to an external load of

FIGURE 8-27
LOAD BALANCING DESIGN OF P.C. BEAMS.

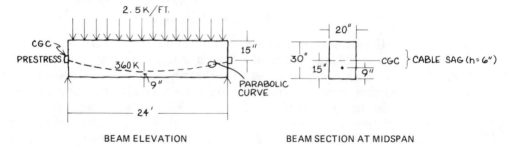

BEAM ELEVATION   BEAM SECTION AT MIDSPAN

2.5 k/ft. (Note that if greater accuracy is desired, the W and the *horizontal* prestress component of 360 k should be used instead of the 360 k, but in most cases the difference will be small.)

When the external load is 3 k/ft (compare with Examples 8-1 and 8-2), the unbalanced load amounts to $3 - 2.5 = 0.5$ k/ft, and its moment at midspan is

$$\frac{wL^2}{8} = \frac{0.5 \times 24^2}{8} = 36 \text{ kip-ft}$$

This can be resisted without cracking because $(K_t/h)(2.5) = (5/6)(2.5) \sim 2.0$ k/ft. The extreme fiber stresses produced by this 0.5 k/ft load are

$$f = \frac{Mc}{I} = \pm \frac{36 \times 12,000 \times 15}{45,000}$$

$$= \pm 144 \text{ psi}$$

or could use $\left(\pm \dfrac{e}{k_t} 600 = \pm 144 \text{ psi} \right)$, where $\dfrac{36 \text{ kip-ft}}{360 \text{ k}} = e = 1.2$ in.

while the resulting stresses, due to combined effect of prestress and an external load of 3 k/ft are

$$f = -600 \pm 144$$

$$= -744 \text{ psi (compression) for top fiber}$$

$$= -456 \text{ psi for bottom fiber}$$

This third concept is often termed *load balancing*. It is a very powerful tool for design and analysis because it cancels out the effect of external loads and arrives at a state of zero deflection under a given loading condition. In many cases it will be sufficient to design to balance the $DL + \frac{1}{2}LL$. Thus it is extremely convenient for the purpose of preliminary design. For a more detailed treatment of prestressing, the reader is referred to *Design of Prestressed Concrete Structures* by T. Y. Lin, which presents this concept in much greater depth and detail.

SECTION 7: Horizontal and Vertical Shear Flow in Beams

As a beam bends, horizontal layers of its material resist slippage between them. This can be visualized by considering a rectangular beam built up of thin strip components as shown in Figure 8-28. Note that if there is no structural interaction between the strip components, they will behave like independent and shallow beams. Each carries an equal part of the total load but the overall deflection will be very large, because the stiffness is reduced for each strip much more than the load. This means that if the beam is to act as a whole and be stiffer, the strips must be structurally connected to resist slippage.

There will be a horizontal component of shear that must match the

FIGURE 8-28
BEAMS MUST BE ABLE TO TRANSFER HORIZONTAL SHEAR STRESSES.

(a) HORIZONTAL STRIP INTERACTION FORCES BEAM TO ACT AS A WHOLE

(b) WITHOUT INTERACTION, EACH STRIP BENDS AS AN INDEPENDENT SHALLOW BEAM AND DEFLECTIONS ARE MUCH GREATER

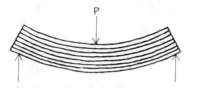
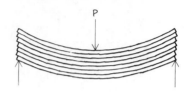

FIGURE 8-29
SHEAR FLOW IN BEAMS

(a) THE VERTICAL SHEAR FLOW (q) DETERMINES THE VERTICAL AND HORIZONTAL INTERACTION STRESSES (v)

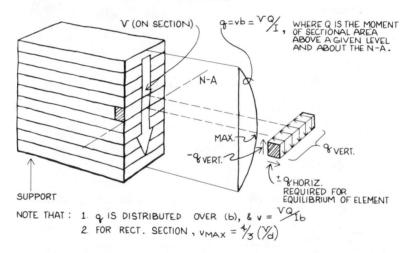

$$q = vb = \frac{VQ}{I},$$ WHERE Q IS THE MOMENT OF SECTIONAL AREA ABOVE A GIVEN LEVEL AND ABOUT THE N-A.

NOTE THAT : 1. q IS DISTRIBUTED OVER (b), & $v = \frac{VQ}{Ib}$

2. FOR RECT. SECTION , $v_{MAX} = \frac{4}{3} \left(\frac{V}{d} \right)$

(b) FOR 'H' SECTIONS

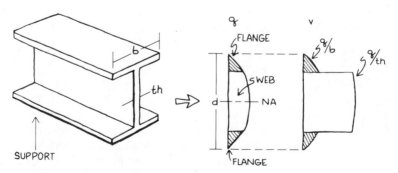

NOTE THAT : 1. WEB CARRIES MOST OF SHEAR & v IS FAIRLY UNIFORM

2. WEB STRESSES CAN BE APPROXIMATED BY $v \sim \frac{V}{th(d_{web})}$

vertical shear across a section of the beam *at a given level* with the shear being maximum at the *N*-*A* and 0 at the top and bottom fibers (Figure 8-29).

The *shear flow* (*q*) at a given level is $q = vb = VQ/I$, where (*q*) represents the total horizontal shear resisting force *across* width (*b*) at a given level (also, the force per unit of length). The *shearing stress* (*v*) is given by $v = VQ/Ib$. *Q* is the statical moment of the area of the beam material above (or below) a given level of the section and about the *N*-*A*. Thus (*v*) varies parabolically in a rectangular section and is maximum at the *N*-*A* of the section.

For the rectangular section (Figure 8-29a), $v_{max} = (3/2)(V/bd)$, and $q_{max} = (3/2)(V/d)$ at the *N*-*A*. For steel wide-flange sections (Figure 8-29b), the shear flow at the flange to web joint is very close to that at the neutral axis, and the value of *v* may be approximated by uniform distribution $v = V/(th)d_{web}$.

With steel and concrete beams, shear resistance is not often as significant a design problem as bending and deflection. But, in the case of timber, it may be useful to know how to provide for shear design when sections are built up to form *T*, box, or glue laminated beams, since the shear strength of nails or glue must be considered (Appendix C-5*a*, *b*, *c* and *d*) to insure that fasteners can resist the shear flow.

Figure 8-31, page 295 shows that stirrups made of re-bars are placed in concrete beams in locations where the actual average shearing stress exceeds the allowable average shearing stress for the concrete. Since the allowable shearing stress $v_c = 2\sqrt{f'_c}$, and $V_c = v_c(A_c)$, concrete can carry some shear on its own. But any excess shear ($V - V_c$) will have to be carried by 'U' or rectangular stirrups often made of no. 4 or smaller re-bars. The spacing of these stirrups is determined by the ACI Code formula relating beam depth (*d*) to spacing (*s*):

$$A_v f_{steel} \times \left(\frac{d}{s}\right) = V - V_c \quad \text{or,} \quad \frac{A_v f_{steel}}{V - V_c} = \frac{s}{d} \quad \text{and,}$$

$$s = \frac{A_v f_{steel}(d)}{V - V_c} \quad \text{(see Ex. 8-5, p. 288).}$$

The relation *s*/*d* is important since a shear crack in concrete can occur at a 45° angle. If $s \geq d$, the crack can occur *between* two stirrups and the beam will fail. Therefore, it is common practice to insure that $s < d$ and if (say for #4 bars) $s \geq d$, use more stirrups of smaller diameter bars to get closer spacing.

SECTION 8: Design Examples

Three design examples are presented here to illustrate the preliminary proportioning of horizontal components in steel, reinforced concrete, and prestressed concrete.

Example 8-4

A typical bay in a story of a high-rise steel *rigid-frame* building is to have wide flange (W) sections as its main girders spaced 20 ft apart along the length of the building. The width of the building is 60 ft, and two alternate layouts are to be considered: a clear span of 60 ft, and two equal spans at 30 ft each.

Due to the stiffness of the columns, the girder moments will be estimated at $wL^2/14$ for the $-M$ at end and $wL^2/16$ for the $+M$ at midspan. Use A-36 steel with allowable stress at 24 ksi. Dead load over the floor is estimated at 80 psf, which includes the girder's own weight; live load is 75 psf reduced to 45 psf for the girders.

   1.   Choose a W section for the 60-ft-span layout and also for the 30-ft-span layout.
   2.   For each case in **1**, compute the live load deflection, using an approximate formula for continuous beams, ($k_3 = 3/384$)

$$\Delta_{LL} = \frac{3wL^4}{384\,EI}$$

       Compare these deflections with an allowable limit of $L/360$.
   3.   On the basis of the above calculations and other pertinent considerations, discuss the economics and desirability of the longer versus the shorter span layout.
   4.   If the lateral resistance of the building is provided by shear walls and utility cores so that rigid-frame action is not required for the steel structure, then the two girders could be simply supported at the ends. Discuss the weight of W sections needed for item **1** above, considering moment and deflection requirements.

Solution: Consider two alternate layouts:

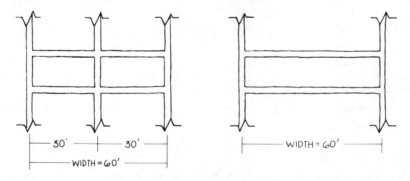

Tributary load @ 20-ft bay, since girders are spaced 20 ft apart:
Estimate column moments = $wL^2/14$ for $-M$ controls girder size.

             Dead load = 80 psf
Reduced Live load = $\dfrac{45 \text{ psf}}{125 \text{ psf}}$

$w_{TOT} = 125 \times 20$ ft $= 2.5$ k/ft of girder length
$w_{LL} = 45 \times 20$ ft $= 0.9$ k/ft of girder length
allowable $f_b = 24$ ksi for A36 steel in bending, $E = 29,000$ ksi.
Section Modulus ($S = I/c$) and, if $f = Mc/I$, S required $= M/f_b$.

1. *Case 1 (30-ft span layout)*

$$-M = \frac{1}{14}(2.5 \times 30^2) = 160 \text{ k ft}$$

$$S_{REQ} = \frac{M}{f_b} = \frac{160 \times 12}{24} = 80 \text{ in.}^3$$

Choose $L/d = 20$, $\dfrac{30 \times 12}{20} = 18$ in.

From Appendix Table C-2, try W18×45:

$$I = 706 \text{ in.}^4, \quad S = 79.0 \text{ in.}^3$$

2. $\Delta_{LL} = \dfrac{3wL^4}{384 \, EI} = \dfrac{3 \times 0.9 \times 30^4 \times 12^3}{384 \times 29,000 \times 706}$

$$= 0.45 \text{ in.}$$

allowable:

$$\frac{L}{360} = \frac{30 \times 12}{360} = 1 \text{ in.} > .045 \text{ in.};$$
$$\text{o.k.}$$

1. *Case 2 (60-ft span layout)*

$$-M = \frac{1}{14}(2.5 \times 60^2) = 640 \text{ k ft}$$

$$S_{REQ} = \frac{640 \times 12}{24} = 320 \text{ in.}^3$$

Choose $L/d = 20$, $\dfrac{60 \times 12}{20} = 36$ in.

From Appendix Table C-2, W36

Sections too heavy; try W30×116

$$I = 4930 \text{ in.}^4 \quad S = 329 \text{ in.}^3$$

2. $\Delta_{LL} = \dfrac{3 \times 0.9 \times 60^4 \times 12^3}{384 \times 29,000 \times 4930}$

$$= 1.12 \text{ in.}$$

$$\frac{L}{360} = \frac{60 \times 12}{360} = 2 \text{ in.} > 1.12 \text{ in.};$$
$$\text{o.k.}$$

3. a. Case 1 would require a lot less steel for the girders and only slightly more steel for the three columns since the vertical force to be transmitted is about the same in both cases.

b. Construction costs would be higher for Case 1 because the additional row of columns requires additional connections, footings, and so on.

c. Case 1 would allow more room to hang ducts and plumbing since the girder depths are considerably less. Holes may have to be cut in the webs of Case 2 girders which would increase fabrication costs.

d. Case 2 would provide slightly more floor space and a much more flexible floor layout.

e. For simple spans $M = wL^2/8$, which is a 75 percent increase over $M = wL^2/14$ for the rigid-frame girder, S_{REQ} would be increased by 75 percent and a heavier section would be needed. The deflection formula for simple spans is $5wL^4/384 \, EI$, which is an increase over $3wL^4/384 \, EI$. Hence, larger I may be needed to control the deflection.

Example 8-5

The 30-ft bay rigid-frame scheme in Example 8-4 is now to be designed in reinforced concrete, using a concrete slab spanning between 30-ft-span girders at 20 ft centers. Dead load consists of the slab (concrete at 150 pcf) and the girders, whose weight is to be estimated. Basic design live load is 75 psf with reductions as permitted by Uniform Building Code.

1. Compute $+M = wL^2/16$ and $-M = wL^2/14$ to determine size and reinforcement of slab and girder. Consider T-beam action at midspan only of girder and use some compression steel over supports if desired. Try to minimize size of girders, particularly their depth. For concrete and for reinforcing steel, $f'_c = 4$ ksi; $/f_y = 40$ ksi. Use ultimate strength method.

2. Check for shear resistance of girders and provide stirrups if necessary. The reader should compare this design with the 30-ft-span steel girders in Example 8-4.

Solution

Estimate dead load = 120 psf for girders

Live Load Reduction

Design live load = 75 psf

Tributary area of girders = $30 \times 20 = 600$ sf

Uniform Building Code Reduction:

1. $R = .08 \times 600 = 48$ percent

2. $R = 23.1\left(1 + \dfrac{D}{L}\right) = 23.1\left(1 + \dfrac{100}{50}\right) = 69$ percent

3. Not to exceed 40 percent reduction; hence, reduced live load = 60 percent (75) = 45 psf

(a) *Design of Slab* (consider 1 ft strip)

For span-depth ratio of [30 + 10 percent (30)] (continuous construction) = 33,

$$t = \frac{20 \times 12}{33} = 7.27; \text{ say, } 7.5 \text{ in.}$$

Use ultimate strength design $1.4D + 1.7L$,

$DL = \dfrac{7.5}{12} \times 150 = 94 \text{ lb/ft} \times 1.4 = 132 \text{ lb/ft}$

$LL = 75 \text{ psf} = 75 \text{ lb/ft} \times 1.7 \quad = 128 \text{ lb/ft}$ (no reduction for slab strip)

$\omega_u = 260 \text{ lb/ft}$

Positive moment $= \dfrac{\omega_u L^2}{16} = \dfrac{260 \times 20^2 \times 12}{(\phi = 0.9)(16)} = 87$ kip-in. (Using strength

reduction factor $\phi = 0.9$)

Compute d': if $\frac{1}{2}$ in. bars are used and $\frac{3}{4}$-in. clear protection

$$d' = 7.5 - (\tfrac{3}{4} + \tfrac{1}{4}) = 6.5 \text{ in.}$$
$$a = 0.9d' = 0.9 \times 6.5 = 5.8 \text{ in.}$$
$$T = \frac{M_u}{a} = \frac{87}{5.8} = 15.0 \text{ k}$$

$A_s = \dfrac{T}{f_y} = \dfrac{15.0}{40} = 0.38 \text{ in.}^2/\text{ft of slab; use no. 4's at 6 in. With } A_s =$
$0.39 \text{ in.}^2/\text{ft} \cdot \text{slab}$

$$\text{Negative moment } M_u = \frac{wL^2}{\phi 12} = \frac{260 \times 20^2 \times 12}{(\phi = 0.9) \times 12} = 116 \text{ kip-in.}$$

$$a = 5.8 \text{ in.}; \quad T = \frac{M_u}{a} = \frac{116}{5.8} = 20.0 \text{ k};$$

$A_s = \dfrac{T}{f_y} = \dfrac{20.0}{40} = 0.50 \text{ in.}^2/\text{ft slab; use no. 4's @ } 4\tfrac{1}{2} \text{ in. with } A_s = 0.52 \text{ in.}^2/\text{ft slab}$
(i.e. 7%, ok).

(**b**) *Girder Design* (T-section for positive moment)

For depth-span ratio of 15, $d = \dfrac{30 \times 12}{15} = 24$ in. let $b = 12$ in. (stem width)

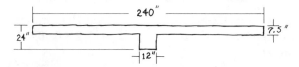

For T-section of this proportion, top slab compression is not a problem.
Design for tension only.

$$DL = \left(20 \text{ ft} \times \frac{7.5}{12} + \frac{12 \times 16.5}{144}\right) 150 \times 1.4 = 2.91 \text{ k/ft}$$
$$LL = 45 \times 20 \times 1.7 \qquad\qquad\qquad = 1.53 \text{ k/ft}$$
$$\omega_u = 4.44 \text{ k/ft}$$

Estimate d': If $1\tfrac{1}{4}$-in. bars are used and $1\tfrac{1}{2}$-in. clear protection of concrete,

$$d' = 24 \text{ in.} - \left(\frac{1.25}{2} + 1.5\right) = 21.9 \text{ in.}$$

$$\text{Positive moment} = \frac{\omega_u L^2}{\phi 16} = \frac{4.44 \times 30^2}{(\phi = .9) \times 16} = 277 \text{ kip-ft}$$

$$a = 0.92d' = 0.92 \times 21.9 = 20.1 \text{ in.}$$

$$T = \frac{M_u}{a} = \frac{277 \times 12}{20.1} = 165 \text{ k}$$

$A_s = \dfrac{165}{40} = 4.1 \text{ in.}^2;$ Use 2 no. 10's and 1 no. 11, $A_s = 4.1 \text{ in.}^2$ (i.e. 0.2% ok)

Negative moment $= \dfrac{wL^2}{\phi 14} = \dfrac{4.44 \times 30^2}{(\phi = .9) \times 14} = 317$ kip-ft (tension at top)

Treat as rectangular beam: $a = 0.9d' = 0.9 \times 21.9 = 19.7$ in.

$T = \dfrac{317 \times 12}{19.7} = 194$ k, $A_s = \dfrac{194}{40} \sim 5.0$ in.^{2} Use 4 no. 10's, $A_s = 5.0$ in.^{2}

Check compression at bottom of girder via percentage of steel (again assume rect. section

$$p = \frac{A_s}{bd} = \frac{5.0}{12 \text{ in.} \times 21.9 \text{ in.}} = 1.9 \text{ percent}$$

which is high but o.k., since usual limits for p is about 1.5 to 2.0 percent for beams. Some increase in stem width (say to 14 in.) or addition of some bottom (compression) bars may be desired.

(c) Check Shear Capacity of Stem

Assume maximum shear $= \frac{1}{2}wL = \frac{1}{2}4.44 \times 30 = 66.6$ k (for shear, use strength reduction factor of $\phi = 0.85$ per ACI Code).

$$\frac{V_u}{\phi} = \frac{66.6}{0.85} = 79 \text{ k}$$

allowable $v_c = 2\sqrt{f_c'} = 2\sqrt{4000} = 126$ psi

allowable $V_c = v_c bd = 0.126 \times 12 \times 21.9 = 33.1$ k < 79 k, indicating need of stirrups. Assuming no. 4 stirrups are used, $A_v = 2 \times 0.20 = 0.40$ in.^{2} and using ACI Code formula,

$$A_v f_y \frac{d}{s} = \frac{V_u}{\phi} - V_c \text{ where } s \text{ is stirrup spacing (} s \text{ should be less than } d\text{);}$$

or, $\dfrac{A_v f_y}{\dfrac{V_u}{\phi} - V_c} = \dfrac{s}{d}$ and $s = \dfrac{A_v f_y d}{\dfrac{V_u}{\phi} - V_c} = \dfrac{0.40 \times 40 \times 21.9}{79 - 33.1} = 7.6$ in.

Use stirrups near support and spaced at 8 in.
As shear V decreases, space farther apart as needed.

EXAMPLE 8-6

A typical bay of the multistory building in Example 8-4 is now to be designed using prestressed concrete girders at 60 ft spans and 20 ft spacing, with prestressed slabs spanning between the girders. Use lightweight concrete (110 pcf) with $f_c' = 5$ ksi, for both girders and slabs. Steel tendons are of (270 ksi) 7-wire strands ($\frac{1}{2}$-in. diameter, $A_s = 0.152$ sq in.), with effective stress at $0.6f_s' = 162$ ksi.

a. Choose a suitable slab thickness and compute the amount of tendons required for each foot width of a typical interior slab span. Balance the dead load and check for live load stresses, using approximate moment coefficients. Discuss end span requirements and possible effect of frictional loss.

b. Suppose the structural depth of the floor is limited to 2 ft 4 in. and the girders are fixed to the supporting columns, compute the amount of prestressing tendons required for the girder. Again balance the dead load and check for live load stresses, using suitable moment coefficients. Assume a reasonable effective flange width for the T-girders.

c. Compute the maximum girder deflection using the approximate formula $k_3 = 3/384$, $\Delta = k_3 \omega L^4/EI$, $E_c = 5000$ ksi; I for T-section as computed on p. 291.

The reader should discuss the feasibility and economics of this prestressed concrete construction, relative to steel or reinforced concrete construction in Examples 8-4 and 8-5.

Solution:

(a) *Design of Slab* (1-ft strip)

Using a span-depth ratio of 44,

$$t = \frac{20 \times 12}{44} = 5\tfrac{1}{2} \text{ in.}$$

Assuming $1\tfrac{1}{4}$ in. to center of gravity of steel, effective sag is

$$h = 5\tfrac{1}{2} \text{ in.} - 2 \times 1\tfrac{1}{4} = 3 \text{ in.}$$

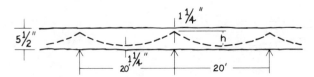

Balanced DL stress:
Assuming $f'_s = 270$ ksi, $f_e = 162$ ksi, $f'_c = 5$ ksi and lightweight concrete at 110 pcf.

$$w_{DL} = \frac{5.5}{12} \times 110 = 50 \text{ lb/ft}$$

Required prestress: use balanced-load design

$$F = \frac{wL^2}{8h} = \frac{0.050 \times 20^2 \times 12}{8 \times 3} = 10 \text{ k/ft}$$

$$A_{S_{req/ft}} = \frac{10 \text{ k}}{162 \text{ ksi}} = 0.06 \text{ in.}^2/\text{ft},$$

$A_{S_{strand}} = 0.152$ in.^{2}, use 1 strand every 30 in. (i.e., $A_s = 0.152/2.5 = 0.06$ in.^{2}/ft).
$f_{avg} = \dfrac{10,000}{12 \times 5.5} = -152$ psi (compression)

which is a low value, since the average precompression for slabs should generally fall between 200 and 400 psi.

LL (Bending) stresses: $LL = 75$ psf $= 75$ lb/ft (no reduction)

$$\text{Negative moment (controlling)} = \frac{wL^2}{12} = \frac{0.075 \times 20^2 \times 12}{12} = 30 \text{ kip-in./ft-slab}$$

$$f = \frac{6M}{bd^2} = \frac{6 \times 30,000}{12 \times 5.5^2} = \pm 495 \text{ psi}$$

Or, you could have used e/k_t, where, for $-M$ at 75 psf, $e = \frac{3}{2}(\frac{2}{3} \cdot h)$.[3]

$f_{\text{top}} = +495 - 152 = +342$ psi (tension)—by adding some reinforcing bars, allowable tension for PC is

$$6\sqrt{f'_c} = 6\sqrt{5000} = 424 \text{ psi} \qquad \text{Hence, o.k.}$$

$f_{\text{bot}} = -495 - 152 = -647$ psi (compression)
which is quite low, since allowable is,

$$0.45f'_c = 2250 \text{ psi; o.k.}$$

Assuming the tendons are anchored at the center of gravity of the slab at the end space, there would be a need for 15 to 20 percent more prestress because the effective sag of the tendon will be somewhat less than 3 in. This can be accomplished by: (1) adding reinforcing steel to the end spans or (2) using more tendons to meet the endspan requirements as well as to meet the required prestress in the interior spans after frictional losses.

(b) *Girder Design*

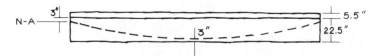

Assume 3 in. protection to center of steel.
Effective sag $= 22.5 - 6$ in. $= 16.5$ in. Assume width of girder $b = 10$ in.

$$DL: \frac{22.5 \times 10}{144} \times 110 + 20 \times \frac{5.5}{12} \times 110 = 1.18 \text{ k/ft, add partition load } 20 \times$$

20 psf $= 0.40$ k/ft.

$$\text{Required prestress:} \quad F = \frac{wL^2}{8h} = \frac{1.58 \times 60^2 \times 12}{8 \times 16.5} = 516 \text{ k}$$

$$A_{s_{\text{req}}} = \frac{516 \text{ k}}{162} = 3.18 \text{ in.}^2$$

$$\text{Number of } \tfrac{1}{2}\text{-in. strands required} = \frac{3.18}{0.152} = 21$$

which can be supplied by 1 or more cables.

[3]That is, for **Dead Load** negative moment, e would be $(\frac{2}{3})h$ and, in this instance, live load is $(\frac{2}{3})DL$.

Live load moments and stresses under *DL + LL*

$$w_{LL} = 0.045 \times 20 = 0.9 \text{ k/ft}$$

$$-M_{LL} = \frac{wL^2}{14} = \frac{0.9 \times 60^2}{14} = -232 \text{ kip ft}$$

$$+M_{LL} = \frac{wL^2}{16} = \frac{0.9 \times 60^2}{16} = 202 \text{ kip ft}$$

Due to Balanced *DL* average pre-compression is

$$f = \frac{516,000}{240 \times 5.5 + 10 \times 22.5} = -335 \text{ psi}$$

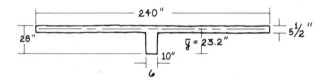

Computation of 'T' section properties (assume *full flange width*):

$$\bar{y} = \frac{5.5 \times 240 \times 25.25 + 10 \times 22.5 \times 11.25}{5.5 \times 240 + 10 \times 22.5} = 23.2 \text{ in.}$$

$$\therefore \quad y_b = 23.2 \text{ in.} \quad y_t = 28 - 23.2 = 4.8 \text{ in.}$$

$$I = \left[\frac{1}{12}(240 \times 5.5^3) + (240 \times 5.5 \times 2.05^2)\right] + \left[\frac{1}{12}(10 \times 22.5^3) + (10 \times 22.5 \times 11.95^2)\right]$$
$$= 50,500 \text{ in.}^4$$

Stresses with *LL* added:

$$f_{top} = f_{av} + \frac{Mc_t}{I} = -335 + \frac{232 \times 12 \times 4.8 \times 1000}{50,500} = -335 + 263 = -72 \text{ psi (no tension)}$$

$$f_{bot} = f_{av} - \frac{Mc_b}{I} = -335 - \frac{232 \times 12 \times 23.2 \times 1000}{50,500} = -335 - 1290 = -1625 \text{ psi}$$

+ Moment

$$f_{top} = -335 - \frac{202 \times 12 \times 4.8 \times 1000}{50,500} = -335 - 230 = -565 \text{ psi}$$

$$f_{bot} = -335 + \frac{202 \times 12 \times 23.2 \times 1000}{50,500} = -335 + 1160 = +825 \text{ psi (tension)}$$

ACI Code: permissible stresses:

Compression: $0.45f'_c = 0.45 \times 5000 = 2250 \text{ psi, o.k.}$

Tension: $6\sqrt{f'_c} = 6\sqrt{5000} = 424 \text{ psi} < 825$

We have exceeded the allowable tensile stress. To correct this without increasing the girder depth, we could balance some of the live load in addition to the dead load, which would mean increasing the prestress force (add more steel). The reader can try with a larger prestress, say 28 instead of $21\text{-}\frac{1}{2}$ in. strands, and check the stresses. Of course one can also derive a formula to find the required prestress, but this will not be attempted here.

(c) *Deflections*

Since the dead load is balanced, $\Delta_{DL} \approx 0$

$$\therefore \ \Delta_{LL} = \frac{3}{384}\left(\frac{\omega L^4}{EI}\right) = \frac{3 \times 0.600 \times 60^4 \times 12^3}{384 \times 5{,}000 \times 50{,}500} = 0.42 \text{ in.}$$

and,

$$\text{Allowable } \Delta_{LL} = \frac{60 \times 12}{360} = 2 \text{ in.; o.k.}$$

Conclusion

One can compare the prestressed concrete design with the reinforced concrete design (Example 8-5) as follows:

1. The PC system uses considerably less (and lighter) concrete for the slab—$5\frac{1}{2}$ in. thick compared with $7\frac{1}{2}$ in., and the PC girder spans twice the distance.
2. The PC system uses less steel for a longer span. But the cost of high-strength steel in place is three to four times that of reinforcing steel. Thus, the total cost of steel may be somewhat higher in the PC system, but it is well justified by the longer span.
3. The cost of formwork is approximately the same for both systems. Precasting may or may not be economical for this case, depending upon several factors. In buildings where many of the structural elements are repeated, precasting becomes economically feasible, particularly in combination with prestressing.

SECTION 9: Connections for Horizontal Components

Connections are a vital part in the design of a building structure. However, they are a part of structural design that is beyond the intended scope of this book. Thus, only a very brief introduction and some general samples of connections will be presented herein. A detailed discussion of connections for structural members of steel, timber, reinforced concrete, and prestressed concrete can be found in various reference books (e.g., the *AISC Manual of Steel Construction* and the PCI Manual on Design of Connections for Precast, Prestressed Concrete).

In the early days, steel members were connected by riveting, which has now been essentially replaced by high-tensile bolts. In addition, welding of steel has developed tremendously and is a common practice, particularly in the fabricating shops, although it is also applied in the field. Some typical connections for steel members using bolting, welding, or a combination are shown in Figure 8-30. In addition, Table 8-1 gives the allowable load in tension and in shear for various sizes of high-tensile bolts. Table 8-2 gives the allowable load per inch of fillet weld. A final design of connections must be made with a real understanding of theory of structures, and mechanics as applied to such connections.

Reinforced concrete members are usually cast monolithic, in which case the problem of joinery is taken care of by proper detailing of the reinforcing bars within the concrete members (Figures 8-31 and 8-32). When hinges or articulations are provided between concrete members, special devices are used. They are somewhat similar to the joinery for precast members (Figure 8-33).

Precast members require careful detailing of their connections. As samples of the type of connections used for precast concrete, whether prestressed or reinforced, some examples are shown in Figure 8-34.

Table 8-1
Connections—Allowable Loads

HIGH-STRENGTH A 325 BOLTS—LOAD IN KIPS

Bolt size diameter inches	$\frac{5}{8}$	$\frac{3}{4}$	$\frac{7}{8}$	1	$1\frac{1}{8}$	$1\frac{1}{4}$	$1\frac{3}{8}$	$1\frac{1}{2}$
Tension, Kips ($F_t = 40$ ksi)	12.3	17.7	24.1	31.4	39.8	49.1	59.5	70.7
Single shear, Kips ($F_v = 15$ ksi)	4.6	6.6	9.0	11.8	14.9	18.4	22.3	26.5
Double shear, Kips	9.2	13.2	18.0	23.6	29.8	36.8	44.6	53.0

Table 8-2
Connections—Allowable Loads

FILLET WELDS—LOAD PER INCH

Weld size, inch	$\frac{3}{16}$	$\frac{1}{4}$	$\frac{5}{16}$	$\frac{3}{8}$	$\frac{1}{2}$	$\frac{5}{8}$
Shear load, Kips ($F_v = 18$ ksi)	2.4	3.2	4.0	4.8	6.4	7.9

FIGURE 8-30
SAMPLE CONNECTIONS FOR STEEL MEMBERS. NOTE THAT CONNECTOR ELEMENTS CAN BE EITHER WELDED AS SHOWN OR BOLTED.

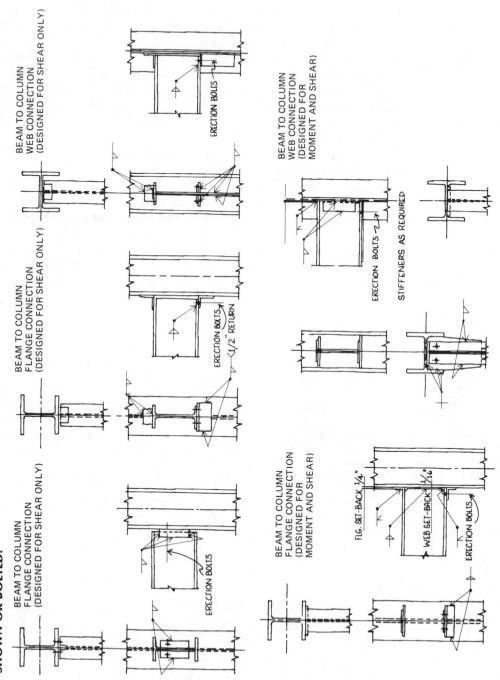

BEAM TO COLUMN
WEB CONNECTION
(DESIGNED FOR SHEAR ONLY)

ERECTION BOLTS

BEAM TO COLUMN
WEB CONNECTION
(DESIGNED FOR
MOMENT AND SHEAR)

ERECTION BOLTS
STIFFENERS AS REQUIRED

BEAM TO COLUMN
FLANGE CONNECTION
(DESIGNED FOR SHEAR ONLY)

ERECTION BOLTS
1/2" RETURN

BEAM TO COLUMN
FLANGE CONNECTION
(DESIGNED FOR SHEAR ONLY)

ERECTION BOLTS

BEAM TO COLUMN
FLANGE CONNECTION
(DESIGNED FOR
MOMENT AND SHEAR)

FIG. SET-BACK 1/4"
WEB SET-BACK 1/16"
ERECTION BOLTS

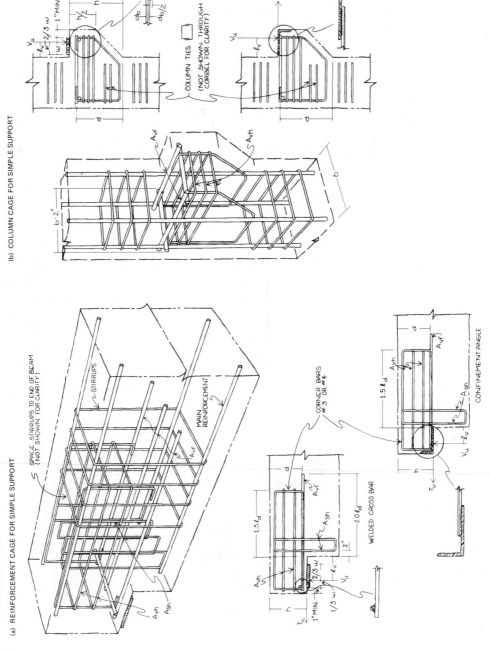

FIGURE 8-31
R.C. BEAM TO COLUMN CONNECTION.

(a) REINFORCEMENT CAGE FOR SIMPLE SUPPORT

(b) COLUMN CAGE FOR SIMPLE SUPPORT

FIGURE 8-32
MORE EXAMPLES OF BEAM-COLUMN CONNECTIONS.

(a) SIMPLE CONNECTIONS

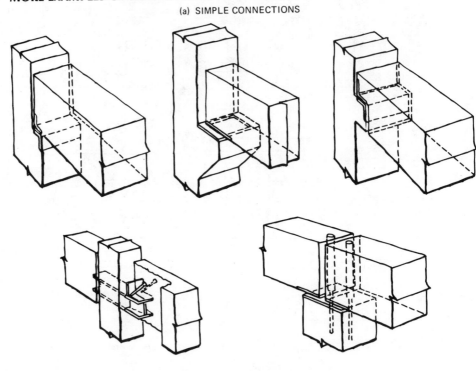

(b) MOMENT RESISTING CONNECTIONS

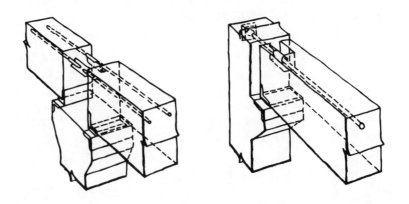

FIGURE 8-33
COLUMN-FOUNDATION ARTICULATIONS.

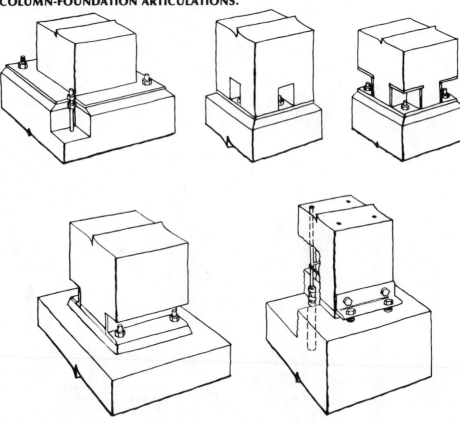

298

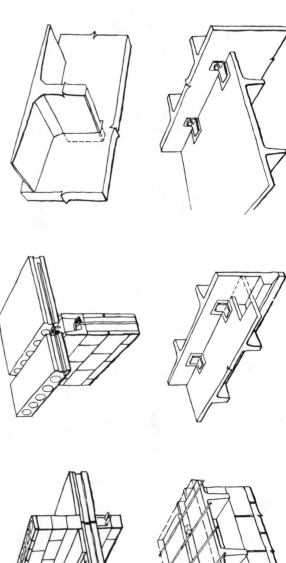

FIGURE 8-34
PRECAST CONCRETE COMPONENT CONNECTIONS.

(a) TYPICAL SLAB TO BEAM CONNECTIONS

(b) TYPICAL SLAB TO WALL CONNECTIONS

SECTION 10: **Trusses**

When dealing with long spans or heavy loads, trusses are more economical than beams and girders, particularly for roof construction. Occasionally, they are used to span long floors and perhaps to serve as transfer floors that carry columns for shorter spans above to longer ones below. In any case, the basic principle of truss action is to resist bending moment by concentrating the moment resisting materials near the extreme fibers so that they will be fully stressed and will act with the maximum lever arm. At the same time, web or diagonal members are provided so that the shear force acting on the truss can also be resisted by direct stress action in these members. Generally, these web members can most economically resist shearing forces when they are orientated at about 45°.

Figure 8-35a shows a truss of rectangular profile but triangulated web, which is a common arrangement. Note that there are many components to be connected. Thus, while a truss is often economical in material, it may not be the cheapest type of construction on account of the cost of joining the members. They become suitable and economical only for relatively long spans or heavy loads, when the savings in material is particularly important; although prefabricated trussed joists are sometimes used for medium span floor systems. Figure 8-35b shows an approximate moment *envelope* for a simple span truss. The maximum moment is indicated $M = wL^2/8 = WL/8$. Figure 8-35c illustrates how the maximum bending moment is resisted by one-half of the truss taken as the freebody. Here, the moment-resisting couple $C–T$ is furnished by the top and bottom chords with a lever arm (a) between them, and since the compression force C in the top chord equals the tension force T in the bottom chord $C–T = M/a$. The area of each chord member can simply be estimated as the force divided by the allowable unit stress (C or $T/F_{allowable} = A_{required}$).

An approximate variation of shear along the length of the span is shown in Figure 8-35d. Note that shear variation will be different for dead load and live load, since live load can act over part of the span whereas dead load always exists over the entire span. Under uniform loading, DL shear diagram is linear, (Figure 8-35e), while live load shear is parabolic (Figure 8-35f), where the midspan shear is obtained by loading half span only.

To design for shear near the support, we take a freebody of the truss, (Figure 8-35g). To resist the shear (V equivalent to the vertical reaction R), the diagonal must possess a vertical force equal to R. Hence, the total force in that diagonal must be $F = V/\text{Sin } \theta$. Having determined the force F, the area required for that diagonal member can be computed similar to the above computation for the chord members ($A = F/f_{allowable}$).

Note that all members in a truss are subjected to *direct* stress action with no bending in them, except some small secondary moments produced by the member's own weight and (if not pin-connected) due to rigidity of the joint. Although pin connections would allow joints to rotate freely, they are

FIGURE 8-35
TRUSSES WITH PARALLEL CHORDS.

(a) TRIANGULAR
TRUSS

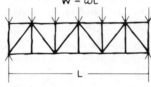

$W = \omega L$

L

(b) APPROXIMATE
MOMENT
ENVELOPE

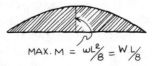

$$MAX. M = \frac{\omega L^2}{8} = \frac{WL}{8}$$

(c) MOMENT-
RESISTING
COUPLE

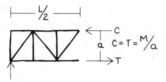

$\frac{L}{2}$

C

a $C = T = \frac{M}{a}$

T

(d) APPROXIMATE
SHEAR
ENVELOPE FOR
DEAD LOAD
PLUS LIVE
LOAD WITH
UNIFORM
INTENSITY w

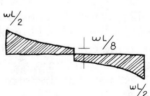

$\frac{\omega L}{2}$

$\frac{\omega L}{8}$

$\frac{\omega L}{2}$

(e) SHEAR DUE TO
UNIFORM DEAD
LOAD w_D

$\frac{\omega_D L}{2}$

$\frac{\omega_D L}{2}$

(f) SHEAR DUE TO
UNIFORM LIVE
LOAD w_L

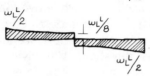

$\frac{\omega_L L}{2}$

$\frac{\omega_L L}{8}$

$\frac{\omega_L L}{2}$

(g) SHEAR-RESISTING
WEB

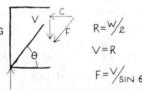

V C

F

θ

$R = \frac{W}{2}$

$V = R$

$F = \frac{V}{\sin \theta}$

often more costly and not used, since members would be conveniently fabricated with a bolted or welded joint at the connections.

In the case of a sloped roof truss (Figure 8-36a), we can also take a half truss section as a free body (Figure 8-36b). The moment M about joint D is carried by horizontal force H with lever arm (a), thus

$$H = M/a.$$

Figure 8-36c shows that there will be a compressive reaction force in the top chord:

$$-F = H/\cos \theta.$$

Similarly, the vertical reaction component for the top chord will act to resist the actual shear V_E across section 1-1:

$$-V_H = F \sin \theta.$$

FIGURE 8-36
TRUSSES WITH SLOPED CORDS.

(a)

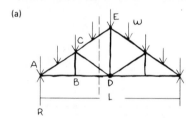

$$R = \frac{\omega L}{2}$$

$$M = \frac{\omega L^2}{8}, \text{ for evenly distributed loading.}$$

(b)

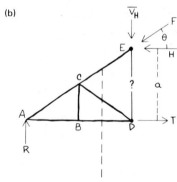

For $\frac{1}{2}$ truss analysis, assume: $H = T = M/a$, points A, C, E carry tributary load, shear increases from V_E at E, to V_C at C, and V_A at A.

$$R = \frac{\omega L}{2} = V_A.$$

(c)

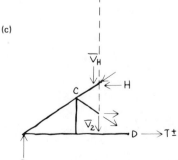

Reactions across section 1-1 must result in a balance between the actual V_E across the section and $-V_H$ supplied by CE.

But if $(-V_H)$ is larger than (V_E), DE will act in tension, as a tie, to transfer the difference (V_2) to the diagonal member CD where,

$$V_2 = V_H - V_E.$$

In this case, there will be a compressive reaction in CD and the tension in the bottom chord member BD is greater than (T).

Stress forces in other members of the truss can be obtained by taking either a section or a joint as a free body. For example, in Figure 8-36c, the force in the vertical member (BC) would be equal to any panel load applied at (B) while the force in member (AB) equals that in BD. The force in BD is given by $T \pm$ any horizontal component of V_2.

In *floor* construction for buildings, triangular trusses (within walls) are oftentimes not desirable in that the diagonal members obstruct necessary openings, such as windows and doors. Therefore, *vierendeel* truss construction may be better (Figure 8-37a), since it eliminates all diagonals. This in fact transforms the truss into a rigid horizontal frame with the

FIGURE 8-37
VIERENDEEL TRUSSES.

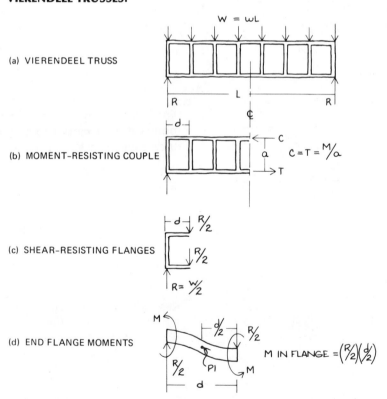

verticals acting as stiff connectors. Note that the moment is still primarily resisted by the chord members as indicated in Figure 8-37b. However, in order to resist the shear without the diagonal members, the chord members act in shear as well as direct compression or tension. Since the flange members are themselves perpendicular to the direction of the shear, they cannot carry shear by *direct stress action*. Instead, the transverse shear produces local moment along the axis of the member, Figure 8-37c. Assuming that the chords resist the shear equally, we have $R/2$ shear in each chord. In actual construction, the two chords will share the shear carrying capacity with respect to their relative stiffnesses and the stiffness of their connecting vertical members.

The chord members can usually resist the transverse shear without too much trouble, but the local moment is another matter (Figure 8-37d). The $(R/2) \times (d/2)$ moment has to be resisted by each chord acting on its own, with a lever arm determined by the depth and shape of the chord member itself. This means much additional material has to be provided in each chord to resist not only the original primary axial force C or T, but also the additional moment. The presence of such moment will not only produce high stresses in the members, particularly at their joints, but will also cause high deflections in vierendeel trusses. Hence, to provide the necessary strength and stiffness, vierendeel trusses will require much more material than a triangular truss. Furthermore, to work at all, their vertical members must be sufficiently stiff and strong to resist rotation and the bending moments at their joints with chord members.

In order to minimize the problem of chord bending in the vierendeel truss, the bow-string truss, Figure 8-38 provides a curved top chord that will resist shear by virtue of its *arch* action. If the curve is such that the moment pressure line will follow the arc of the top flange, then no bending

FIGURE 8-38
BOW STRING TRUSSES.

BOW-STRING TRUSS

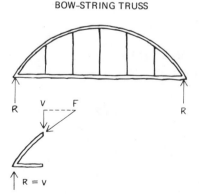

will be introduced in the chord (to be explained in Chapter 11). However, this is not always possible, particularly on account of the variation in live loading positions. Nevertheless, practically all the dead load stresses can be carried in the arc, without causing chord member bending. This truss can be easily applied to roof construction.

Figure 8-39 shows a trussed tied arch for a long roof span. When spans are long and wind load may produce high bending moments in the chord, it may be economical to truss the arch so that it can take a certain amount of bending. Similarly, for very long span roofs using cable construction, the horizontal component may be built of truss members to strengthen its moment capacity and to increase its rigidity (Figure 8-40).

It is sometimes economical to truss a vertical shaft so as to increase its strength and rigidity against lateral force (Figure 8-41). Assuming a building or a shaft with height, h, acted on by a horizontal load of w lb per linear foot, then the total load acting horizontally would be $W = wh$. Under the action of these lateral forces, the moment and the shear diagrams are shown in Figures 8-41b and 8-41c. Note that this is a vertical cantilever supported at the bottom but free to move at the top. Therefore, the moment and shear are different from those in simple beams. Having computed the moment and the shear, the design of the members can be accomplished. For example, the two flange members on the first floor will

FIGURE 8-39
TRUSSED TIED ARCH

TRUSSED TIED-ARCH

FIGURE 8-40
SUSPENDED ROOF WITH STIFFENING TRUSS.

SUSPENDED ROOF WITH
STIFFENING TRUSS

FIGURE 8-41
TRUSSED WALLS OR SHAFTS.

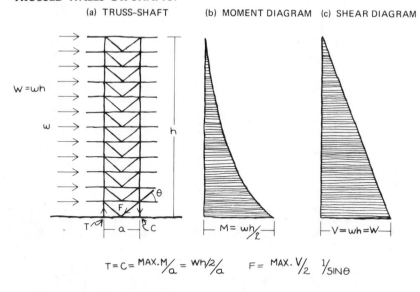

(a) TRUSS-SHAFT   (b) MOMENT DIAGRAM   (c) SHEAR DIAGRAM

$$T = C = \frac{MAX.M}{a} = \frac{wh/2}{a} \qquad F = \frac{MAX.V}{2} \cdot \frac{1}{SIN\theta}$$

be stressed by a force.

$$T = C = \max M/a$$

In other words, $T = C = \dfrac{w(h/2)}{a}$.

The size of the members can be estimated as shown earlier in this section.

The shear in this case is also maximum at the ground level and equals the total horizontal force W. Since we use a K frame in this case, there are two diagonals, both resisting the shear. If they resist shear equally, then the force can be shown to be

$$F = \frac{\text{maximum shear}/2}{Sin\ \theta,} = \frac{V/2}{\sin \theta}$$

with θ being the angle shown in Figure 8-41.

Truss members can be built of steel, timber, reinforced concrete, or prestressed concrete. Generally, the computation of the main stresses is a relatively simple matter, whereas secondary stresses and detailing can be a tedious problem. For timber and concrete construction, the joint design for the transmission of tension force is relatively complicated. Prestressing steel through RC joints to supply the necessary tensile force can be a very economical solution. Steel members, on the other hand, can be conventionally bolted or welded at the joints, and they can also extend continuously through these joints with splices made away from the joints. The choice will depend upon the requirements of fabrication, transportation, and erection.

9
Vertical Linear Components

SECTION 1: Ties, Hangers, and Tension Members

A tie, hanger, or tension member is a straight member subjected to two direct (axial) *pulling* forces applied to its ends. It is an efficient and economical member because it utilizes the full area of the material effectively at a uniform tensile stress. As examples, ties can be used to tie the abutments of an arch together, hangers can be used to hang lower floors from top ones or from outrigger trusses, and tension members exist in a truss where some of the members are in compression and others in tension.

Timber is capable of resisting tension but is not an economical material for this purpose because of the difficulty in developing connections at the ends of the member to transfer tensile forces to other components. Reinforced concrete is not a good material for that purpose because concrete will be subjected to cracking under tensile stresses and frequent cracks in the member would appear unsightly and may cause corrosion in the reinforcing bars due to moisture and oxygen penetration.

Two ideal materials for tension are structural steel and prestressed concrete, with structural steel being the most common material in contemporary use.

In general, there are four groups of structural steel tension members: wires and cables, rods and bars, single structural shapes and plates, and built-up members. It is not the purpose of this text to discuss each of these in detail. Therefore, only a brief discussion will follow. For a fuller discussion of these members, refer to Chapter 7 of *Design of Steel Structures*, second edition, by Bresler, Lin, and Scalzi, John Wiley, New York, 1968.

Small tension members are often made from square or round rods, or flat bars. The main disadvantage of these members is their inadequate stiffness, resulting in noticeable sag under their own weight if placed in a nonvertical position. When some amount of rigidity is desired, structural shapes, either single or built-up, can be employed (Figure 9-1). The simplest and most commonly used shape as a tension member is the angle. Angles are considerably more rigid than wire ropes, rods, and bars, but for long members and members subjected to bending, other structural shapes or built-up shapes are desired.

The design of these members is a straightforward matter, using the

FIGURE 9-1
SIMPLE AND BUILT-UP STEEL TENSION MEMBERS.

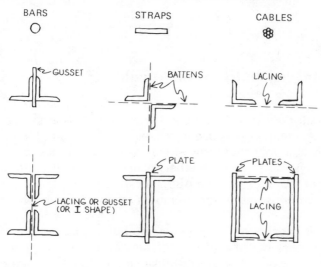

familiar formula $f = P/A$, where f is the unit stress, P is the total load, and A is the cross-sectional area. Using working load design, f would be the working stress, and P the working load. If ultimate design is used, f will be the yield point (perhaps slightly reduced), while P will be the factored ultimate load.

If the tension member is additionally subjected to a bending moment, M, it is then necessary to supply an additional area of cross section so that the moment of inertia provided by that added material will yield the following stress: $f_{allowable} = Mc/I$, where $c =$ the distance of the extreme fiber from the centroidal axis and $I =$ moment of inertia of this added cross-section about the centroidal axis.

When deflection or elongation of the tension member is under question, it will be desirable to check the amount of elongation by the following conventional formula: $\Delta = (P/AE)(L)$, where $\Delta =$ the deflection, $P =$ the tensile load, $A =$ cross-sectional area of the member, $E =$ the modulus of elasticity of the material, and $L =$ the length of the member.

In a prestressed concrete tension member, the concrete would be precompressed by high-strength steel tendons (Figure 9-2). If the steel has an area A_s and is tensioned against the concrete to a stress f, then the total tensile force in that steel is $F = fA_s$. This will place a compression in the concrete equal to F, and therefore the unit stress in the concrete will be $f_c = F/A_c$. It is obvious that if an external axial force, P, is applied to that member, then when $P = F$ the concrete will be subjected to zero stress.

Since it is not desirable to place the concrete under tension or to subject it to cracking, the externally applied load P should not be greater than the internally applied prestress, F. To supply additional ultimate strength, it may be desirable to apply a prestress, F, somewhat greater than P. Or else, some conventional reinforcing bars may be placed along the member to minimize the cracking of concrete and to increase the ultimate strength so as to provide for the possibility that an externally applied load may be higher than the design value of P.

FIGURE 9-2
P.C. TENSION MEMBERS.

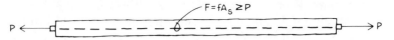

Prestressed concrete tension members can be quite desirable and economical for special tension members in a building, such as ties and hangers. By applying the prestress, perhaps in stages, the externally applied tensile force, stresses, and strains in the tension member as well as the structure as a whole can be controlled. This was designed into the Exhibit Hall of the Moscone Convention Center in San Francisco, where each of the eight pairs of concrete arches are tied below the ground with tendons exerting a force of 8000 kps for each arch (see Photo 9-1).

PHOTO 9-1
ARTIST'S RENDERING OF THE 1000-FT-LONG MOSCONE CENTER EXHIBIT HALL, SAN FRANCISCO. POSTTENSIONED TIES BELOW THE FLOOR RESIST AND CONTROL THE 280-FT ARCH THRUST. THESE ARCHES SUPPORT 3 FT OF SOIL FOR A CITY PARK ABOVE. (H. O. K, ARCHITECTS; T. Y. LIN, INTERNATIONAL STRUCTURAL ENGINEERS.)

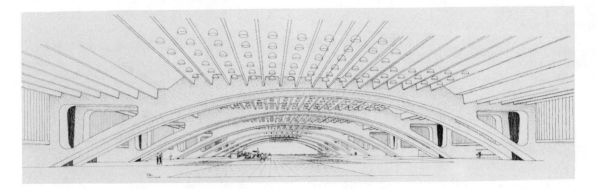

SECTION 2: Short Columns—Axially Loaded

Most vertical components in a building are columns and walls under direct compression due to the fact that a building is supported from the bottom up. Furthermore, the gravity loads on these members accumulate from the top of the building down to its foundation. Compared with tensile members, columns must be stiffer in order to resist a tendency to buckle under load. Depending on the slenderness of a column, the design load may be less than could be allowed for a tensile element of the same section.

When a column is "short," its proportions are such that it has little or no tendency to buckle and it is easier to design these columns as compared to long ones. Therefore they will be discussed first.

Although it is difficult to exactly define the term "short" as applied to columns, a "short" column may be described as one whose buckling effect is negligible. This can be expressed in terms of L/b, where (L) is the unsupported height or length of the column and (b) is the least dimension of the column section. But the ratio varies for different materials.

A timber column may be considered as "short" if the L/b ratio is less than (5 or 8). A reinforced concrete column may be termed as short if the L/b ratio is less than (8 or 12). In the case of steel, the ratio 10 applies to an unfavorable section such as an unsymmetrical angle, whereas the ratio 20 is more applicable to a favorable section such as a hollow box.

Some typical sections for steel columns are shown in Figure 9-3. The most

FIGURE 9-3
BUILT-UP STEEL COLUMN SECTIONS.

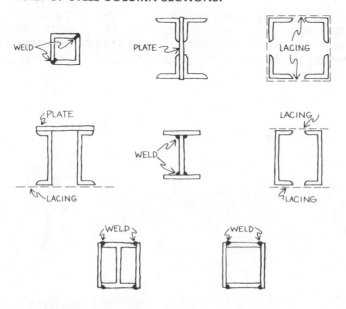

common section is the *W* (or *H*) shape (see Appendix C-2), since there are many available sizes and proportions of this type of rolled section. Other sections shown in Figure 9-1 (for tension members) can also be used as columns. However, in order to obtain sufficient stiffness and to avoid buckling, shapes shown in Figure 9-3 are often preferred.

When *steel columns* are "short," they can be simply designed by the formula: $A = P/f$, where A is the cross-sectional area required, P is the axial compression load applied, and f is the allowable compressive stress, which is generally taken as $0.6f_y$, where $f_y =$ the yield point of the steel. Some of the more common values of f are listed in Appendix B-2.

Short *timber columns* are designed in a similar manner, using the formula $A = P/f$, where f values for various types of wood are usually given in timber handbooks. The normal range for structural wood varies, say, between 1.2 ksi and 2.0 ksi. For purposes of quick approximation, you may use a value of 1.5 ksi.

Concrete columns are usually reinforced with steel bars. Plain concrete columns are seldom used except when the columns are really short, say, with L/b ratios less than (2 or 3). Furthermore, in addition to supporting loads, reinforcement in concrete helps to provide resistance to bending transferred from other members or due to eccentric loads. It also controls the cracks in concrete that might take place as a result of shrinkage and temperature changes.

Reinforcement along the length of RC columns is designated as longitudinal steel and ranges between 1 and 8 percent of the column concrete section. A minimum amount is desired for crack control and to resist unexpected bending. A maximum amount is specified to avoid the difficulty of placing concrete around the reinforcing steel and obtain an efficient interaction of the two materials under ultimate conditions of stress. Since concrete is weak in tension and reinforcing bars are slender, horizontal *ties* or *spirals* are provided to apply lateral restraint to these longitudinal bars in order to prevent their possible buckling. In addition, for fire and corrosion protection, a minimum of concrete cover (on the order of $1\frac{1}{2}$ in.) is usually required.

In tied columns the longitudinal bars are restrained by independent horizontal ties of $\frac{1}{4}$ in. to $\frac{3}{8}$ in. diameter placed at regular intervals, say, at 12 in. c.c. They are often spaced closer near the more critical top and bottom sections of a column (say, four intervals at 3 in. centers). Usually, tied *RC* columns are square or rectangular in section (Figure 9-4). Specific rules for determining the size and spacing of these ties are given in textbooks or building codes (*UBC*) on reinforced concrete design.

Spiral reinforcement is more closely spaced and acts *not only* to prevent buckling of steel *but also* to confine the enclosed concrete (Figure 9-5). As a result, spiral columns are generally stronger than tied ones. It is commonly used for circular or octagonal columns, but can also be used for square and rectangular sections. These spirals are made of small diameter wire and are

FIGURE 9-4
R.C. COLUMNS WITH TIES.

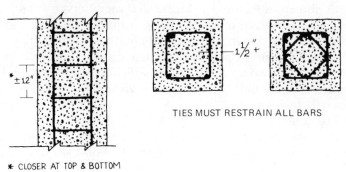

TIES MUST RESTRAIN ALL BARS

* CLOSER AT TOP & BOTTOM

FIGURE 9-5
R.C. COLUMNS WITH SPIRALS.

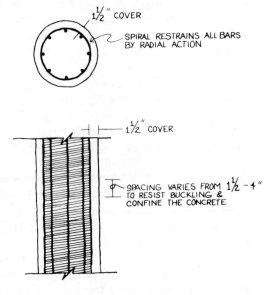

placed as a continuous coil, running the entire length of a column. The wires are of about $\frac{1}{8}$-in. diameter up to about $\frac{1}{4}$-in., and the coils are stretched like a spring to achieve a spacing of between $1\frac{1}{2}$-in. and 4-in. centers generally. Since spirals tend to confine the enclosed concrete, the usable strength of both reinforcing bars and concrete is increased, as compared with tied columns. Again, specific rules for the size and spacing of spirals are to be found in any standard treatise on reinforced concrete design.

Reinforced concrete columns can be designed either at working loads and working stresses, or at ultimate loads and ultimate stresses. Using working load design, $P = f_c A_c + f_s A_s$, where f_s and f_c = allowable working stresses in steel and concrete, A_s and A_c = areas of steel and concrete. For tied columns, f_c is taken as $0.20\ f'_c$ and f_s is taken as $0.32\ f_y$; for spiral columns, f_c as $0.25\ f'_c$, and f_s as $0.40\ f_y$. Thus we obtain the following two formulas for allowable axial load P_a:

Tied column $\qquad\qquad\qquad P_a = 0.20\ f'_c\ Ag + 0.32\ f_y A_s$

Spirally reinforced column $\qquad P_a = 0.25\ f'_c Ag + 0.40\ f_y A_s,$

where f'_c = 28-day cylinder strength of concrete, psi; Ag = gross cross-sectional area of concrete, in.^{2}, including the part occupied by steel bars; f_y = yield point of steel, psi; A_s = gross cross-sectional area of steel bars. Appendix Table B-3 gives some typical bar sizes. Note that the allowable stresses are lower for tied columns because of possible brittle failures, since the failure of concrete in compression would result in a sudden collapse of the column.

Spiral columns behave differently. First, the concrete cover for the column outside of the spiral would spall off under the ultimate load, then the core concrete, contained by the closely spaced spiral, will continue to carry load and to shorten appreciably before a total collapse. This means that spiral columns will be able to resist ultimate loads with greater ductility, therefore giving early warning of impending collapse and at the same time permit load redistribution to other supporting members of the structure.

EXAMPLE 9-1: Column Design

A 20 in. × 20 in. reinforced concrete column is to support an axial load of 400 k. If the concrete has compressive strength f'_c = 3000 psi, determine the amount of longitudinal steel required for both a tied and a spiral column. Show a suitable arrangement of steel bars (number of bars, size, and location). The steel is of intermediate grade, having f_y = 40,000 psi.

(a) TIED COLUMN

(a) *Tied Column*

$400\ \text{k} = 0.20 \times 3\ \text{ksi} \times 400\ \text{in.}^2 + 0.32 \times 40\ \text{ksi} \times A_s$

$A_s = \dfrac{400 - 240}{12.8} = 12.5\ \text{in.}^2$

$\dfrac{12.5}{400} \times 100 \sim 3$ percent; o.k.

Use 8 no. 11 bars at 1.56 in.^{2}

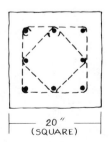

20″
(SQUARE)

(b) *Spiral*

$$400 \text{ k} = 0.25 \times 3 \text{ ksi} \times 400 \text{ in.}^2 + 0.40 \times 40 \text{ ksi} \times A_s$$

$$A_s = \frac{400 - 300}{16.0} = 6.25 \text{ in.}^2$$

$$\frac{6.25}{400} \times 100 \sim 1.5 \text{ percent, o.k. but low}$$

Use 8 no. 8 bars at 0.79 in.^{2}
(Or, could reduce Concrete area)

27″
(DIAMETER)

When using *ultimate strength design*, the ultimate load of the column is computed with the following formula for both tied and spiral columns:

$$P_{ult} = Cf'_c \cdot A_c + f_y \cdot A_s$$

C relates the strength of concrete in a short column to compressive strength f'_c of a standard 6 ft × 12 ft concrete cylinder, and is usually approximated by a constant = 0.85; A_c is the net cross-sectional area of concrete. The ACI Building Code specifies that this ultimate strength is further reduced by a strength reduction factor $\phi = 0.7$ for tied columns and 0.75 for spiral columns, to insure a sufficient safety factor.

The above ultimate strength of the column is then compared with the ultimate load on the column which is computed by multiplying the working loads with load factors of 1.4 for dead load and 1.7 for live load.

SECTION 3: Long Columns

Long columns are more slender in L/b proportion and tend to fail by buckling at loads below the yielding of steel or the crushing of concrete. This is known as the buckling instability of slender columns. Photos 9-2a and *b* show some unusual columns during and after erection. These pretensioned concrete columns were 90 ft long, precast at Basalt Rock Co., Napa, California, and trucked 50 miles to the job site. After completion, these columns are stabilized and greatly stiffened by the precast walls. In practice, most concrete columns are poured in-place, while steel columns are prefabricated and erected into position.

Due to variation in the properties of materials, initial crookedness in the fabrication of the column, or some eccentricity in the load application, there always exists a tendency for columns to bend sideways. When a slight lateral deflection takes place for a relatively short and stout column, a sufficient

PHOTO 9-2
NINE-STORY OFFICE BUILDING FOR UNIVERSITY OF CALIFORNIA, DAVIS (T. Y. LIN INTERNATIONAL STRUCTURAL ENGINEERS). (a) ERECTION OF 90-FT HIGH PRETENSIONED CONCRETE COLUMNS. (b) ERECTION OF PRECAST FLOOR MEMBERS.

(b)

(a)

internal resisting moment would be created to resist the small eccentricity produced between load and column axis. However, when a column is relatively slender, any such lateral deflection would be relatively large, and the internal resisting moment so created would be small. At some point the increase in moment produced by the sideways deflection is greater than the increase of internal resisting moment created by bending, and "buckling" occurs. That is, the deflection will continue until collapse occurs.

For the same amount of lateral deflection, a column with higher E (due to material properties) and I (due to sectional shape) would generate a higher internal resisting moment, the resistance to buckling will be higher. On the other hand, if the length, L, of the column is large relative to (b), a larger deflection will be produced in the column and the tendency to buckle would occur sooner, other things being equal. For a column with both ends pinned and subjected to a direct load, a theoretical relationship has been derived to compute the critical buckling load: $P_{cr} = \pi^2 EI/L^2_{effective}$ where, in *this* case, $L_{eff} = L$.

Figure 9-6 illustrates how a change in the end restraint conditions of a column changes its effective length. Since this must be taken into consideration when designing a long column to resist buckling, a coefficient K is added in the above formula (i.e., $L_{effective} = KL$) to reflect the end condition effect on the overall column behavior in buckling. Figure 9-6a shows a common situation with the two ends being pin-connected so that they are free to rotate but not free to move sideways. This is the standard case where $K = 1.0$. Figure 9-6b shows a column supported at the bottom but free to deflect sideways at the top. To compare this column to a simple column like Figure 9-6a, it can be seen that it would have deflection

FIGURE 9-6
THE EFFECTIVE LENGTH OF SLENDER COLUMNS VARIES WITH FIXITY AT COLUMN ENDS.

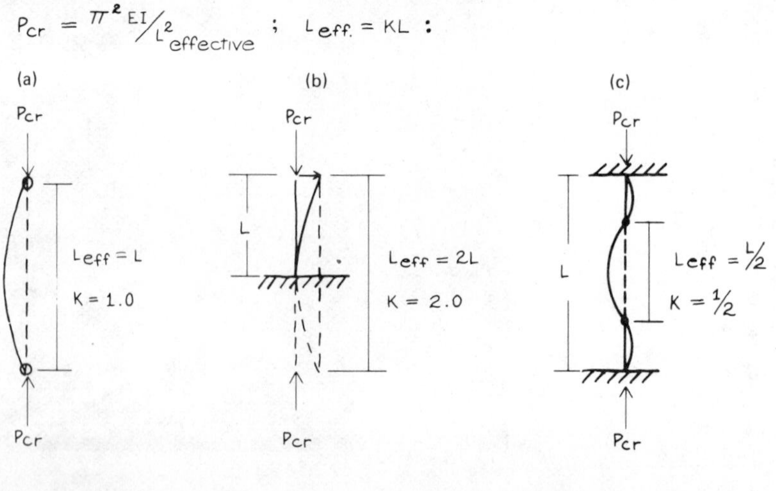

$$P_{cr} = \pi^2 EI/L^2_{effective} \quad ; \quad L_{eff.} = KL :$$

(a)

P_{cr}

$L_{eff} = L$

$K = 1.0$

P_{cr}

(b)

P_{cr}

L

$L_{eff} = 2L$

$K = 2.0$

P_{cr}

(c)

P_{cr}

L

$L_{eff} = L/2$

$K = 1/2$

P_{cr}

characteristics similar to a simple column with twice the length L. Thus the coefficient $K = 2.0$. As a contrast to Figure 9-6b, Figure 9-6c shows a column of the same length L, but with both ends fixed against rotation and translation. In this case, the effective length insofar as buckling is concerned is only $L/2$. Hence the critical buckling load for Figure 9-6b is only $\frac{1}{4}$ that of Case a, and $\frac{1}{16}$th that of Case c. This illustrates how important the end conditions can be insofar as they affect buckling of columns.

The above three cases and some additional ones are summarized in Table 9-1. This table recommends practical K values to be used in the previous formula. They not only represent the theoretical conditions of end restraint, but also other considerations which tend to modify the theoretical values to a certain degree. The critical load P_{cr} at which buckling takes place can now be expressed by the following formula:

$$P_{cr} = \pi^2 EI/(KL)^2$$

where K = coefficient reflecting the end conditions of the member; E = elastic modulus of the material; I = moment of inertia of the cross section; L = unsupported length of member.

Now, let us look at E in the previous formula. E, the elastic modulus, varies with different materials, and approximate values are listed below:

Steel	29×10^6 psi
Aluminum	10×10^6 psi
Concrete	3 to 5×10^6 psi
Wood	1.2 to 1.7×10^6 psi

note that the E value for concrete varies from 3 to 5 million psi, according to the strength of concrete, and is conveniently taken at approximately

Table 9-1
A Comparison of Theoretical and Practical K Values for Design.

	PIN ... PIN	FIXED ... FIXED	PIN ... FIXED	FIXED (WITH SIDESWAY) ... FIXED	FIXED ... PIN / PIN ... FIXED (WITH SIDESWAY)
BUCKLED SHAPE					OR
THEORETICAL K	1.0	0.50	0.7	1.0	2.0
RECOMMENDED FOR DESIGN	1.0	0.65	0.8	1.2	2.0

$1000 \times$ the ultimate stress of concrete, or $(10^3 \times f'_c)$, although there are other better approximations.

The actual value of E for concrete also changes with the duration of the load application. For an instantaneous load (occurring in a fraction of a second), the E value of concrete can be 50 percent higher than given above. On the other hand, if the load is sustained for months or years, then concrete tends to creep and shorten continuously, and the E value to be used should be perhaps one-half to one-third those indicated above.

The moment of inertia (I) of the section cannot be obtained unless a section has been chosen or designed. When proceeding with a preliminary dimensioning of the member, the I value can be guessed at. To simplify this matter, it is convenient to express $I = Ar^2$, where r^2 is a shape property called the radius of gyration. The r about an axis represents the distance from the axis where the entire sectional area can be concentrated for the purpose of computing I. Thus, $r^2 = I/A = d^2/z$; for a rectangular section $z = 12$ (from Chapter 3), and $r = \sqrt{1/12} \ (d) \simeq 0.29d$.

Table 9-2 shows the approximate radii of gyration for some selected sectional shapes. If this radius of gyration is known for a given section, then the moment of inertia can be manipulated by varying the sectional area (A). (A) can be guessed at when the load P is known and the allowable unit stress (f) assumed. Then one can compute $I = Ar^2$ with r approximated from Table 9-2. More exact r values can be obtained from design manuals.

Consider a rectangular section with depth d_1, and thickness d_2. That section may bend or buckle about either axis. If the member is supported in the same manner along both axes, then the weaker or smaller I or r will control. For circular or square sections, the I and r about both axes are the same.

Although the above formula for P_{cr} will yield the buckling load, obviously a factor of safety must be placed on this load in order to arrive at an allowable load. This factor is on the order of 2 for metals such as steel or aluminum, but higher for nonhomogeneous materials such as concrete or timber, perhaps 2.5 to 3.

For the purposes of preliminary design of slender concrete columns, the following *rule-of-thumb* is often used. For *reinforced concrete* columns with L/d under 10, no allowance or reduction need be made for the effect of buckling, and short column strength will apply. Reinforced concrete columns should preferably not have an L/d ratio in excess of 20. For L/d between 10 and 20, one may deduct 3 percent of the short column load for each additional unit increase of L/d. Thus, for $L/d = 20$, the strength of the column can be approximately reduced by 30 percent relative to its short column strength. For reinforced concrete columns with L/d in excess of 20, special and careful analysis should be made before adopting such a design. For *timber columns*, we may place a limit of L/d at 15, in excess of which a reduction of perhaps 7 percent for each unit increase of L/d up to 25. For $L/d > 25$, refer to Figure 9-7a. *Steel design* will be discussed below.

Table 9-2
Approximate Values for Radius of Gyration
for Some Section Shapes.

TYPE OF SECTION	APPROXIMATION OF r VALUE		APPROXIMATE $I = Ar^2$
	$r_X = 0.29 d_1$	$r_y = 0.29 d_2$	
	$r_X = 0.40 d_1$	$r_y = 0.23 d_2$	
	$r = 0.25 d$	——	
	$r = 0.35 d$	——	
	$r_X = 0.41 d_1$	$r_y = 0.41 d_2$	
	$r_X = 0.21 d_1$	$r_y = 0.21 d_2$	
	$r_X = 0.36 d_1$	$r_y = 0.28 d_2$	

As is seen from the above formula for P_{cr}, it should be noted that the buckling stress for a long column of any material is dependent only on its E value, but independent of its yield point or ultimate stress. On the other hand, for short columns the ultimate strength will be dependent on the yield strength, since buckling will not take place.

FIGURE 9-7
ALLOWABLE STRESS FOR AXIALLY LOADED WOOD AND STEEL COLUMNS.

(a) WOOD COLUMNS

(b) STEEL

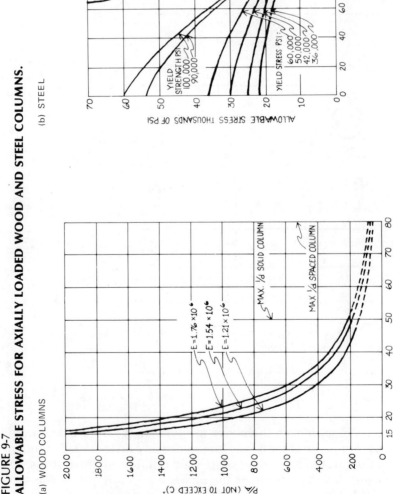

1 "C" IS ALLOWABLE COMPRESSIVE STRESS PARALLEL TO THE GRAIN
FROM NATIONAL DESIGN SPECIFICATION FORMULA:

SOLID COLUMNS $P/A = \left(\dfrac{0.30E}{(l/d)^2} \right)$

SPACED COLUMNS   CONNECTORS WITHIN 1/20 FROM COLUMN END; MULTIPLY P/A BY 2.5
CONNECTORS FROM 1/20 TO 1/10 FROM COLUMN END; MULTIPLY P/A BY 3.0
d = THICKNESS OF INDIVIDUAL MEMBER

One can also express the limitations on steel design by considering buckling in terms of allowable buckling stress. Figure 9-7*b* shows buckling stress of steels of different yield strengths in terms of the ratio KL/r. It also shows the allowable stress for steel columns possessing various yield strengths. For really short columns, an allowable value of 60 percent of the yield strength is permitted. As soon as the column is not really short, it will be affected by excentric loads and certain preexisting internal stresses in rolled or welded steel sections. These stresses will tend to reduce the strength of steel columns for slenderness ratio between 10 or 20 and up to 60 or 80. For slenderness ratios exceeding 60 or 80, steel columns will be controlled by buckling and independent of the yield stress.

Note the set of allowable stress curves for steel columns as shown in Figure 9-7*b*. These are based on a factor of safety of 1.67 for short columns, 1.92 for long columns. Also note that for *primary column* members of steel, it is not desirable to use too high a slenderness ratio, 120 being the usual limit, and 200 being the absolute maximum. For secondary columns, a smaller safety factor is permitted when failure under buckling controls. The reason is that when secondary members start to buckle, they will still supply the necessary bracing effect and any additional load can be redistributed to other members. Secondary members are those members that may be omitted without affecting the static stability of the structure as a whole. In other words, secondary members are usually bracing members.

When designing a steel column for a given design load (P), one may use the following abbreviated procedure: Assume a column size based on $A_s = P/0.6f_y$ or somewhat more. Pick a shape and estimate KL and r. Then pick an allowable stress from Figure 9-7 for the corresponding steel. Again compute required A_s of column. Repeat process until sufficiently accurate answer is obtained ($A_s = P/f_a$).

A more elaborate explanation is included in the following summary of how to design steel columns, which is usually done by a trial-and-error procedure, as follows:

1. Determine axial load P to be carried by the column under consideration (use tributary area).
2. Estimate effective column length KL. For a building column L, unsupported length is usually the floor-to-floor height. For first trial, K may be taken as unity for columns braced at top and bottom.
3. Select steel grade to be used, i.e., f_y = yield point of steel. Note: For heavy column loads and small slenderness ratio (say less than 90), some advantage is obtained by selecting $f_y > 36$ ksi (say $f_y = 50$ ksi). For light columns and slenderness ratio larger than about 100, little advantage is obtained by using steels with $f_y > 36$ ksi, since the value of E_s controls anyway.
4. Guess allowable stress f_a. A value of $f_a = 15$ ksi may be assumed for first trial, unless experience of judgment suggests a better value.

5. First trial area of column section required is calculated;
$$A_1 = P/f_a$$

6. For first trial area A_1, select an appropriate steel section (see Appendix B or *AISC Handbook*). Decide shape of section, that is, *WF*, pipe, box, built-up column, or other section, then select standard shape having area A closest to A_1.

7. Determine values of r_x and r_y for the shape selected. Then calculate effective $(KL/r)_{max}$, using r_{min} and appropriate K (usually, $K = 1$ for preliminary design). Based on $(KL/r)_{max}$, obtain from Figure 9-7 corresponding allowable stress f_{a_2}. Calculate $P_a = f_{a_2} \times A_1$.

8. Compare the value of P_a computed in Step 7 with the actual load P in Step 1. If the difference is significant, repeat Steps 4 through 7 until P_a is close to P and an economical section is selected.

EXAMPLE 9-2:   Design of Slender Steel Columns

A steel column in a building is to carry an axial load of 250 k, without any bending. The column is 12 ft high, pin-ended at top and bottom. Choose a W section, A-36 steel.

Assume $f_a = 15$ ksi

$$A_s = \frac{250}{15} = 16.7 \text{ in.}^2$$

Try 12 W 58

$$A_s = 17.1 \text{ in.}^2$$
$$r = 2.51 \text{ in.}$$
$$\frac{KL}{r} = \frac{12 \text{ ft} \times 12 \text{ in.}}{2.51} = 57$$

From curve (Figure 9-7),

$$\text{allow } f_a = 18 \text{ ksi}$$

Hence, revised area of steel needed is,

$$A_s = \frac{250}{18} = 14 \text{ in.}^2$$

One can try 12 W 50 and carry on to obtain a final section.

SECTION 4:   Columns Under Bending—Steel

In addition to carrying axial load and resisting buckling, columns are frequently subjected to bending. Bending may result from an eccentric load on the column, that is, the load not being applied axial to the centroid of the column section. When the axis of resistance and the line of external force are eccentric, a moment arm exists and bending results. Further, when columns are rigidly connected to girders or beams, these girders may tend to bend the columns and moment will develop at the end and along the columns. As we have seen in Chapter 3, when a multistory rigid-frame is acted on by lateral loads, the horizontal shear in each story will produce moments in the columns. Furthermore, there may even be local forces such as wind and earthquake applied along the height of the column when it helps to transfer local horizontal loads to the floor diaphragms. In fact, the majority of columns are subjected to both axial load and bending. When the amount of bending cannot be accurately computed, it is often assumed that the axial load will have at least a certain amount of eccentricity, equal to 5 or 10 percent of the width of the column, to compensate for this uncertainty. In any case, column bending will reduce the axial load a given column can carry.

The behavior of a steel column subjected to bending is quite different from a concrete column. Compression in a steel column always reduces its capacity to carry bending because steel is almost equally strong in compression and in tension. Therefore, any axial compressive load will reduce the capacity of a column in carrying bending and vice versa.

On the other hand, concrete is strong in compression and weak in tension. Although an axial load on a concrete column does decrease its compressive fiber capacity for bending, it increases its tensile fiber capacity because cracking in concrete will be delayed as a result. If the tension side controls the ultimate failure, then the addition of axial compressive load may increase the capacity of the column. Only when a concrete column fails in compression will the addition of compressive load to bending hurt its capacity to carry bending. Hence, this section deals with steel. Concrete columns under bending will be discussed in Section 5.

The proportioning of *steel columns* under axial load and bending moment acting simultaneously may be done by an "interaction equation":

$$(P/P') + (M/M') \leqq 1,$$

where P = applied axial load in the member, M = applied maximum bending moment in the member, P' = allowable axial load in the absence of bending, and M' = allowable bending moment in the absence of axial load. From this, another form of interaction equation, expressed in terms of stresses can be

$$\frac{f_P}{f'_P} + \frac{f_M}{f'_M} \leqq 1$$

where $f_P = P/A$ and $f'_P = P'/A$, A = cross-sectional area of structural member, $f_M = M/S$ and $f'_M = M'/S$, S = section modulus of structural member. Either can be represented graphically by a straight line (Figure 9-8).

FIGURE 9-8
**THE RELATIONSHIP BETWEEN AXIAL
LOADING AND BENDING
MOMENTS ON STEEL COLUMNS.**

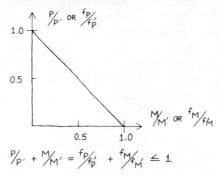

$$\frac{P}{P'} + \frac{M}{M'} = \frac{f_P}{f'_P} + \frac{f_M}{f'_M} \leq 1$$

To initiate design, the above formula can be modified as follows:

$$\left(\frac{f_P}{f'_P} + \frac{f_M}{f'_M}\right) = \frac{1}{A}\left(\frac{Af_P}{f'_P} + \frac{AM}{Sf'_M}\right) = \frac{1}{A}\left(\frac{P}{f'_P} + \frac{BM}{f'_M}\right) \leq 1$$

From this modified form, the required cross-sectional area A may be found as:

$$A \geq \left(\frac{P}{f'_P} + \frac{BM}{f'_M}\right)$$

where $B = (A/S) = $ (area/section modulus) is often designated as the bending factor. It is, in fact the reciprocal of the kern distance $k_d = S/A$ (in inches) at which point, if a load is applied, the bending tensile stress neutralizes the axial compressive stress at one extreme fiber. Thus, for a rectangular cross section,

$$A = bd, \quad S = \frac{bd^2}{6} \quad \text{and} \quad B = \frac{A}{S} = \frac{bd.6}{bd^2} = \frac{6}{d} = \frac{1}{k_{in.}}$$

The actual value for (B) will grow larger as the actual value of $(d$ or $k)$ decreases. And, for W shapes used as columns, the following range of $B_x = (A/S_x)$ and $B_y = (A/S_y)$ values are obtained in (inches)$^{-1}$:

$$0.18 < B_x < 0.36, \quad \text{and} \quad 0.49 < B_y < 1.5$$

Note that the product (BM) yields an equivalent force to be resisted by the stress (f'_M) in the above formula. For preliminary design, a rough value of $B_x = 0.25$ or $B_y = 0.7$ may be used as illustrated in Example 9-3.

EXAMPLE 9-3: Design of Steel Columns Under Bending

Select proportions of a structural steel column, in a rigid-frame building, with axial load $P = 250$ k, bending moment $M = 75$ kip-ft, and effective length of column $(KL) = 12$ ft.

COLUMN-BEAM FIXITY

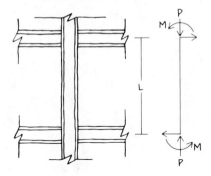

COLUMN SECTION

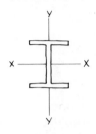

For bending about x-x axis ($B_x = 0.25$): Estimate cross-sectional area:

$$A = \frac{P}{f'_P} + \frac{BM}{f'_M}$$

$f'_M = 24$ ksi, and Estimate $f'_P = 17$ ksi, then:

$$A = \frac{250}{17} + \frac{0.25 \times 75 \text{ kip-ft} \times 12}{24} = 14.7 + 9.38 = 24.08 \text{ in.}^2$$

(for preliminary design).
Try W sections with $A = 24 \pm \text{in.}^2$

$$14 \ WF \ 78 - A = 22.94 \text{ in.}^2, \ S_x = 121.1 \text{ in.}^3, \ r_y = 3.0$$

Then:

$$f_P = \frac{P}{A} = \frac{250}{22.9} = 10.9 \text{ ksi}$$

$$f_M = \frac{M}{S_x} = \frac{75 \times 12}{121.1} = 7.4 \text{ ksi}$$

f'_P depends on

$$(KL/r) = \frac{12 \times 12}{3.0} = 48$$

from Figure 9-7 (for allowable stress, steel $f_y = 36$ ksi)

$$f'_P = 18.5 \text{ ksi.}$$

Then: rechecking 14 WF 78

$$\frac{f_P}{f'_P} + \frac{f_M}{f'_M} = \frac{10.9}{18.5} + \frac{7.4}{24} = 0.59 + 0.31 = 0.90 < 1.0$$

The section is safe, because sum of

$$\frac{f_P}{f'_P} + \frac{f_M}{f'_M}$$

is less than unity. Probably a lighter section can be used. (One may try to verify 14 W 74, but note that for preliminary design 14 W 78 is sufficiently close to final design.)

SECTION 5: Columns Under Bending—Concrete

Proportioning of reinforced concrete columns under combined axial compression and bending is a complicated problem. One might consider using an interaction formula similar to that for steel:

$$\frac{P}{P'} + \frac{M}{M'} \leq 1.0$$

where P = applied axial load; P' = axial load capacity of the column; M = applied moment; M' = moment capacity of the column section.

But this interaction formula would yield a very gross approximation of the behavior of reinforced concrete columns under combined compression and bending.

As stated in the previous section, a certain amount of axial load may help to increase the bending capacity for concrete columns, and concrete columns may fail either in tension or in compression. For tension failure, axial compressive load helps to increase the bending capacity because cracking of concrete is delayed. Hence we need a different approach, which will be presented as follows (note the similarity to idea of k_b and k_t).

First, ascertain the relative significance of the axial load P and the bending moment M on the column section by computing the eccentricity:

$$e = \frac{M}{P}$$

Then compare that e-value with the assumed column dimension (d) in the direction of the moment M by computing e/d. A critical point occurs when e/d is about $\frac{1}{4}$ for rectangular R.C. columns *with a normal amount of steel* (say 2–3 percent), at which point the column can be simply designed for axial load $2P$ to take care of both P and M. When e/d is greater than about $\frac{1}{2}$, tension failure may control the strength of the column.

If e/d is between 0 and $\frac{1}{4}$, design the column for an axial load interpolated between P and $2P$. If e/d is between $\frac{1}{4}$ and $\frac{1}{2}$, interpolate between $2P$ and $3P$. For e/d above $\frac{1}{2}$, design the column like a beam acted on by moment M, because bending or tension controls the section. This approach is illustrated in Example 9-4.

EXAMPLE 9-4: Design of Concrete Columns Under Bending

Select proportions of a square tied reinforced concrete column under combined axial load P and bending moment M.

$$P = 250 \text{ k}, \ M = 75 \times 12 = 900 \text{ kip-in.}$$

$$\text{use } f'_c = 5 \text{ ksi, } f_y = 60 \text{ ksi.}$$

Solution

Assume d to be 18 in. in the direction of the moment,

$$e = \frac{M}{P} = \frac{900}{250} = 3.6 \text{ in., and } \frac{e}{d} = \frac{3.6}{18} = 0.2$$

Alternatively, one could shoot for $\sim 1.8P$ by setting $(d \approx 5e)$ to start. Interpolating 0.2 between 0 and 0.25, the effect of M is equivalent to $0.8P$, thus design for

$$P + 0.8P = 1.8P = 1.8 \times 250 = 450 \text{ k.}$$

Although allowable concrete stress is $0.2f'_c = 1$ ksi, *we can allow 1.5 ksi to include effect of steel.*
$A_g = 450/1.5 = 300 \text{ in.}^2$
Try 18 in. $\times$ 18 in., $A_g = 324 \text{ in.}^2$

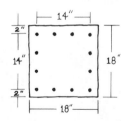

P taken by concrete $= 324 \times 1 \text{ ksi} = 324 \text{ k.}$
Remaining $450 - 324 = 126$ k to be taken by steel at $0.32 \ f_y = .32 \times 60 = 19.2$ ksi., $A_s = 126/19.2 = 6.6 \text{ in.}^2$

Use 8 no. 8 bars, $A_s = 6.4$ in.^{2} $P_s = A_s/A_g = 6.4/324 = 2$ percent which is o.k. for preliminary design.

EXAMPLE 9-5: Design of Concrete Columns With Large Moments

If the 18 in. × 18 in. column in the previous example is subjected to $P = 250$ and $M = 2250$ kip-in., check the size and the approximate reinforcement.

Solution

$$\frac{M}{P} = \frac{2250}{250} = 9 \text{ in.}$$

$$\frac{e}{d} = \frac{9}{18} = 0.5$$

The design is on the borderline between column action and beam action. As a column, design for $3P = 3 \times 250 = 750$ k. Concrete carries 324 k, steel to carry $750 - 324 = 426$ k

$$\frac{426 \text{ k}}{19.2 \text{ ksi}} = 22.1 \text{ in.}^2,$$

$$P_s = \frac{22.1}{324} = 6.8 \text{ percent which is high for columns } \textit{under bending.}$$

Compare the above with design as a beam (but using column rather than bending stress): Design for $M = 2250$ kip-in.,

$$a = 18 \text{ in.} - (2 \times 2) = 14 \text{ in.}$$

$$T = \frac{2250}{14} = 160 \text{ k}$$

$$A_s = \frac{160}{19.2} = 8.4 \text{ in.}^2 \text{ required for } \textit{each side.}$$

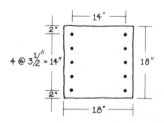

The total A_s of 16.8 in on 5.2 percent A_g still may be high, but try to use 5 no. 11 each side to resist tension. Since there should be $1\frac{1}{2}$ in. clear (or 1.5 diameters c-c) between bars, it appears that the column size may not be sufficient to accommodate 5 no. 11 bars each side, and a larger size may be needed. Note this is *not* an attempt at a final design, but it does give an idea of the dimensioning and proportioning problems.

SECTION 6: Examples for Seismic Design of Columns

Two Examples are presented here illustrating preliminary seismic design for a first-floor column in a 13-story building (same building in Example 7-5), first using a concrete column and then a steel column. The same procedure can be applied when designing for wind resistance.

EXAMPLE 9-6: Concrete Column

For the 13-story building in Example 7-5, make a preliminary design of a typical column on the first floor, as follows:

1. Determine the size of the interior column on the first floor to carry axial load only, using a tied reinforced concrete column. Assume the column to carry all loads within a tributary area of 20 ft × 30 ft, with dead load of 120 psf and reduced live load at 37.5 psf of floor area. The column is to be 20″ wide on its exterior face and may have any depth.

2. Modify the above design (by adding more rebars but keeping the same column size), taking into account the live load moment produced by the 60′ girder at $wL^2/12$ (which is assumed to be equally shared by the columns above and below the girder). Since the dead load is balanced by posttensioning, neglect its moment.

3. Check the column in **2.** for seismic effects, Zone 3. Assume the frames to carry 25 percent of the total shear (which is primarily carried by the core) and further assume each *interior* frame to share $\frac{1}{8}$ of the total frame shear. Inflection points of the columns are assumed to be at their midheight. Allow $\frac{1}{3}$ increase in stresses.

Solution

(a) *Size column for axial load only*:

Dead load = 120 psf   Tributary area = 20 ft × 30 ft

Reduced live load = 37.5 psf

Axial load $P = \dfrac{20 \times 30(120 + 37.5)}{1000} \times 13 = 1230$ k

For medium reinforcement: Assume $p = 2$ percent *and find* A_g.

$$\therefore \ A_s = 0.02 A_g$$

$$P = P_{conc} + P_{steel}$$

$$P_{conc} = 0.20 f'_c A g$$

$$P_{steel} = 0.32 f_y A_s$$

$$1230 \ k = 0.20 \times 4 \ ksi \times A_g + 0.32 \times 40 \ ksi \times 0.02 A_g$$

$$A_g = 1165 \ in.^2$$

$$d = \frac{1165}{20 \ in.} = 58.25 \ in., \text{ say } 60 \ in.$$

$\therefore$ Make column 20 in. $\times$ 60 in. $A_g = 1200 \ in.^2 \ A_s = .02 \times 1200 \approx 24 \ in.^2$

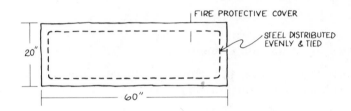

(b) *Add more steel to resist axial load + live load moment:*

Since the second and first floor columns are assumed to share the girder moment equally, $M = \frac{1}{2} \times wL^2/12$

$$w_{LL} = 20 \ ft \times 37.5 = 0.750 \ k/ft$$

$$M_{LL} = \frac{1}{2} \times \frac{0.75 \times 60^2}{12} \times 12 = 1350 \ \text{kip-in.}$$

$$e = \frac{M}{P} = \frac{1350}{1230} = 1.1 \ in.$$

$$\frac{e}{d} = \frac{1.1}{60} = 0.02,$$

which indicates the column can be designed for

$$P\left(1 + \frac{.02}{.25}\right) = 1230 \times 1.08 = 1330 \ k \text{ and using } f'_c = 4 \ ksi$$

$$P_{conc} = 0.20 \times 4 \ ksi \times 1200 \ in. = 960 \ k$$

$$P_{steel} = 1330 - 960 = 370 \ k,$$

$$f_s = .32 \times 40 = 12.8 \ ksi$$

$$A_s = \frac{370}{12.8} = 29 \ in.^2$$

Compared to case (a), very little *additional* steel is needed. Use 24 no. 10 bars, $A_s = 30$ in.^{2}

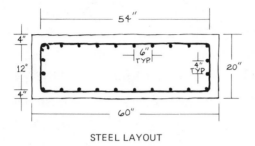

STEEL LAYOUT

(c) *Check column for seismic effects*:

Assume overall base shear $= 644$ k $\times 0.25$ (frames take 25 percent) $= 161$ k,
Overall Base moment $= 67,000$ k $\times 0.25$ (frames take 25 percent) $= 16,750$ kip-ft,
Interior frame takes $\frac{1}{8}$ of frame shear;
Moment due to column shear $= \frac{1}{8} \times \frac{1}{2} \times 161 \times 12$ ft/2 $= 60$ kip-ft.

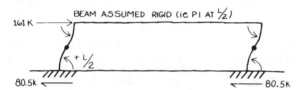

Total axial force in nine columns due to overturning moment:

$$T = C = \frac{M}{a} = \frac{16,750}{60} = 280 \text{ k}, \quad P_{EQ} = \frac{280}{9} = 31 \text{ k/column}$$

Totals for each column,

$$P_{\text{Total}} = 1230 + 31 = 1260 \text{ k, (compare to } P = 1230 \text{ k)}$$

$$M_{\text{Total}} = 1350 + (60 \times 12) = 2070 \text{ kip-in., (compare to } M = 1350 \text{ kip-in.)}$$

Note the increase in steel requirement due to seismic effect is small for both P and M, is less than the $\frac{1}{3}$ increase allowed for. Hence gravity load controls the design, and no revision is needed for seismic effects.

EXAMPLE 9-7: Steel Column

If steel rigid frames are used for this building, instead of reinforced concrete, and the central core is made up of non-load-carrying partitions, then seismic forces are resisted entirely by the frames. Using simplifying assumptions, make a preliminary design for a typical column, using A-36 steel. Proceed in a manner similar to Parts (a), (b), and (c) in Example 9-6.

Solution

For system with ductile moment resisting frame:

Assume $Z = 1$, $K = 0.67$ (from Chapter 5)

Fundamental period $T = 0.10N$ (where N = number of stories)

$$T = 0.1 \times 13 = 1.3 \text{ secs.}$$

$$C = \frac{0.05}{\sqrt[3]{1.3}} = 0.5$$

$W = 16,100 \text{ k}$ (from Example 7-5),

$V = ZCKW = 0.05 \times 0.67 \times 16,100 = 540 \text{ k}$ (Note that $V \sim 0.03W$).

Assuming triangular distribution of lateral loads along height of building,

$$\text{Base moment} = M = \left(\frac{2}{3} \times 156\right) \times 540 = 56,200 \text{ kip-ft.}$$

(a) Axial load, $P = 1230 \text{ k}$ (from Example 9-6).
For building frames, it is common and conservative practice to assume columns as pin-ended with, $K = 1.0$
Guess $f'_p = 17 \text{ ksi}$, $A_1 = \dfrac{P}{f'_p} = \dfrac{1230}{17} = 72 \text{ in.}^2$
Choose W 14×246, $r_y = 4.12 \text{ in.}$, $A = 72.3 \text{ in.}^2$, $B_x = 0.18$
$$\frac{KL}{r_{min}} = \frac{1.0 \times 12 \times 12}{4.12} = 35$$
From Figure 9-7, $f'_p = 19.6 \text{ ksi}$

The column above was designed *for axial load only*. The intention was to give us better approximations for B_x and f'_p to be used in Part (b).

(b) For axial load plus moment due to live and dead load:
Dead load = 120.0 psf
Live load = ___37.5__ psf (reduced)
157.5 psf $\times$ 20 ft = 3.15 k/ft
$$M = \frac{1}{2} \times \frac{wL^2}{12} = \frac{3.15 \times 60^2}{2 \times 12} \times 12 = 5670 \text{ kip-in (i.e. } \tfrac{1}{2} \text{ girder } M)$$
$$A_{req} = \left(\frac{P}{f'_p} + \frac{B_xM}{f'_M}\right) = \frac{1230}{19.6} + \frac{0.18 \times 5670}{24} = 105 \text{ in.}^2$$
Choose W 14×370, $A = 109 \text{ in.}^2$, $r_y = 4.27 \text{ in.}$
$$\frac{KL}{r_y} = \frac{1.0 \times 12 \times 12}{4.27} = 34, \quad S = 608 \text{ in.}^3, \quad B_x = 0.180$$
From chart, $f' = 19.65 \text{ ksi}$

Check interaction formula:

$$\frac{f_P}{f_P'} + \frac{f_M}{f_M'} \leq 1.0$$

where

$$f_P = \frac{P}{A} = \frac{1230\ k}{109} = 11.3\ k/in.^2$$

$$f_M = \frac{M}{S} = \frac{5670}{608} = 9.32\ ksi$$

$$\frac{11.3}{19.65} + \frac{9.32}{24} = .96 < 1;\ o.k.$$

(c) For E.Q. design, $V = 540\ k$; $M = 56{,}200$ kip-ft.

Shear on 1 interior column $= (540/8)(\frac{1}{2}) = 34\ k$, producing column moment

$$M = \frac{34 \times 12 \times 12}{2} = 2450\ \text{kip-in.}$$

Due to overturning

$$P = \frac{56{,}200}{60' \times 8} = 117\ k$$

Hence, for seismic effects

$$f_P = \frac{117\ k}{109} = 1.07\ ksi,\ f_M = \frac{2450}{608} = 4.03\ ksi$$

and the interaction formula control factor increases from $1 \rightarrow \frac{4}{3}$ for E.Q. or Wind.

Check interaction (Cases $b + c$):

$$\frac{f_P}{f_P'} + \frac{f_M}{f_M'} \leq 1.33$$

$$\frac{11.3 + 1.07}{19.65} + \frac{9.32 + 4.03}{24} = 1.19 \leq 1.33;\ o.k.$$

Hence, no increase in column size is required for seismic resistance.

10
High-Rise Buildings

SECTION 1: Introduction

It is difficult to define a high-rise building. One may say that a low-rise building ranges from 1 to 2 or 3 stories. A medium-rise building probably ranges between 3 or 4 stories up to 10 or 20 stories. Thus a high-rise building is probably one that has at least some 10 stories or more.

Although the basic principles of vertical and horizontal subsystem design remain the same for low-, medium-, or high-rise buildings, when a building gets high the vertical subsystems become a controlling problem for two reasons. Higher vertical loads will require larger columns, walls, and shafts. But, more significantly, the overturning moment and the shear deflections produced by lateral forces are much larger and must be carefully provided for.

The vertical subsystems in a high-rise building transmit accumulated gravity load from story to story, thus requiring larger column or wall sections to support such loading (Figure 10-1a). In addition, these same vertical subsystems must transmit lateral loads, such as wind or seismic loads, to the foundations. However, in contrast to vertical load, lateral load effects on buildings are not linear and increase rapidly with increase in height (Figure 10-1b). For example, under wind load, the overturning moment at the base of a building varies approximately as the *square* of the height of the building, and the lateral deflection at the top of a building may vary as the *fourth power* of building height, other things being equal. Earthquake produces an even more pronounced effect.

When the structure for a low- or medium-rise building is designed for dead and live load, it is almost an inherent property that the columns, walls, and stair or elevator shafts can carry most of the horizontal forces. The problem is primarily one of shear resistance. Moderate additional bracing for rigid frames in "short" buildings can easily be provided by filling certain panels (or even all panels) without increasing the sizes of the columns and girders otherwise required for vertical loads.

Unfortunately, this is not so for high-rise buildings because the problem is primarily resistance to moment and deflection rather than shear alone. Special structural arrangements will often have to be made and additional structural material is always required for the columns, girders, walls, and slabs in order to make a high-rise building sufficiently resistant to much higher lateral loads and deformations.

FIGURE 10-1
**HIGH-RISE DESIGNS REQUIRE MORE MATERIAL TO RESIST BOTH VERTICAL
AND HORIZONTAL LOADS.**

(a) FOR EQUIVALENT FLOOR AREA PER BAY, A STACKED SCHEME WILL REQUIRE $(\frac{n+1}{2})$
     TIMES MORE (VERTICAL LOAD) COLUMN UNITS THAN A SINGLE STORY SCHEME

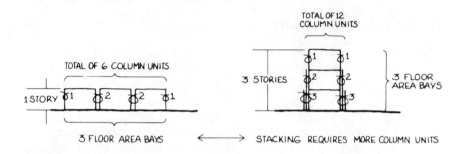

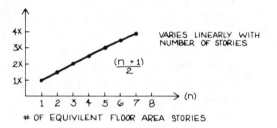

(b) THE INCREASE FOR HORIZONTAL LOAD RESISTANCE IS NOT LINEAR

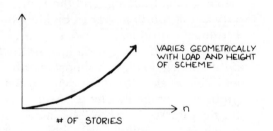

As previously mentioned, the quantity of structural material required per
square foot of floor of a high-rise building is in excess of that required for a
low-rise building. The vertical components carrying the gravity load, such as
walls, columns, and shafts, will need to be strengthened over the full height
of the building. But quantity of materials required for resisting lateral forces
is even more significant. The graph shown in Figure 10-2 illustrates how the
weight of structural steel in pounds per square foot of floor increases as the

FIGURE 10-2
**STRUCTURAL REQUIREMENTS FOR RESISTANCE TO LATERAL LOADS
CAN BE VERY IMPORTANT.**

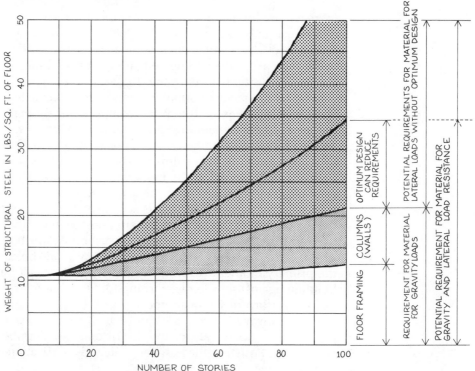

TAKEN FROM "OPTIMIZING STRUCTURAL DESIGN IN VERY TALL BUILDINGS", ARCHITECTURAL
RECORD, AUGUST, 1972

number of stories increases from 1 up to 100. Note that by using *optimum
structural systems* with suitable width and arrangement, the additional
material required for lateral force resistance can be controled such that,
even for buildings of 100 stories, the total structural weight of steel may be
only about 34 pounds per square foot, whereas it is noted previously
(Figure 5-1), some buildings quite a bit shorter require much more structural
steel.

   With reinforced concrete, the quantity of material also increases as the
number of stories increases. But here it should be noted that the increase in
the weight of material added for gravity load is much more sizable than for
steel, whereas for wind load the increase for lateral force resistance is not
that much more since the weight of a concrete building helps to resist
overturn. On the other hand, the inherently greater mass of a concrete
building can aggravate the problem of design for earthquake forces.
Additional mass in the upper floors will give rise to a greater overall lateral
force under the action of seismic effects.

In the case of either concrete or steel design, there are certain basic principles for providing additional resistance to lateral forces and deflections in high-rise buildings without too much sacrifice in economy:

1. Increase the effective width of the moment-resisting subsystems. This is very useful because increasing the width will cut down the overturn force directly and will reduce deflection by the third power of the width increase, other things remaining constant (Figure 10-3). However, this does require that the vertical components of the widened subsystem be suitably connected to actually gain this benefit.

2. Design subsystems such that the components are made to interact in the most efficient manner. For example, use truss systems with chords and diagonals efficiently stressed, place reinforcing for walls at critical locations, and optimize stiffness ratios for rigid frames.

3. Increase the material in the most effective resisting components. For example, materials added in the lower floors to the flanges of columns and connecting girders will directly decrease the overall deflection and increase the moment resistance without contributing mass in the upper floors where the earthquake problem is aggravated.

4. Arrange to have the greater part of vertical loads be carried directly on the primary moment-resisting components. This will help stabilize the building against tensile overturning forces by precompressing the major overturn-resisting components.

5. The local shear in each story can be best resisted by strategic placement of solid walls or the use of diagonal members in a vertical subsystem. Resisting these shears solely by vertical members in

FIGURE 10-3
THE EFFECTIVE WIDTH OF THE MOMENT-RESISTING SUBSYSTEM IS IMPORTANT TO EFFICIENCY AND STIFFNESS.

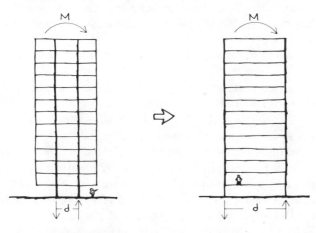

PHOTO 10-1

339

SIXTY-FIVE STORY P.C. OFFICE BUILDING DESIGNED FOR KAISER CEMENT AND GYPSUM CORPORATION. (STRUCTURAL ENGINEERING BY T. Y. LIN; ARCHITECTURE BY IRVING D. SHAPIRO & ASSOCIATES, A.I.A.)

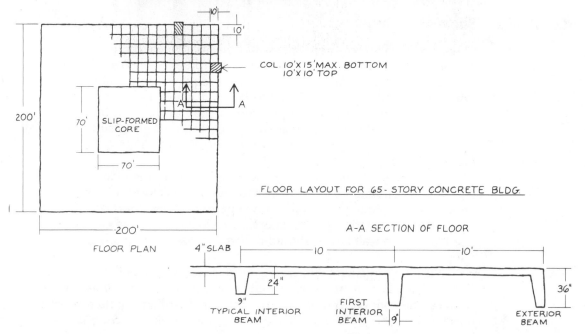

COL. 10' X 15' MAX. BOTTOM
10' X 10' TOP

SLIP-FORMED CORE

70'

70'

200'

200'

FLOOR PLAN

FLOOR LAYOUT FOR 65-STORY CONCRETE BLDG.

A-A SECTION OF FLOOR

4" SLAB

10

10'

24"

9"

TYPICAL INTERIOR BEAM

FIRST INTERIOR BEAM

9"

36"

EXTERIOR BEAM

bending is usually less economical, since achieving sufficient bending resistance in the columns and connecting girders will require more material and construction energy than using walls or diagonal members.

6. Sufficient horizontal diaphragm action should be provided at each floor. This will help to bring the various resisting elements to work together instead of separately.

7. Create mega-frames by joining large vertical and horizontal components such as two or more elevator shafts at multistory intervals with a heavy floor subsystem, or by use of very deep girder trusses. Photo 10-1 shows the design of a 65-story building, using eight massive columns and three horizontal transfer floors to form a giant earthquake-resistant cage, which is further stiffened by individual posttensioned concrete waffle floors.

Remember that all high-rise buildings are essentially vertical cantilevers which are supported at the ground. When the above principles are judiciously applied, structurally desirable schemes can be obtained by walls, cores, rigid frames, tubular construction, and other vertical subsystems to achieve horizontal strength and rigidity as introduced and illustrated in Chapter 7. Some of these applications will now be described in subsequent sections of this chapter.

SECTION 2:   Shear-Wall Systems

When shear walls are compatible with other functional requirements, they can be economically utilized to resist lateral forces in high-rise buildings. For example, apartment buildings naturally require many separation walls. When some of these are designed to be solid, they can act as shear walls to resist lateral forces and to carry the vertical load as well. For buildings up to some 20 stories, the use of shear walls is common. If given sufficient length, such walls can economically resist lateral forces up to 30 to 40 stories or more.

However, shear walls can resist lateral load *only* in the plane of the walls (i.e. not in a direction perpendicular to them). Therefore, it is always necessary to provide shear walls in two perpendicular directions, or at least in sufficient orientation so that lateral force in any direction can be resisted. In addition, the wall layout should reflect consideration of any torsional effect.

As illustrated in Figure 7-5, two or more shear walls can be connected to form *L*-shaped or channel-shaped subsystems. Indeed, four internal shear walls can be connected to form a rectangular shaft that will then resist lateral forces very efficiently. If all external shear walls are continuously connected, then the whole building acts as a tube, and is excellent in resisting lateral loads and torsion, as will be discussed in Section 4.

Whereas concrete shear walls are generally of solid type with openings when necessary, steel shear walls are usually made of trusses (Figure 10-4). These trusses can have single diagonals, "X" diagonals, or "K" arrangements. A trussed wall will have its members act essentially in direct tension or compression under the action of lateral forces. They are effective from a strength and deflection-limitation point of view, and they offer some opportunity for penetration between members. Of course, the inclined members of trusses must be suitably placed so as not to interfere with requirements for windows and for circulation and service penetrations through these walls.

As stated above, the walls of elevator, staircase, and utility shafts form natural tubes and are commonly employed to resist both vertical and lateral forces. Since these shafts are normally rectangular or circular in cross-section, they can offer an efficient means for resisting moments and shear in all directions due to tube structural action. But a problem in the design of these shafts is to provide sufficient strength around door openings and other penetrations through these elements as illustrated in Example 10-1. For reinforced concrete construction, special steel reinforcements are placed around such openings. In steel construction, heavier and more rigid connections are required to resist racking at the openings.

In many high-rise buildings, a combination of walls and shafts can offer excellent resistance to lateral forces when they are suitably located and connected to one another. It is also desirable that the stiffness offered by these subsystems be more-or-less symmetrical in all directions.

FIGURE 10-4
SHEAR WALLS IN STEEL AND CONCRETE.

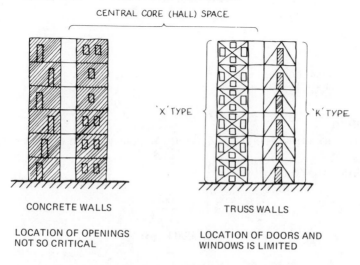

CENTRAL CORE (HALL) SPACE

'X' TYPE   'K' TYPE

CONCRETE WALLS

TRUSS WALLS

LOCATION OF OPENINGS
NOT SO CRITICAL

LOCATION OF DOORS AND
WINDOWS IS LIMITED

EXAMPLE 10-1: Design of Shear Wall Building

A typical reinforced concrete shear wall in a 30-story apartment building is shown, with openings through the corridors at every floor so that the two half-walls are connected only by beams above the corridors. These shear walls are spaced 30-ft centers and will be designed to resist a wind load of 40 psf on the building. The connecting beams are 8 ft long, limited to 30 in. deep × 8 in. wide. Make an approximate calculation to determine the adequacy of these beams.

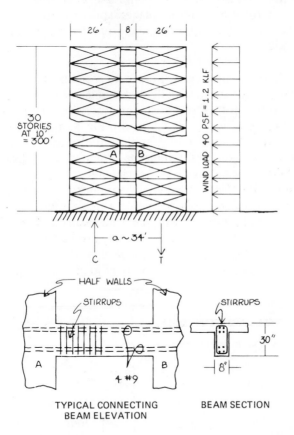

TYPICAL CONNECTING   BEAM SECTION
BEAM ELEVATION

Solution

Wind load 40 psf acting on wall with tributary width of 30 ft = 1.2 klf.

$$\text{Moment at base of wall} = \frac{wL^2}{2}$$

$$= 1.2 \times \frac{300^2}{2}$$

$$= 54{,}000 \text{ kip-ft}$$

which is resisted by a tension force T at base of one half-wall and a compression force C at base of the other half-wall, as shown. These two forces act with a lever arm (a) between them, which can be conservatively estimated to be 34 ft—the distance between the centers of the two half-walls. Thus

$$T = C = \frac{54,000}{34} = 1590 \text{ k}$$

This is also the total vertical shear transmitted through the 30 connecting beams. Assume the beams to carry equal shears, the shear in each beam is

$$V = \frac{1590}{30} = 53 \text{ k}$$

With the point of inflection at midspan of each beam, the moment at each end of the beam is

$$53 \times \frac{8}{2} = 212 \text{ kip-ft}$$

For a beam depth of 30 in., the resisting arm for either the top or the bottom reinforcing bars can be approximated at 24 in. or 2 ft, and the tension force required for the bars is

$$T = \frac{212}{2'} = 106 \text{ k}$$

Using an allowable stress of 20 ksi for the bars, plus $\frac{1}{3}$ increase permitted under wind load, we would need a steel area of

$$106/(20 \times 4/3) = 4.0 \text{ sq in.}$$

which represents a steel ratio in the beam of

$$4.0/(8 \text{ in.} \times 27 \text{ in.}) = 1.8 \text{ percent (for } T \text{ force only),}$$

which is reasonable. *Since we have an equal amount of steel at top and bottom of beam, 2 or even 3 percent can be allowed.*

Note that these bars are needed both at top and at bottom of each beam since wind can act in either direction.

Suppose we provide four no. 9 bars for $A_s = 4$ sq in., they have to be arranged in two layers in the 8-in. thick beam.

The average shearing stress in the beam is

$$\bar{v} = \frac{V}{bd'}$$

$$= \frac{53,000}{8 \times 24 \text{ in.}}$$

$$= 123 \text{ psi}$$

which can be resisted with sufficient stirrups (Note that Table B-2 in the Appendix allowable $v = 100 \sim 200$ psi even for 3 ksi concrete).

The above indicates that the beams can be properly designed to resist the shear and moment produced by the wind force, although they will need to be carefully detailed.

SECTION 3: Rigid-Frame Systems

As discussed in Chapter 7, rigid-frame systems for resisting vertical and lateral loads have long been accepted as an important and standard means for designing buildings. They are employed for low- and medium-rise buildings up to high-rise buildings perhaps 70 or 100 stories high. When compared to shear-wall systems, these rigid frames provide excellent opportunity for rectangular penetration of wall surfaces both within and at the outside of a building. They also make use of the stiffness in beams and columns that are required for the building in any case, but the columns are made stronger when rigidly connected to resist the lateral as well as vertical forces through frame bending.

Frequently, rigid frames will not be as stiff as shear-wall construction, and therefore may produce excessive deflections for the more slender high-rise building designs. But because of this flexibility, they are often considered as being more *ductile* and thus *less* susceptible to catastrophic earthquake failure when compared with (some) shear-wall designs. For example, if over stressing occurs at certain portions of a steel rigid frame (i.e., near the joints), ductility will allow the structure as a whole to deflect a little more, but it will by no means collapse even under a much larger force than expected on the structure. For this reason, rigid-frame construction is considered by some to be a "best" seismic-resisting type for high-rise steel buildings. On the other hand, it is also unlikely that a well-designed shear-wall system would collapse.

In the case of concrete rigid frames, there is a divergence of opinion. It is true that if a concrete rigid frame is designed in the conventional manner, without special care to produce higher ductility, it will not be able to withstand a catastrophic earthquake that can produce forces several times larger than the code design earthquake forces. Therefore, some believe that it may not have the additional reserve capacity possessed by steel rigid frames. But modern research and experience has indicated that concrete frames can be designed to be ductile, when sufficient stirrups and joinery reinforcement are designed into the frame. Modern building codes have specifications for the so-called ductile concrete frames. However, at present, these codes often require excessive reinforcement at certain points in the frame so as to cause congestion and result in construction difficulties. Even so, concrete frame design can be both effective and economical.

Of course, it is also possible to combine rigid-frame construction with shear-wall systems in one building. For example, the building geometry may be such that rigid frames can be used in one direction while shear walls may be used in the other direction.

EXAMPLE 10-2: Design of Rigid-Frame Building

A 20-story office building has its steel frames spaced at 25-ft centers. Each frame has two columns and 60-ft girders. It is required to estimate the effect of seismic forces on the size of the second floor girders. The total dead load on the building is estimated at 80 psf of floor area, the dead plus live load on the girders estimated at 120 psf, and the seismic lateral load is given as 4 percent of the dead load $(0.04W_{DL})$.

(a) STEEL FRAME ELEVATION

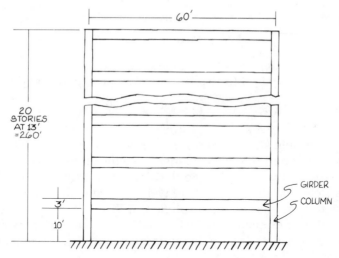

(b) FORCE DIAGRAM ON GIRDER

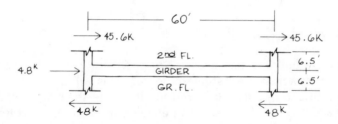

Solution

The tributary dead load on each frame is

$$80 \text{ psf} \times 25 \text{ ft} \times 60 \text{ ft} = 120 \text{ kips per floor}$$

which means a lateral force of,

$$0.04 \times 120 = 4.8 \text{ kips per floor.}$$

Thus the total shear across the ground floor is,

$$20 \times 4.8 = 96 \text{ kips}$$

and across the second floor is

$$19 \times 4.8 = 91.2 \text{ kips.}$$

If we assume point of inflection at column's midheight, we have a force diagram for the girder as shown. Using center-line dimensions, the moment at the end of the girder is

$$45.6 \times 6.5 + 48 \times 6.5 = 295 + 312 = 607 \text{ kip-ft}$$

The vertical load on the girder is estimated

$$120 \text{ psf} \times 25 \text{ ft} = 3 \text{ k/ft of girder.}$$

Assuming girder ends more or less fixed due to the relative stiffness of the columns compared to that of the girder. The girder end moment is

$$-M = \frac{wL^2}{12} = \frac{3 \times 60^2}{12} = 900 \text{ kip-ft}$$

The allowable stress in the steel can be increased by only one-third when considering seismic loads. Since the added seismic moment of 607 kip-ft is greater than one-third of the vertical load moment of 900 kip-ft, it is clear that the seismic load will have a definite effect on the girder design. Note that the final solution of this problem involves other considerations such as sufficient stiffness of the columns to minimize lateral sway and to delay plastic hinge formation in the columns. But they are beyond the scope of this treatise.

SECTION 4:   Tubular Systems

High-rise buildings in excess of 30 or 40 stories may be best designed using tubular systems to resist lateral forces. This will give the building greater strength and rigidity compared to either the shear-wall or the rigid-frame system. By effective use of the enveloping material, maximum lever arm between the resisting forces is achieved.

A natural way to build a tubular system would be to connect the exterior walls to form an overall tube structure. The tube can be rectangular, circular, or some other fairly regular shape. The exterior walls may be penetrated with holes which form round or rectangular windows. An example is the newly completed Hong Kong high-rise of 50 or 60 stories and round windows (Photo 10-2).

If it is desired to have rectangular window frames on the outside of the high-rise building, these frames can be integrated into a frame-tube design

PHOTO 10-2
CONNAUGHT BUILDING, HONG KONG, 586-FT HIGH.

using rather heavy spandrel beams, to connect the closely spaced columns or heavy window mullions. However, it should be noted that when a frame-tube bends, as a vertical cantilever supported at the base, the frame racking effect may result in a significant *shear lag* between the supporting columns. As a result, the stress distribution would not be linear and those

FIGURE 10-5

SHEAR LAG REDUCES THE EFFECTIVENESS OF SOME COLUMNS AS COMPARED WITH AN IDEALIZED (LINEAR) ASSUMPTION OF THE DISTRIBUTION OF AXIAL FORCES.

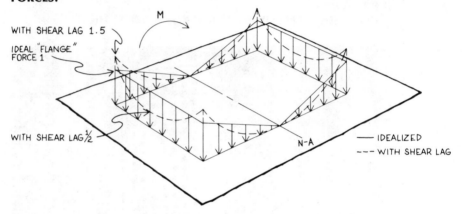

(NOTE: IN EITHER CASE, OVERALL FLANGE EFFECTIVENESS IS ABOUT 3 TIMES THAT OF WEB COLUMNS)

FIGURE 10-6

AS COLUMNS AND SPANDRELS ARE WIDENED, FRAME ACTION WILL MORE CLOSELY APPROXIMATE WALL TUBE ACTION.

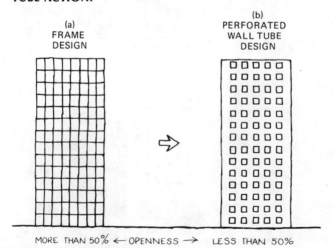

MORE THAN 50% ← OPENNESS → LESS THAN 50%

columns away from the N-A will be stressed quite a bit higher than would be expected for a linear assumption, as shown in Figure 10-5. Shear lag in tube design can be fairly well analyzed using modern computer programs. But, for schematic purposes, the figure suggests that the effect on corner columns can be approximated at 1.5 times greater than indicated by the linear assumption. The objection to these frame tubes is the necessity for fairly heavy spandrel girders. Of course, when the girders are so deep and the columns so wide that the openings are very small (i.e., ~30 percent), the frame tube will degenerate into a perforated wall tube (Figure 10-6).

Although wall tubes with small windows are generally of concrete construction, the frame tubes can be either of concrete or steel. For steel, the trussed tube is often used (Photo 10-3). By trussing the exterior columns

PHOTO 10-3
**THE JOHN HANCOCK CENTER,
CHICAGO. (SKIDMORE, OWINGS, AND
MERRILL, ARCHITECTS AND ENGINEERS).**

and forming them into a tube, they are very efficient in resisting lateral forces because they utilize overall dimensions for overturn resistance and efficient truss members in direct stress (rather than in bending) for shear resistance. The shape and size of windows are influenced by the location of diagonals, but a larger percentage of opening is possible in comparison to a concrete wall tube.

It is also possible to develop a concrete trussed tube design. This is accomplished by blocking certain panels at different levels so that these blocked panels along an inclined line form an inclined member of the truss. This apparently can be an economical solution for concrete high-rise buildings.

EXAMPLE 10-3: Design of A Round Tube Building

A 40-story reinforced concrete round building utilizes its exterior walls to carry $\frac{3}{8}$ of the floor loads and resist wind forces. The diameter is 100 ft and the walls are 8 in. thick with window openings for about 50 percent of the surface. It is required to investigate the strength of the walls for a wind load of 50 psf.

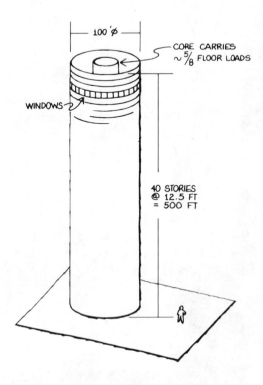

Solution

Wind force per foot of building height, using a shape factor of 0.6 for this round building, is

$$50 \text{ psf} \times 0.6 \times 100 \text{ ft} = 3 \text{ k/ft}$$

The overturning moment at the base of the building is

$$\frac{wL^2}{2} = \frac{3 \times 500^2}{2} = 375,000 \text{ kip-ft}$$

The moment of inertia of a circle 8 in. (or 0.67 ft) thick with a diameter of 100 ft is,

$$I = \pi r^3 t = \pi 50^3 \times 0.67 = 262,000 \text{ ft}^4$$

The tensile or compressive stress at the extreme fibers is

$$f = \frac{Mc}{I} = 375,000 \times \frac{50}{262,000} = 71 \text{ ksf}$$

Now let us compute the approximate vertical loading on the walls. The wall's own weight averages about 80 psf of wall surface (including the effect of window openings), hence its weight at base *per foot of wall perimeter* is

$$80 \times 500 \text{ ft} = 40 \text{ k}$$

The dead weight of the floors, assuming a tributary floor area of ~15 sq ft per ft of wall perimeter, and an average dead load of 120 psf, is

$$15 \times 120 \times 40 \text{ floors} = 72 \text{ k}$$

Hence, total dead weight on wall base is,

$$40 + 72 = \frac{112 \text{ k}}{\text{ft}}$$

Thus the maximum and minimum stresses per foot of wall are

$$112 + 71 = 183 \text{ k comp.}$$
$$112 - 71 = 41 \text{ k comp.}$$

which indicates that there will be no tension under wind load and the wall thickness is governed by compression. Since only 50 percent of the wall area is effective across window openings, the 8-in. wall has an effective concrete area of $4 \times 12 = 48$ in.^{2} per foot of perimeter.

The average unit compressive stress is

$$\frac{183,000}{48} = 3810 \text{ psi}$$

which is high, indicating that the 8-in. wall thickness is not sufficient for the base and should be increased. If 5 ksi concrete is used with allowable stress of about 2 ksi under wind (allowing for proper reinforcement), the wall thickness will need to be doubled at the base of the building. In any case, the calculation indicates that careful design and analysis will be needed.

The tube-in-tube concept offers another excellent approach (Figure 10-7). The exterior tube with its larger width can resist overturn forces very efficiently. But, the openings required in this tube may reduce its capacity to resist shear, particularly at the lower floors. On the other hand, an inner tube can better resist the story shear, being more solid than the exterior tube. But inner tubes will not be as effective in resisting the overturning moment since they will be quite slender in comparison to the outer tube. This tube-in-tube combination can be applied to either a steel building or a concrete building, or even used to combine steel and concrete design. For example, a concrete shaft as an inner tube in combination with a steel rigid-frame exterior tube can be very efficient in resisting the bending moment and in giving sufficient shear rigidity to the structure as a whole.

The bundled-tube concept is a new one. Thus far, it has been applied only to the Sears Building in Chicago (Photo 10-4). The 110-story Sears Tower in Chicago, 1450 ft high, is now the tallest building in the world. The design calls for a steel frame of nine tubes, 75 ft × 75 ft square, that make up the 225 ft sq steel building. The nine tubes have common

FIGURE 10-7
A TUBE-IN-TUBE SCHEME COMBINES THE FLOOR-TO-FLOOR STIFFNESS OF A MORE SOLID CORE TUBE WITH THE OVERALL AXIAL STIFFNESS OF AN OUTER FRAME TUBE.

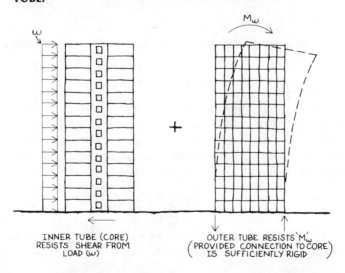

INNER TUBE (CORE)
RESISTS SHEAR FROM
LOAD (w)

OUTER TUBE RESISTS M_w
(PROVIDED CONNECTION TO CORE
IS SUFFICIENTLY RIGID)

PHOTO 10-4
THE SEARS TOWER, CHICAGO, 110-STORY OFFICE BUILDING (SKIDMORE, OWINGS AND MERRILL, ARCHITECTS AND ENGINEERS).

columns that link their faces to make up two exterior and two interior diaphragms in each direction. The building plan is thus trisected in two directions and the result is a greatly stiffened overall structure. This building has a steel weight of only 33 lb per sq ft, whereas with a traditional rigid-frame system, the weight might have reached 60 to 70 lb per sq ft.

The bundle approach is also interesting in that only two of the building's 75 ft-square modules rise to the full 1450-ft height. Two corner modules drop off at the fiftieth floor, two more at the sixty-sixth, and three at the ninetieth floor. This arrangement satisfies the needs of large amount of office space on the lower floors required for the owners' offices, and it has the advantage of reducing mass in the upper floors and of renting entire floors to tenants who want exclusive occupancy.

Dropping off the upper modules will also reduce wind sway by reducing exposed surface area and by breaking up the flow of wind. In other words, the resulting turbulence may tend to minimize deflection oscillations. The building's exterior columns are built up of wide flange sections on 15-ft centers. The corner columns of each module are heavier at 39 in. × 39 in. built-up sections. The other columns measure 39 in. from flange to flange, but have a width from 24 to 30 in., since under horizontal load they will be loaded less than the corner columns due to shear lag.

Example 10-4 illustrates a simple calculation for the columns of the Sears Building:

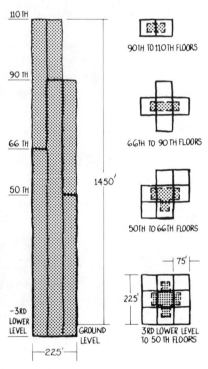

EXAMPLE 10-4: Sears-Roebuck Building, Chicago

For a simple calculation of column size at base:

Plan Modules by storeys:

$2 \times 20 = 40$
$5 \times 24 = 120$
$7 \times 16 = 112$
$9 \times 50 = 450$
$\overline{722} \div 9 =$ an average of 80 modules for each tube

$$h_{(averaged)} = \frac{80}{110} \times 1450 \text{ ft} = 1060 \text{ ft} \sim 1100 \text{ ft.}$$

$$w_h = 50 \text{ psf} \times 225 \text{ ft} = 11.5 \text{ k/ft, assume (a)} \sim \frac{1400 \text{ ft}}{2},$$

$$M = 11.5 \text{ k/ft} \times 1100 \times \frac{1400}{2} \simeq 9{,}000{,}000 \text{ kip-ft}$$

Assume flange diaphragms take $\frac{2}{3}M \simeq 6{,}000{,}000$ kip-ft

For flange M-load, $M \div 225 \text{ ft} \simeq 24{,}000 \text{ k}$

Distribute flange load over 225 ft/15′ c.c. = 16 columns:

$$24,000 \text{ k} \div 16 = 1500 \text{ k/column (average)}$$

Vertical load: 80 stories $\times$ [(225^2 $\times$ 100 psf) = 5000 k/floor] = 400,000 k

For Cols: $W_{\text{Total}} = 400,000 \div 112$ cols. [i.e. (16 $\times$ 4) + (1)] = 3800 k/col.

$$W_{\substack{\text{flange} \\ \text{columns}}} = 3800 \text{ k} + 1500 \text{ k (Flange } M\text{-load)} = 5300 \text{ k}$$

$$A_{\substack{\text{s-Flange} \\ \text{columns}}} = \frac{5300 \text{ k}}{20 \text{ ksi}(\frac{4}{3})} = 200 \text{ in.}^2$$

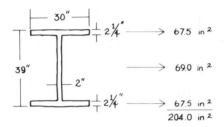

SECTION 5:   Special Systems

Although the previous three systems (shear-wall, rigid-frame, and tubes) are probably the most commonly used systems for high-rise buildings, the emergence of other so-called special systems is possible.

For high-rise buildings requiring long spans, it is often desirable to use the *staggered truss* (or wall) arrangement shown in Figure 10-8.

The staggered arrangement was initially developed for steel truss construction. By providing deep trusses that connect and support two floors, but are vertically staggered, long-span construction can be obtained for medium- or high-rise buildings. In a long-span high-rise building, the exterior columns will carry many stories of loading and are usually quite heavy. They are therefore able to absorb considerable bending moment. Since the staggered truss can serve as a very stiff connection between these heavy exterior columns, it can provide additional strength as well as rigidity. This is because, although the trusses are staggered, the columns will not have to carry the story shear. Shear is essentially resisted by the trusses since they are linked by the diaphragm action of the floor planes. However, it should be noted that staggered systems will take care of the lateral force in the direction of the truss, *but not perpendicular to them* unless lateral trusses are also employed.

It is also possible to employ a system of staggered or alternate story wall arrangements for concrete construction. Instead of the diagonal trusses we

FIGURE 10-8
**A STAGGERED TRUSS SCHEME WILL BE VERY STIFF,
WHILE ALLOWING LONG SPANS FOR INTERIOR
SPACES.**

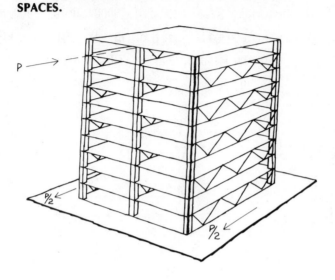

FIGURE 10-9
**A VIERENDEEL TRUSS IS ACTUALLY A FRAME WITH THE OVERALL
RESISTING MOMENT BEING PRIMARILY SUPPLIED BY AXIAL FORCES
IN THE TOP AND BOTTOM CORDS.**

VIERENDEEL TRUSS

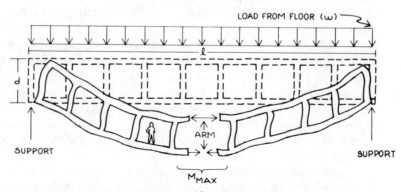

[MAX. C & T FORCES WILL BE ROUGHLY M_{MAX}/d PROVIDING THAT THE VERTICAL
MEMBERS ARE VERY STIFF]

will have the walls acting as very deep girders, or possibly as vierendeel trusses (Figure 10-9). These walls may be reinforced, or they can be posttensioned with draped tendons buried in the walls (or the vierendeel chords). Evidently this can be an exceedingly economical and functional arrangement. Photo 10-5 illustrates a *PC* vierendeel student design for floor-high trusses.

When buildings get 50 or 60 stories or higher, it will be worthwhile to investigate the use of super-frames or megastructures. Let us say megaframes. The megaframe approach takes advantage of the fact that mechanical floors will be provided every 15 or 20 stories to facilitate air-conditioning and other service equipment. The entire 1- or 2-story depth of mechanical floors can be utilized to build up a strong and stiff horizontal subsystem. This allows use of very large girders or space trusses connected to large outside columns to form rigid megaframe action over 15 to 20 floors (Figure 10-10). The megaframe may be filled in with a secondary (and much lighter) frame of standard design.

The megaframe approach was used in the design of the First National Bank of Chicago (Photo 10-6*a* and *b*). In this case, a tapering form, 800 ft high, rises in a sweeping curve from a 55,000 sq ft base to a 29,000 sq ft tower at the top. With 60 stories, the building encloses 2,000,000 sq ft. The design makes possible a clear-span banking floor which has no columns interior to the entire 55,000 sq ft base.

Other special approaches include curved-shell systems. The Toronto City Hall building, for example, makes use of its curved shape to provide

FIGURE 10-10
WHERE PRACTICAL, SHAFTS ON THE PERIPHERY OF A HIGH-RISE PLAN CAN BE CONNECTED TO ACHIEVE MEGAFRAME ACTION.

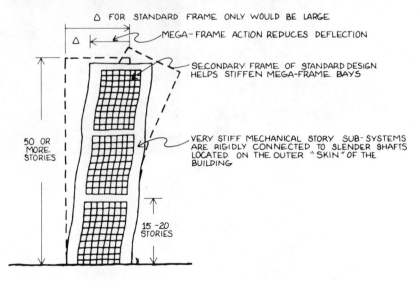

△ FOR STANDARD FRAME ONLY WOULD BE LARGE

MEGA-FRAME ACTION REDUCES DEFLECTION

SECONDARY FRAME OF STANDARD DESIGN HELPS STIFFEN MEGA-FRAME BAYS

VERY STIFF MECHANICAL STORY SUB-SYSTEMS ARE RIGIDLY CONNECTED TO SLENDER SHAFTS LOCATED ON THE OUTER "SKIN" OF THE BUILDING

50 OR MORE STORIES

15-20 STORIES

PHOTO 10-5
**FLOOR DEEP *PC* VIERENDEEL TRUSSES ON ALTERNATE FLOORS TIE 12 COLUMNS
TOGETHER IN A 60-STORY STUDENT ARCHITECTURAL DESIGN PROJECT AT UCB.
(a) FORM MODEL E ELEVATION (b) GROUND-LEVEL AND CORNER PERSPECTIVE
(ON FACING PAGE) (c) VIERENDEEL LAYOUT (d) ANALYSIS OF DL AND EQ
BEHAVIOR.**

(a)

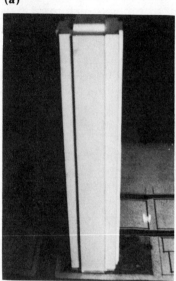 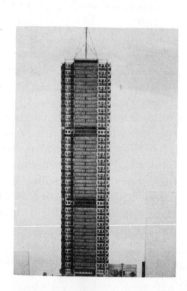

(b)

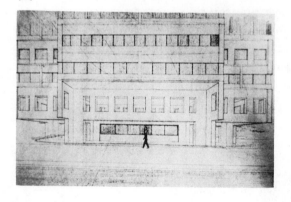 

(c)

(d)

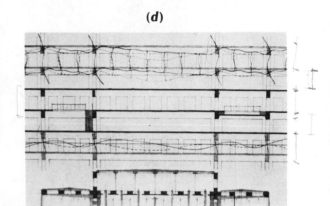

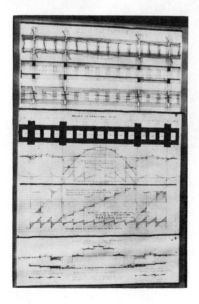

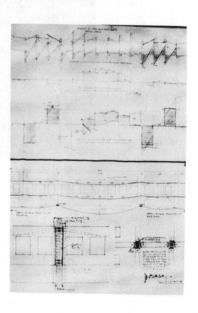

PHOTO 10-6
(*a*) **FIRST NATIONAL BANK BUILDING,
CHICAGO (COURTESY OF HARR, HEDRICH-
BLESSING, CHICAGO).**
(*b*) **(ON FACING PAGE) SCHEMATIC OF
STRUCTURE.**

(b) <u>SCHEMATIC OF STRUCTURE</u>

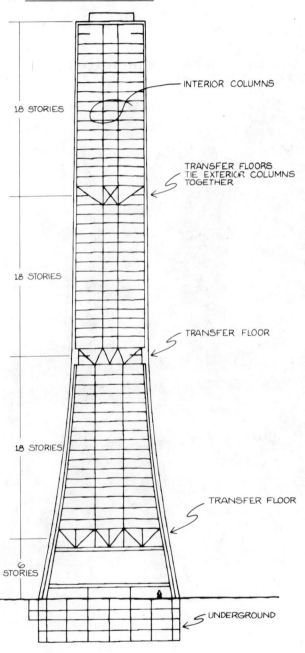

18 STORIES

INTERIOR COLUMNS

TRANSFER FLOORS
TIE EXTERIOR COLUMNS
TOGETHER

18 STORIES

TRANSFER FLOOR

18 STORIES

TRANSFER FLOOR

6
STORIES

UNDERGROUND

PHOTO 10-7
**TORONTO CITY HALL. (COURTESY OF THE TORONTO
DEPARTMENT OF THE CITY CLERK).**

additional strength and rigidity (Photo 10-7). The basic design concepts are presented in an excellent article by Fazler Kahn of S.O.M., "The Future of High-Rise Structures," (*Progressive Architecture*, October 1972), which is excerpted here:

Future forms

Concrete is a moldable material and its character leads to many unusual forms, while retaining high efficiency in construction. It is through these newer forms that concrete probably will find its use in future ultra high rise buildings for offices and housing in those urban areas of the world where steel is relatively expensive. Form may follow function, but certainly it can also give strength through shaping the entire building into structurally efficient and stable overall shapes.

Take, for example, a typical apartment building in the U.S. that is generally about 70 to 90 ft wide and possibly as long as 200 or 300 ft. The width of each building is naturally limited because of the normal requirement of windows in each room. A double-loaded corridor plus typical living and dining rooms would result in a maximum width of about 90 ft. To build a tall apartment building, therefore, one has to face the reality of the height-to-width ratio. If the height of the building is more than eight times the width of the building, there is a good chance that the building will be too flexible no matter how it is constructed. Lateral sway under

(a) STABILITY OF VARIOUS WALL SHAPES WOULD
    STABILIZE TALL, THIN BUILDINGS

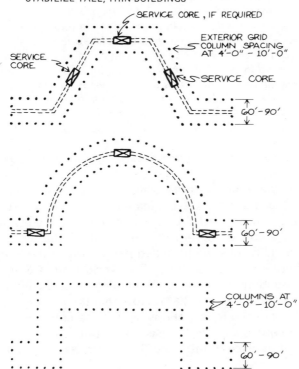

(b) PROPOSED SERPENTINE SHAPED APARTMENT PROJECT

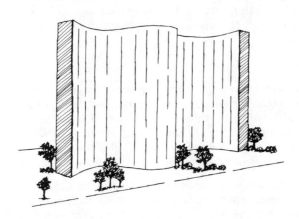

wind load may be perceptible too often to the dwellers and, as a result, such buildings may be structurally adequate but otherwise unlivable. The maximum height of these buildings therefore would be limited to about 700 ft.

How can we then, if need be, extend the height to 1000 ft or more? One simple, possible way to circumvent this limitation is to curve the entire shape of the building in the form of a folded plate, a channel or a fluted shape (a). This is, in fact, a simple concept used for sheet metal floor and roof decks or in concrete shells. The trick is to create an equivalent thin plate out of each exterior face of the building. This can be done with closely spaced columns on each face of the building, connected by relatively deep spandrel beams running along the facade. A somewhat romantic version of this structural concept was proposed by one of the graduate students at IIT, Alfonso Rodriguez, also under architectural advisorship of Myron Goldsmith (b). In reality, these shapes could be arranged in many possible forms and proportions as long as the overall section property of the equivalent thin plates on each face of the building provides the required stiffness, rigidity and strength against the overturning forces caused by wind or earthquake.

Although the components of most high-rise buildings are commonly constructed in place of either steel elements or reinforced concrete, the use of properly designed prefabricated steel or precast concrete *subassemblies* may cut down the cost of high-rise buildings. But either way, sufficient strength and ductility must be provided particularly at the joints and at connections between subassemblies. The World Trade Center and Standard Oil of Indiana Building (Chicago, 1971) employed steel columns and spandrel subassemblies Figure 10-11. Note that the exterior wall sections are designed to serve as the fenestrations, and as the columns and spandrel beams. Precast concrete could be used as spandrels and as forms for columns poured in place to integrate the precast exterior walls into a frame tube.

In another case, the Concrete Technology Corporation of Tacoma,

FIGURE 10-11
**COLUMN-SPANDREL SUBASSEMBLY OF 3-STORY SECTION
FOR EXTERIOR WALL FOR STANDARD OIL OF INDIANA
BUILDING, CHICAGO, 1971.**

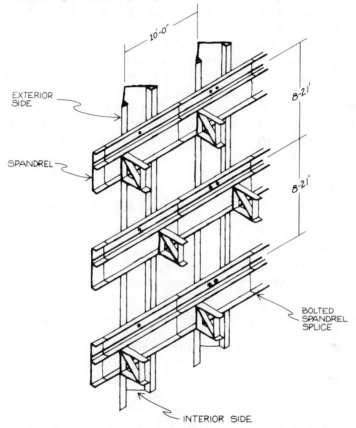

SOURCE: *ENGINEERING JOURNAL*/AISC. Second quarter 1975.

Washington, proposed the use of a circular floor plan, with two concentric
rings of columns to resist lateral loads following the tube-in-tube principle
previously mentioned. The column rings are connected by radial rectangular
prestressed concrete beams. Solid 3 in.-thick prestressed concrete planks,
topped by a 2 in. cast-in-place slab, span between them.

SECTION 6: Floor Systems

In general, floor systems for high-rise buildings are the same as for low- and
medium-rise buildings (Chapter 6). However, there are several features

which magnify themselves in the case of high-rise buildings, and they will be discussed here.

Floor systems in a high-rise building repeat themselves dozens of times. Therefore, a careful design becomes very important because of their effect on the vertical and lateral load-carrying systems. The structural depth of a high-rise floor system is very significant. Each foot of depth required for a floor accumulates to increase the overall height of a building (i.e., 1 ft × 60 floors = 60 ft of height). This means that the effect of wind and earthquake would be seriously increased for a design which calls for a 30 in. as opposed to an 18 in. floor design. Of course, the cost of the elevators, the wall cladding, and other subsystems would also be increased.

The weight of the floor systems will also add up and can affect the cost of the foundation very seriously. Furthermore, heavier floor systems require heavier columns or walls, which again adds to the material and placement cost of the superstructure as well as the weight and cost of the foundation. For heavier floor systems, inertia forces in upper floors during an earthquake are seriously increased and again affect the structure of the building.

In any high-rise building, the overturning moment becomes a significant factor. Thus it is important to design the floor systems so that a major part of the weight of the building will be carried onto the more exterior resisting elements in order to stabilize them by precompression against tensile overturning force requirements. This usually means the elimination of as many interior columns as possible, and the use of long-span floor subsystems.

Long-span floor systems also serve as horizontal diaphragms. They tie the vertical elements together and may have to be strengthened so as to effectively carry out this mission. On the other hand, if the vertical resisting elements are more-or-less evenly and symmetrically distributed throughout the plan of the building, then the floor system does not need much strengthening to perform its diaphragm action. Again note that every pound saved on one floor is multiplied many times throughout the height of the building.

In a steel building of high-rise construction, the most common floor system is the corrugated metal deck topped with a thin layer of lightweight concrete (Figure 6-20). This provides a lightweight floor that is also capable of providing for small service ducts throughout the floor for distribution of communication and power transmission wires.

In a concrete building, it will often be economical to use lightweight structural concrete construction (say, 100–115 psf). This would indicate a thin slab of 3 or 4 in. supported on a system of beams and girders. Occasionally, and particularly for apartment buildings, a prestressed flat slab design would do very well for spans not exceeding 25 or 30 ft. In such cases, the slab need be only 6 to 7 in. or at most 8 in. thick, supported directly on columns, with no beams. The weight of such a flat slab would be almost the same as slab-on-beam construction. A flat slab with no supporting beams allows minimum structural depth and also facilitates the connection and arrangement of conduits below the slab for offices. Of course, if services are not suspended, floor-to-floor height can be minimized and the slab can serve as the ceiling for floors below and therefore is economical for apartment designs.

For long-span office designs and where square floor panels are possible, the grid system for concrete floors can be very economical. Such a system can be built with a minimum quantity of concrete, but it does require a floor depth greater than the others. For example, slab-on-beam-and-girder construction allows the designer to pass the ducts in one direction between beams and then connect with other floor sections via holes in girders running in the other direction (Figure 10-12). Such ducts must be hung below in grid construction, since generally the grid beams are of equal depth in both horizontal directions. However, it is possible for air-conditioning and other ducts to be passed through holes provided through the grid beams to permit such passage in both directions. But this may be a costly arrangement to construct unless carefully incorporated in the overall design.

In rigid-frame construction, the beams and girders spanning in both directions are often required to be relatively stiff so as to be able to minimize the drift of the high-rise building floors. This means additional

FIGURE 10-12
A SLAB ON BEAM AND GIRDER LAYOUT ALLOWS MAJOR DUCTING TO PASS THROUGH A GIRDER BETWEEN BEAMS TO KEEP OVERALL FLOOR-TO-CEILING DEPTH SMALL.

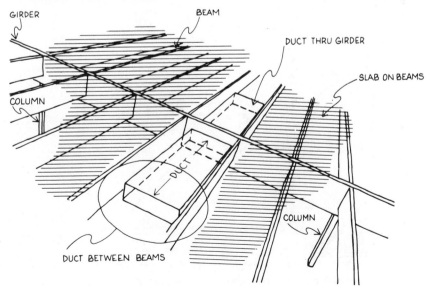

FIGURE 10-13
SEARS TOWER FLOORING SUBSYSTEM.

(a) BUNDLED FRAMED "TUBE" TRUSS FLOOR LAYOUT

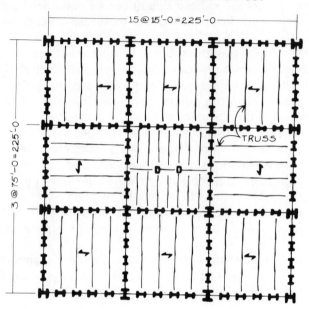

TRUSS

15 @ 15'-0 = 225'-0

3 @ 75'-0 = 225'-0

(b) COMPOSITE TRUSS ASSEMBLY

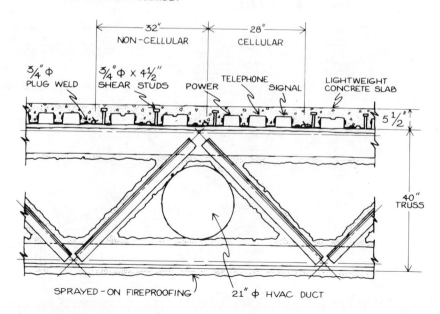

32"
NON-CELLULAR

28"
CELLULAR

3/4" φ
PLUG WELD

3/4" φ x 4 1/2"
SHEAR STUDS

POWER

TELEPHONE

SIGNAL

LIGHTWEIGHT
CONCRETE SLAB

5 1/2"

40"
TRUSS

SPRAYED-ON FIREPROOFING

21" φ HVAC DUCT

material for the beams and girders, unless the floor drift is controlled by some other vertical elements such as shear wall, shafts, and so on.

For long-span designs, trussed joist or beams can be economical for high-rise buildings, even though they require greater structural depth. The trick lies in passing mechanical ducts through the web areas of the trussed joists so that no additional depth is required beneath the bottom cord of the truss. This has been done within the nine-plan modules of the Sears Tower in Chicago Figure 10-13.

SECTION 7: Deflections, Vibrations, and Strength

Lateral deflections and vibrations become exceedingly significant because of their greater magnitude as the height of the building increases. Two main causes for lateral deflections and vibrations are wind load and seismic forces. A third factor is the temperature differential, between the shady and sunny sides on the inside and outside of a building.

The nature of deflections and vibrations as produced by wind is shown in Figure 10-14. Under a steady wind, the building deflects statically to some degree as shown, the amount depending on the wind force and the overall stiffness of the building. Then, due to wind gusts, the building will oscillate as shown in the Figure. Smaller modal deflections also cause vibrations in a building. For both perceptual and operational reasons, such deflections, oscillations, and vibrations must be limited. Too large deflections may cause elevators to be out of plumb or building floors to slope excessively. Hence, a certain amount of stiffness or rigidity is important. Practical rules limit the drift of a building on each story to a certain ratio of its height, such as 1 : 1000.

FIGURE 10-14
DEFLECTIONS DUE TO WIND LOADS.

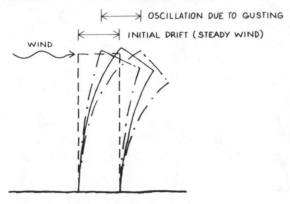

Small oscillations may not cause any mechanical disturbance, still, they may cause insecure or uncomfortable feelings among the occupants. Although it is difficult to predict human response to such oscillations, it is generally desirable to provide for stiffness in a building so that such oscillations will not be noticeable, as discussed in Chapter 8.

Seismic movements in a building are different from those produced by wind. Basic limitations have to do with allowable stresses in the structure of a building under the so-called code-specified earthquake forces. However, under catastrophic earthquakes, a building will deflect much more and, may deflect in any random direction. Then the problem is to avoid movements large enough to produce collapse. Predictions of such building movements in a high-rise structure is a complicated matter, since there are several modes of vibration. The first and second modes are shown in Figure 10-15

It should also be noted that the requirements of a high-rise building to resist wind and seismic forces may contradict. A stiff building will react favorably to wind because the amplitude of vibration is small. On the other hand, for better seismic behavior, a flexible building is often desired so that it will be out of resonance with seismic disturbances and will not result in excessive stresses. The dominant periods of vibration of earthquakes are on the order of fractions of a second, whereas a flexible high-rise building will have a period of several seconds. Hence the building period is very much out of phase with the force period, preventing a force buildup. When the fundamental period of vibration of a building gets to be several seconds, even the higher modes induced by earthquakes will not get into resonance and therefore the seismic response is limited.

The above explanation is probably oversimplified but may help to explain

FIGURE 10-15
**BENDING DUE TO SEISMIC LOADS BEGINS WITH
THE FIRST MODE BUT INCLUDES A WHIPLIKE
SECOND, THIRD, ETC.**

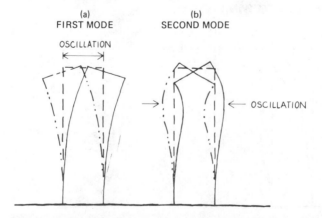

why a high-rise building cannot be *easily* designed to resist both wind and seismic forces in an optimum manner. But note that buildings *can* be designed to be rigid under wind action *and* to escape damage under code-specified earthquake forces. *In order to resist catastrophic earthquakes*, certain parts of the structure may be allowed to yield or crack in local areas, then the building's period of vibration would be lengthened and its damping increased. Thus the building will be able to resist strong seismic action without a structural failure.

In addition to the above, the so-called ductility requirement for

PHOTO 10-8
BANCO DE AMERICA BUILDING, MANAGUA, NICARAGUA, SURVIVED THE 1972 EARTHQUAKE. (T. Y. LIN INTERNATIONAL, STRUCTURAL ENGINEERS).

earthquake design, means that a building should have reserve plastic strength, beyond its elastic behavior limits, so that it could sway with the earthquake but not fail in any significant way. The design may provide for local points of yielding but will not damage the overall structural integrity.

Thus, to design a high-rise building against both wind and seismic loads, it may be desirable to devise a structural system that will be stiff under wind load or code earthquakes, but that may change by controlled yielding into a more ductile response if the earthquake forces become extraordinarily large. This was in fact accomplished by the authors in the case of the Banco de America Building, which survived the 1972 Managuan quake "with flying colors," as commented on by *Engineering News-Record* of January 11, 1973 (see Photo 10-8). This will be briefly described as follows.[2]

An intense earthquake hit Managua, Nicaragua, December 23, 1972. The major part of its downtown area, 511 blocks, was devastated with a death toll estimated at over 10,000. Managua's tallest building, the 200-ft. Banco de America, a reinforced concrete tower, was situated at epicenter, with a fault of $\frac{1}{2}''$ occurring just outside along the street. However, structural damages were limited to mid-span cracks in the beams connecting the shaft walls and they could be easily repaired. The main architectural damage was the falling of marbles from the staircase and elevator core walls, indicating high stresses in these walls, although no structural cracks were discernible. This means that seismic shears and moments were resisted by these walls within the elastic range.

The quake produced at least 0.35g horizontal force on this building which was designed in 1963 using the then existing UBC Zone 3 requirements with a horizontal force of 0.06g. Thus, the core shafts were stressed to six times the design earthquake and survived well. As expected, the weakest links were the connecting beams with large openings at center where spalling of concrete occurred.

This 200-ft.-high reinforced concrete building was designed with the central core to serve as the main earthquake resisting element. Its 38-ft.-square core is made up of four smaller cores, each only 15-ft.-square, connected to one another with beams at every floor.

The four 15-ft.-square cores were designed to act both together and separately. When acting together as one core 38-ft. square, it possesses great stiffness, having an aspect ratio of about 5. In order that each small core can effectively function as a unit by itself, all walls of each core were reinforced almost equally.

Thus a fail-safe feature was provided in that the lateral load resisting capacity of the assembly of cores was not much decreased should the connecting beams totally fail. In addition, when plastic hinges formed at the center of these beams, the dynamic response of the building to the earthquake was greatly decreased thus insuring stability and better behavior.

[2]*Engineering Bulletin*, T. Y. Lin International, 1973.

If the four cores were designed to act independently without the connecting beams, greater deflections would have occurred causing cracking and failure of other architectural elements. On the other hand, if the entire core was designed only as one piece, the failure of the connecting beams could have resulted in an overstressed structure.

A dynamic analysis of this building was made by Professor Vitelmo Bertero, University of California, Berkeley, after the earthquake. The following table summarizes the various responses of the building to the Managuan earthquake corresponding to two different structural behaviors:

	4 Shafts Together	4 Shafts Acting Separately
Period of Vibration, seconds	1.3	3.3
Base Shear, tons	2,700	1,300
Overturning Moment, ton meters	93,000	37,000
Displacement at Top of building—cm . .	12	24

It is noted from the above table that seismic forces induced in the structure were greatly reduced if the four shafts acted independently.

The preceding excerpt illustrates the possibility of designing a high-rise building to meet the requirements both of stiffness under wind and of strength under large seismic disturbances. In this connection, it may be mentioned that Professor Muto of Japan has devised a "slit-wall" system whereby shear walls are slit with weakened vertical separations so that sliding can take place along these controlled lines in case of severe earthquakes. Readers interested in drift and deflections of high-rise buildings are referred to two excellent papers: "Drift in High-rise Steel Framing" by John B. Scalzi for *Progressive Architecture* and "Deflections of High-rise Concrete Buildings," by Mark Fintel for *ACI Journal*.

SECTION 8: Weight of Structural Materials

For high-rise buildings, the total dead-load weight of structural steel or concrete will of course vary with the height of the building and the horizontal and vertical systems chosen. It can be generalized that each floor increment of the horizontal subsystem will require a certain amount of structural material to carry gravity load. This amount is more-or-less independent of the building height, but varies greatly with the span between vertical supports. For vertical subsystems, columns and walls, the dead load material will vary with the height of the building since a greater number of stories is carried by the lower columns of a high-rise building.

FIGURE 10-16

**A COMPARISON OF THE ADDED COST PAID FOR WIND
LOAD RESISTANCE IN A HIGH-RISE STRUCTURE RELATIVE
TO AN IDEALIZED DESIGN FOR VERTICAL LOAD ONLY.**

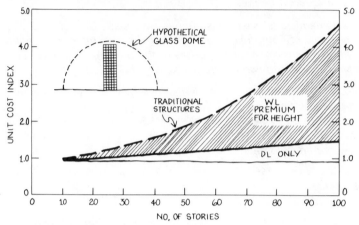

PREMIUM FOR HEIGHT DUE TO WIND LOADS

FROM FAZLUR KAHN'S ARTICLE "THE FUTURE OF HIGH RISE STRUCTURES", PROGRESSIVE
ARCHITECTURE; OCTOBER, 1972

   Then there is the additional material required to carry lateral loads and to
limit deflections. This can become very significant as the building height
increases. For example, in a 20-story building, such material may amount to
an additional 10 percent over that required for gravity load, while for 60 or
80 stories additional material for lateral-load-carrying purposes can be 80 to
100 percent or more (Figure 10-16). Of course, this is even more significant
for higher buildings, say 100 stories or more.

   Therefore, in high-rise building design, it is very important to minimize
the extra amount of material required for lateral load resistance. If the
overall shape and structural subsystems are optimized, additional material
requirements can be made more reasonable, say, to one-half or one-third of
the values indicated above. This was shown previously, in Figure 10-3, for
structural steel, and also referred to in the graph in Figure 8-1.

11
Arch, Suspension, and Shell Systems

SECTION 1: Introduction

Long-span structures can be built using flat construction such as heavy girders, trusses, space-frames, as discussed in Chapter 6. However, for spans in excess of 100 ft, it is often more economical to build a system made up of curved members, in the form of *arches*, *suspension cables*, and *thin shells*.

Philosophically, all structural designs, whether for long or short spans, are based on the same general concepts. But in practice, the controlling proportions of short-span structures are often determined by minimum available dimensions and other nonstructural requirements. Hence, such structures can often be designed by inspection or by comparison with existing buildings, without refined analysis, and incorporating only straight members. On the other hand, when one deals with long-span structures, the relationship between dead weight, strength, and proportions becomes critical, and the design effort needs to be far more rational in order to arrive at safe, economical, and aesthetic solutions.

In order to optimize dead-load efficiency, long-span structures should have their shapes approximate a natural line of pressure, such as a parabolic arch (for evenly distributed load—Figure 11-1), or a catenary (the self-weight curve of a chain). When this is done, the shear- and moment-resisting force always acts at the c.g. Hence, the use of curved forms is often efficient as well as appealing. Since there is (theoretically) only direct axial force *in* the system and the material requirements are minimized, such curved systems are efficient because they yield overall structural depth for long spans without increasing sectional depth.

In contrast, long-span design of a beam would require much more sectional depth along its entire length, or at least much of it, because the overall moment increases with L^2, deflection with L^4. Therefore, ordinary long-span beams need a lot of additional material to maintain a flat surface. With curved structures, the required moment-resisting forces (*C-C* or *T-T* in Figure 1-1) can be controlled by increasing the rise or the sag of the system *as a whole* rather than by increasing the cross-sectional depth. As the overall structural depth is maximized, the thrust of an arch or the tension in a cable, as well as local bending requirements, will be minimized.

The main objection to curved systems is that man cannot function as easily along curved surfaces as along flat surfaces. Thus, where habitable

FIGURE 11-1

**THE MOMENT-RESISTING FORCES IN AN ARCH OR SUSPENSION
SYSTEM ARE LOWER AS COMPARED WITH A BEAM SYSTEM
BECAUSE THE OVERALL DEPTH (*h*) IS GREATER.**

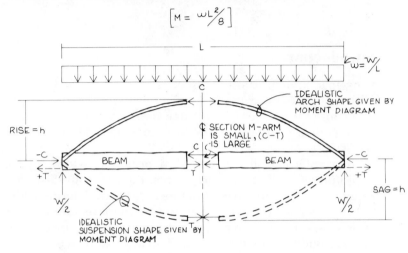

$$\left[M = \frac{wL^2}{8} \right]$$

NOTE THAT THE "IDEAL" SHAPE FOR AN ARCH OR SUSPENSION
SYSTEM IS EQUIVILENT TO THE DESIGN LOAD MOMENT DIAGRAM

surfaces are desired, an auxiliary flat system must be supported or
suspended from a primary curved structural system.

In this chapter we will discuss three types of curved systems: (1) arch
systems, (2) suspension systems, and (3) shell systems. Shell systems will be
discussed last because they employ many of the principles that will be
discussed for arch and suspension systems, and they take many different
shapes. Thus Sections 4 through 7 will deal separately with folded-plate,
cylindrical, dome, dish, and hyperbolic parabaloid shells.

SECTION 2: Arch Systems

Since the beginning of history, mankind has tried to span distances using
arch construction. Essentially this was because an arch required materials
resisting only compression, and large quantities of materials like stone or
mud for bricks were readily available. Later, fired brick, concrete, and steel
were produced and utilized.

The basic issues of statics in arch design are illustrated in Figure 11-2. A
uniform load, (*w*) units per linear foot, is supplied along the projected
horizontal length of the arch. Due to symmetry, the vertical component of
the end reactions is $V = wL/2$ (Figure 11-2a). Note that this load reaction is

FIGURE 11-2
THE STATICS OF A THREE-HINGE ARCH.

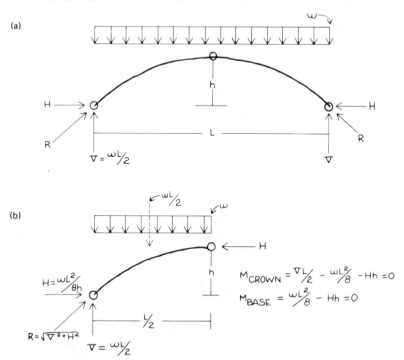

the same as for a simple beam and, similarly, there is no shear across the midspan of the arch, as illustrated in the one-half arch freebody (Figure 11-2*b*). Taking moment about the crown, we obtain

$$M_{resist} = Hh = \frac{wL}{2}\left(\frac{L}{2} - \frac{L}{4}\right) \qquad \text{and} \qquad H = \frac{wL^2}{8h}$$

where (H) is analogous to the *C-T* forces in a beam and (h) is the overall height of the arch. Obviously, (h) is much larger than (d) for beam design and $H \ll C\text{-}T$.

Since equilibrium requires that H be constant across the arch, a parabolic curve would theoretically result in no moment on the arch section. The resultant follows the *natural line of pressure*, and the reaction at the supports is given as $R = \sqrt{H^2 + V^2}$.

Figure 11-2 illustrates a symmetrical three-hinged arch, with one hinge at midspan and two at the ends. Under these conditions, there can be no moment at the hinges and the analysis previously presented is rigorously correct. If the arch has only two hinges, or if it has no hinge, as illustrated in Figures 11-3 and 11-4 respectively, then bending moments may exist

FIGURE 11-3

A TWO-HINGED ARCH WOULD CARRY SOME LOAD IN BENDING AND SOME IN ARCH ACTION.

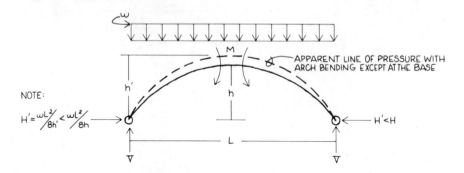

FIGURE 11-4

A NO-HINGE ARCH WOULD CARRY EVEN MORE LOAD IN BENDING DUE TO MOMENTS AT THE BASE.

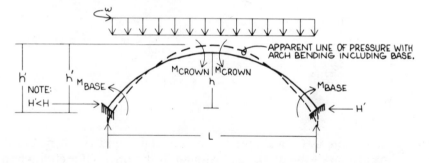

either at the crown or at the ends of the arch, or at both places. In these cases, the horizontal reactions will be somewhat different from a three-hinged arch, even though the shape is parabolic due to deflection causing bending. However, because the section depth is usually small relative to the overall depth, the deflected difference will usually be small enough so that, for a preliminary design, the calculations for reactions in Figure 11-2 can still be applied. For rectangular sections, a rough comparison of bending versus arch action is indicated by $2h^2/d^2$ where, h = arch rise and d = thickness.

Since the force H varies inversely to the rise h, it is obviously desirable to use as high a rise as feasible. For a combination of aesthetic and practical considerations, a span/rise ratio ranging from 5 to 8, or perhaps as much as 12, is frequently adopted. As the ratio goes higher, there is a local buckling problem because the sections required for axial resistance of H forces will normally be very slender relative to the arch length (i.e., ~1/50). As a result, higher ratios would require an increased sectional depth, and the arch advantage diminishes.

We have said that the ideal shape of an arch should be such that it will follow a natural line of pressure *as determined by the vertical load distribution*. If so, the arch would theoretically sustain no bending moment and would be in direct compression along its entire arc. As a start, it can often be assumed that the statical load on the arch is uniform and thereby arrive at a shape of a parabola. This is why, in theory, a three-hinge parabolic arch under uniform load would be assumed to have no bending moment along its entire length. Only if the arch has less than three hinges would small local bending moments exist as mentioned above.

However, in practice, it should be noted that an arch is not necessarily subjected to a uniform load. First, there is usually a difference in the *depth* (and thus the weight) of the arch per foot of span. Then the weight per foot varies again because the *inclination* of the arch rib itself varies. Furthermore, live load may act on portions of the arch so that the line of pressure will not conform to any predetermined shape that considers only the dead load. Therefore, in final design, it is always necessary to design the arch to carry a certain amount of bending moment to allow for a changing live load if nothing else.

The design of an arch to resist bending moment is a complicated problem, but it can be approximated, particularly if the dead load is heavy and the live load relatively light. For example, Figure 11-5 shows a pressure line produced by the combination of dead and live load. This line deviates from the arch axis itself and, *in comparison to h*, the amount of deviation is a measurement of the relative moment to be taken in bending. Note that, due to its curved shape, moments in an arch of any shape will always be much less than the moment that would develop in a simple beam spanning the same distance.

Since the greatest total force in the arch (R) is at its support, the section of the arch is often largest at that point. The crown requires a smaller section because it is subjected essentially to the horizontal force H, which is smaller than the total reaction $(R = \sqrt{H^2 + V^2})$ at the support. When the force is known, for a given arch profile, the cross-sectional area of the arch

FIGURE 11-5
LINE OF PRESSURE AND ARCH MOMENT.

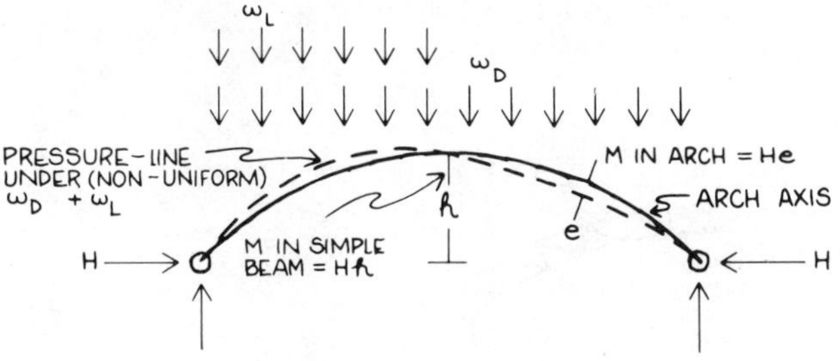

is obtained by dividing that force by an average unit stress applicable to that arch material. However, for arch design, the allowable unit stress is set at a rather low value so as to allow for any possible moment that might exist at such a section. Thus, *when the moments in the arch are not considered* (i.e., design for axial forces only), a nominal factor of safety of five or more is usually assigned to the ultimate compressive strength of the material used. When the bending moments are carefully considered, taking into account various loading conditions, then the allowable fiber stresses can be much higher, say with a factor of safety of perhaps only two or three, as in flexural design for beams.

In modern arch construction, materials such as reinforced concrete or steel are used. Since these materials can resist tension as well as compression, they can carry sizable local moments and arch ribs can be quite slender compared with the span. But it is often necessary to stiffen the ribs against overall buckling in the vertical *and horizontal* planes.

A *steel* arch can be vertically stiffened with a *girder* or trussed arch design (Figure 11-6). Figure 11-7 (elevation) illustrates that vertical stiffening can also be achieved by use of a girder or truss along the deck level. The structural depth of a stiffening *truss* ranges between $\frac{1}{40}$ and $\frac{1}{80}$ of the arch span L, whereas a stiffening girder can be somewhat shallower. Note that the Figure 11-7 arch ribs must be stiffened laterally by some kind of a curved truss serving as a diaphragm, along the plane of the ribs, or by rigid framing across sections perpendicular to the rib (Figure 11-7). To resist the horizontal thrust at the ends of an arch, heavy abutments on solid ground or pile foundations must be built. As an alternative solution, the two ends can be tied with a horizontal member as indicated in Figure 11-7.

To provide a long-span building with an arch roof, a barrel arch system is sometimes constructed (Figure 11-8). As we shall see in Section 4, a static

FIGURE 11-6
ARCH RIB STIFFENED WITH GIRDER OR TRUSS.

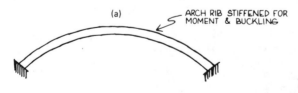

(a)   ARCH RIB STIFFENED FOR MOMENT & BUCKLING

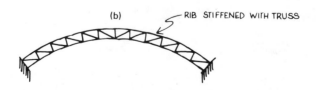

(b)   RIB STIFFENED WITH TRUSS

FIGURE 11-7
ARCH RIB HORIZONTALLY STIFFENED BY BRACING.

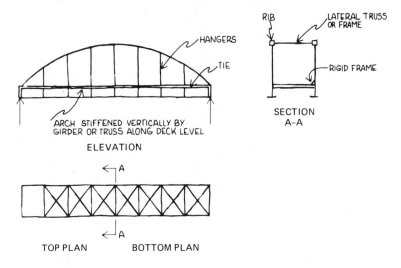

HANGERS

TIE

ARCH STIFFENED VERTICALLY BY
GIRDER OR TRUSS ALONG DECK LEVEL

ELEVATION

RIB

LATERAL TRUSS
OR FRAME

RIGID FRAME

SECTION
A-A

TOP PLAN   BOTTOM PLAN

PLAN

FIGURE 11-8
ARCH BARRELS.

(a) LONG BARREL   SHORT BARREL   (b) MULTIPLE BARREL

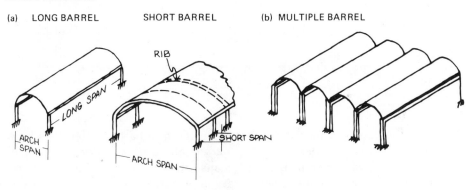

RIB

LONG SPAN

ARCH
SPAN

SHORT SPAN

ARCH SPAN

(c) STIFFENING MEMBERS REQUIRED AT ENDS AND EDGES

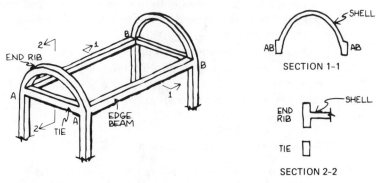

END RIB

A

B

B

TIE

EDGE
BEAM

A

SHELL

AB   AB

SECTION 1-1

END
RIB

SHELL

TIE

SECTION 2-2

FIGURE 11-9
MULTIWAVE ARCH SHELL.

MULTIWAVE SHELLS

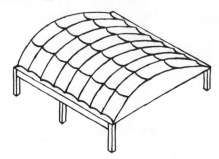

FIGURE 11-10
SCHEMATIC OF GARAGE AND HOTEL BUILDING SUSPENDED OVER MULTILANE HIGHWAY.

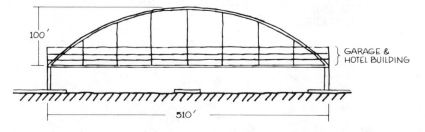

analysis for the schematic design of a barrel system can be made similar to that of *a beam for long barrels* and a *rib arch for short barrels*. This is the case for short barrels because the overall bending forces in the short direction are of secondary importance where the opposite may be true for long barrels. Short barrels are essentially like a continuous arch supported at the edges. When a short barrel arch span becomes long, it is possible to minimize buckling effects by providing a grooved section for the arch (Figure 11-9). This type of construction becomes economical for spans of 200 ft and perhaps up to 1000 ft.

Arch design concepts discussed above can be applied to building design in many ways. Figure 11-10 illustrates a proposed arch design for a megastructure using air rights over a multi lane highway. Example 11-1 will now show how to make an approximate design for the arch components.

Example 11-1:   Design of an Arch

A long arch 100 ft high and spanning 510 ft is to be designed for a garage and hotel building, using air rights over roads and highways (Figure 11-10).

It is necessary to determine preliminary dimensions for the size of the arch section. The arches are spaced 60 ft on centers and carry four-story loading totaling 27 k/ft length along each arch.

Solution

$DL + LL$ for all floors estimated on a psf basis at	27 k/ft
Arch own weight estimated as a start at	6 k/ft
(~ 25 percent since loads are heavy and span is long)	
Total load	$\overline{33 \text{ k/ft}}$

$$H = \frac{wL^2}{8h} = \frac{33 \times 510^2}{8 \times 100} = 10{,}700\ k$$

$$V = \frac{wL}{2} = \frac{33 \times 510}{2} = 8{,}400\ k$$

$$R = \sqrt{H^2 + V^2} = 13{,}600\ k$$

Use concrete-filled steel pipe for arch section:
Try $\frac{1}{2}$-in. steel plate at 6 ft ϕ, $A_s = 72\pi \times \frac{1}{2}$ in. $= 113$ in.^{2}
$A_c = \pi 3^2 \times 144 = 4070$ in.^{2}
For a first approximation and assuming arch is braced by floors, allow $f_s = 20$ ksi and $f'_c = 5$ ksi; $f_c = 2.5$ ksi (including effect of confinement within pipe and the addition of some re-bars). Note that the *strength of confined concrete* can be two to three times that indicated by the f'_c value.

$$\begin{aligned}
\text{Capacity } A_s &= 113 \times 20 \text{ ksi} &= 2{,}260\ k \\
\text{Capacity } A_c &= 4{,}070 \times 2.5 \text{ ksi} &= \underline{10{,}180\ k} \\
& \text{Total} &= \overline{12{,}440\ k}
\end{aligned}$$

which is o.k. for 10,700 k at crown, but not quite for abutments at 13,600 k. This process of trial and error will be continued with changes in the size of the arch diameter, strength of steel and concrete, a new estimate of arch weights, and the like until a reasonable answer is obtained as required for the particular level of the design effort.

SECTION 3: Suspension Systems

A suspension cable is the reverse of an arch rib, using materials under tension instead of compression. When materials taking tension are available, it is often more economical to use a suspension instead of an arch system. The suspension system has the advantage of not being subject to buckling, and overall span-to-depth ratios of up to about 10 are possible. However, it does have to be stiffened to avoid excessive flexibility under dynamic and partial loadings.

Up to now the longest spans are for bridges built of high-strength steel cables, because erection methods have been worked out for suspension bridges to eliminate the need for falsework. Here, loads are heavy and the span-to-depth ratio tends to be lower than 10 to reduce anchor forces. But, there have been innovative applications to buildings such as the Minneapolis Federal Reserve Building (See Photo 11-1) and (Figure 11-11). In buildings where the use of suspension systems is limited to roof structures the ratio can be higher because the loads are much less and falsework is often possible.

Figure 11-12a illustrates a typical suspension bridge. If we isolate one-half of the cable in the main span (L) as a free body (Figure 11-12b), we observe that the horizontal component in the cable can be approximated by using the arch formula:

$$H \times h = \frac{wL^2}{8} \quad \text{and} \quad H = \frac{wL^2}{8h}$$

PHOTO 11-1
STRUCTURAL DESIGN OF THE FEDERAL RESERVE BANK BUILDING, MINNEAPOLIS. (GUNNER BERKERTS AND ASSOCIATES, ARCHITECTS; SKILLING, HELLE, CHRISTIANSEN, ROBERTSON, STRUCTURAL CONSULTANTS).

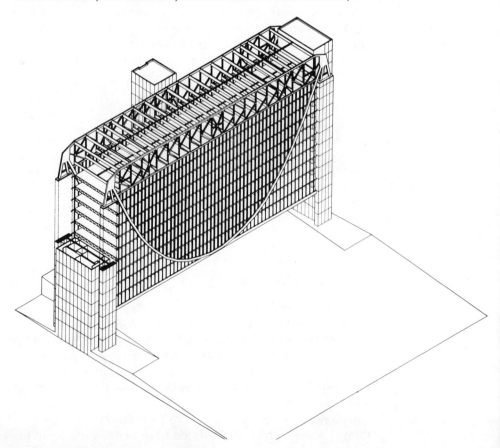

FIGURE 11-11
FEDERAL RESERVE BANK BUILDING IN MINNEAPOLIS AS CONCEIVED FOR POSSIBLE EXPANSION BY USING AN ARCH TO CARRY ADDITIONAL FLOORS.

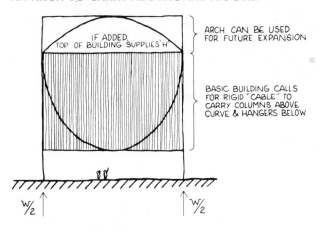

FIGURE 11-12
TYPICAL SUSPENSION BRIDGE SYSTEMS.

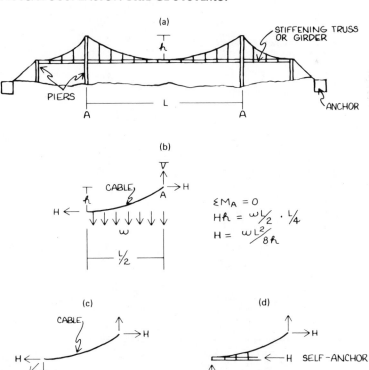

Thus for *both* an arch or suspension system, the horizontal component varies inversely with the rise or sag *h* and increases directly with the uniform load intensity *w* and the second power of the span length *L*. The main difference between a suspension layout and an arch is that an arch may be able to take some bending moment along its length by virtue of the stiffness of the thicker arch rib, whereas the very slender suspension cable adjusts itself to the loading condition and affords practically no bending resistance. Thus if no stiffening is given to the cable, it will change its curvature and move with changes in the loading conditions. In order to avoid that, a stiffening girder or truss is usually provided along the length of

FIGURE 11-13
AN INCLINED ARCH CAN CARRY A SUSPENSION CABLE ROOF SUBSYSTEM.

(a)

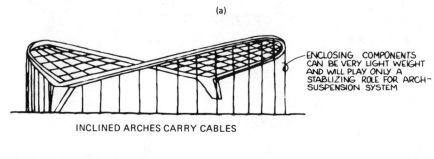

ENCLOSING COMPONENTS CAN BE VERY LIGHT WEIGHT AND WILL PLAY ONLY A STABILIZING ROLE FOR ARCH-SUSPENSION SYSTEM

INCLINED ARCHES CARRY CABLES

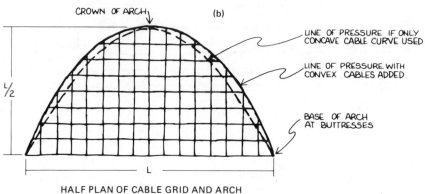

CROWN OF ARCH

(b)

LINE OF PRESSURE IF ONLY CONCAVE CABLE CURVE USED

LINE OF PRESSURE WITH CONVEX CABLES ADDED

BASE OF ARCH AT BUTTRESSES

$L/2$

L

HALF PLAN OF CABLE GRID AND ARCH

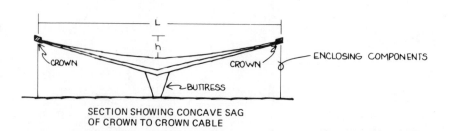

L

h

CROWN

CROWN

ENCLOSING COMPONENTS

BUTTRESS

SECTION SHOWING CONCAVE SAG OF CROWN TO CROWN CABLE

a bridge deck. It can also be provided along the cable itself, although this is seldom done because it is not as economical or as easily erected.

The horizontal component in the cable must be anchored or resisted by some other components of the structure. For example, Figure 11-12c indicates that the cable at the end span can be balanced over a pier but must finally be anchored into the ground to resist both the vertical and the horizontal components of the force in the cable. In order to compromise the problems of tower height and anchor requirements, bridges normally have a span-to-sag ratio of about 7 to 10.

Figure 11-12d illustrates that the horizontal force can also be returned into the deck itself to put it under compressive force H. This is known as a self-anchored suspension span and is sometimes used for bridges. The Minneapolis building uses self-anchoring into a compression and stiffening truss at the top of the building.

When there is a *two-way network* of cables, it is possible to carry the horizontal component of the suspension cables into an arch system (Figure 11-13). Here the horizontal and (possibly) the vertical forces from the cables can be transferred to produce a compression in the inclined arches that is eventually carried into two abutments. Grid design will be discussed further in Section 7.

The cable-stayed system, Figure 11-14, may be considered as a variation of the regular (curved) suspension system in that the cables are straight and the deck is utilized to span between the stays and carry the compression, similar to a self-anchored suspension span. Here the overall structure behaves very much like a cantilever. The basic calculations for such a system can be approximated as in Figure 11-14b. If the total tributary load to a given cable is W, the horizontal component and the stress in the cable can be determined by θ as shown ($H = V \cot \theta$). This will yield an approximate size for the cables and also the compressive force to be resisted by the deck. The concept can be extended as illustrated by the Ruck-a-Chucky Bridge designed for the Auburn Dam Project, California (Photo 11-2) and (see Appendix G-2).

FIGURE 11-14
TYPICAL CABLE STAYED ROOF SYSTEM.

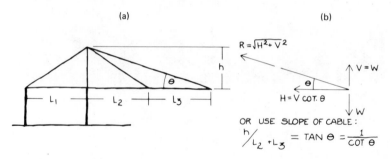

(a)   (b)

$$R = \sqrt{H^2 + V^2}$$

$$V = W$$

$$H = V \cot \theta$$

$$W$$

OR USE SLOPE OF CABLE:

$$\frac{h}{L_2 + L_3} = \tan \theta = \frac{1}{\cot \theta}$$

PHOTO 11-2

THE RUCK-A-CHUCKY BRIDGE, A 1300-FT SPAN ON A 50° CURVE IS HUNG FROM THE MOUNTAIN SIDES WITH STAY CABLES THAT EXERT DIRECT COMPRESSION FORCES INTO THE BRIDGE DECK BY THEIR POSITIONING AND TENSIONING PROCEDURES. (T. Y. LIN INTERNATIONAL, STRUCTURAL ENGINEERS; HANSON ENGINEERS, GEOTECHNIC CONSULTANTS; SKIDMORE, OWINGS & MERRILL, ARCHITECTURAL CONSULTANTS. RENDERING BY POPULAR SCIENCE MAGAZINE).

EXAMPLE 11-2: Design of a Suspension Subsystem

The Federal Reserve Building in Minneapolis (Figure 11-11), an 11-story
building, has to span 273 ft, supported on columns at the ends only. A load
of 30 kips per linear foot is carried by each hanging cable. The cable has a
rise of 150 ft as indicated by the proportions of the building. It is necessary
to compute the maximum force in the cable to determine the cross-
sectional area required for the cable and for the top steel truss.

Solution

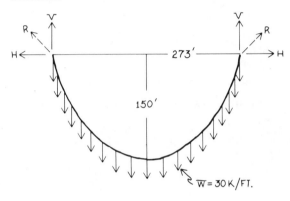

$$H = \frac{wL^2}{8h} = \frac{30 \times 273^2}{8 \times 150} = 1870 \text{ k}$$

$$\bar{V} = \frac{wL}{2} = \frac{30 \times 273}{2} = 4100 \text{ k}$$

$$R = \sqrt{H^2 + V^2} = 4500 \text{ k}$$

Use cables at 240–270 ksi ultimate strength.

If not prestressed, $f_s = 60$ ksi (because a cable is very flexible and, at high
stress, deflections large),

$$A_s \text{ required} = \frac{4500}{60} = 75 \text{ in.}^2$$

If a prestressed Concrete Catenary were used, $f_s = 150$ ksi (much stiffer
because concrete acts to make deflections smaller),

$$A_s \text{ required} = \frac{4500}{150} = 30 \text{ in.}^2$$

Note H is resisted by steel truss at that level, and eventually could be
counteracted by thrust from added arch on top floors (Figure 11-11).

EXAMPLE 11-3: Design of a Cable Stayed Roof Subsystem

A cable-stayed steel roof has typical interior bents as shown, spaced at 40-ft
centers. Steel joists span between the bents. Obtain a 36-in. W section
suitable for the bottom chord of member (C–B).

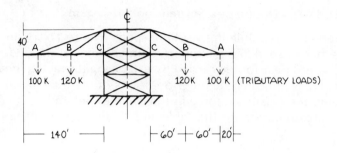

Solution

Assume: $DL = 30$ psf

$\underline{LL = 20\text{ psf}}$

$D + L = 50$ psf, $\times$ 40 ft width = 2 k/ft of cantilever.

Axial load: at A, 2 k/ft $\times$ (30 ft + 20 ft) = 100 k, $H = 100 \times \dfrac{120}{40} = 300$ k

 at B, 2 k/ft $\times$ (30 ft + 30 ft) = 120 k, $H = 120 \times \dfrac{60}{40} = \underline{144 \text{ K}}$

 Axial compression in BC = 444 k.

If $f_s = 18$ ksi, A_s required $= \dfrac{444}{18} = 25$ in.^{2}

Bending: Assume maximum $-M = \dfrac{wL^2}{12} = \dfrac{2 \times 60^2}{12} = 600$ kip-ft

To resist $-M$, assume depth of C–B is 1/20 of span (arm = 2.8 ft), and $\dfrac{5}{6}$ of

600 kip-ft = 500 kip-ft taken by flanges,

$$T_{\text{flange}} = \frac{500}{2.8} = 180 \text{ k, at 20 ksi, } A_s = \frac{180}{20} = 9 \text{ in.}^2$$

Select section:

Two flanges $A_s = 18$ in.^{2} $\Big\}$ for M, (note ratio is 2 : 1)

Web, assume $= 9$ in.^{2}

Axial load $= 25$ in.^{2}

Total $A_s = \overline{52 \text{ in.}^2}$

52 $\times$ 3.4 lb/in^2 ft = 177 lb/ft, try W 36 $\times$ 182.

SECTION 4: Folded-Plate and Cylindrical Shells

As illustrated in the previous sections and elsewhere, it is often economical to increase the overall structural depth of load-carrying systems. For a roof system, it is possible to fold a thin flat plate and develop a currugated section (Figures 11-15a and 11-15b).

FIGURE 11-15
**OVERALL DESIGN OF FLAT PLATE AND CYLINDRICAL SHELL IS
ANALOGOUS TO THAT OF A BEAM.**

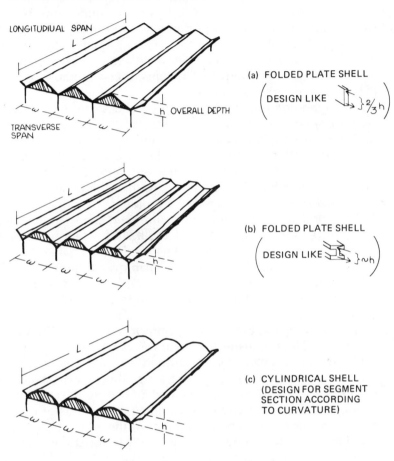

LONGITUDIUAL SPAN

L

(a) FOLDED PLATE SHELL

$\left(\text{DESIGN LIKE} \quad \right\} \tfrac{2}{3}h\right)$

h OVERALL DEPTH

TRANSVERSE
SPAN

(b) FOLDED PLATE SHELL

$\left(\text{DESIGN LIKE} \quad \right\} \sim h\right)$

(c) CYLINDRICAL SHELL
(DESIGN FOR SEGMENT
SECTION ACCORDING
TO CURVATURE)

A number of folded-plate configurations are possible. Essentially it is
necessary to get an optimum *overall* structural depth, to span the long
distance (*L*) between supports, and at the same time have short transverse
distances for the flat plates that span between the folds. In this manner, the
plate can be quite thin relative to the overall span, whether made of steel,
timber, or concrete. Essentially, longitudinal design is analogous to that of a
narrow beam. Transverse design is analogous to a continuous flat slab
supported *at the folds* or a series of arches spanning between the valleys.

Figure 11-15c shows a cylindrical shell. This is better than folded plates in
that a thin curved shell can span a longer transverse distance, allowing
fewer vertical supports; and it yields greater depth, which enables it to also
span the longer longitudinal distance between supports. The reason for this
is that the lever arm is longer for longitudinal design (see Appendix C-E on

shell-arch profile properties), and there will be less transverse moment due to curved arch action.

When folded-plate or cylindrical shells are made of concrete, their thickness is usually 3 or 4 in. This is the minimum allowable for on-site construction to give sufficient protection to the reinforcing or prestressing steel. This is usually sufficient for very long spans, since the primary (membrane) stresses are generally small and a thicker shell is not necessary. As a result, concrete thin shell systems of considerable span are relatively light and therefore economical.

Although folded-plate and cylindrical shells can theoretically span up to 200 ft or more, the actual spans are usually limited to about 100 ft or 150 ft. Roughly speaking, a conventionally reinforced concrete folded plate can easily span 50 ft, but when stretched over a span of 80 ft or more, deflections may become objectionable. On the other hand, if such a shell is prestressed to balance its own weight, then the deflections can be controlled and may no longer be a problem.

Cylindrical shells are inherently stiffer and can span longer, say up to 100 ft for reinforced concrete and about 200 ft when prestressed. The span/depth ratio will vary depending on the material and the method of construction. Generally speaking, a span/overall depth ratio should not exceed 10, or at most 15, as recommended for ordinary reinforced concrete construction. But a larger ratio—say, up to 20—can be used for prestressed construction.

There should always be *end diaphragm action* to stiffen the plate or shell sections and help transfer the shear forces (Figure 11-5a, b, c). These forces act both transversely and vertically and tend to cause flattening of the end sections if there are no lateral restraints. This may be achieved by employing solid diaphragms, rigid frame, or tied arched rib design over the supports. As long as sufficient transverse stiffening at the ends of these plates and shells is achieved, the choice is open. At the edge of these shell systems, there may also be a need for edge beams to resist the arch action thrust, which also tends to cause flattening of the shell.

The principles of design for prestressed concrete plates and shells are illustrated in Figure 11-16. Essentially the cables are draped in the plane of the plate or shell and prestressed to balance the weight of the shell roof. Taking each shell as a whole, we can again use the simple formula $F = wL^2/8h$, where the height (h) is measured vertically from the centroid of the steel to the centroid of the concrete section, as illustrated.
As a rough check on the sufficiency of the concrete section to sustain the prestress force F, one may assume no moment exists and obtain the average prestress, $f_{avg} = F/A_c$. For concrete construction, if this average value does not exceed 500 psi, it is considered a good solution. Often this average stress is as low as 200 or 300 psi and occasionally can be allowed to go up to about 700 psi.

Cylindrical shells can have transverse sections that are concave or convex and may not require edge beams, although often they will. This detail

FIGURE 11-16
PRESTRESSING FOLDED PLATES AND CYLINDRICAL SHELLS OF SIMPLE SPANS.

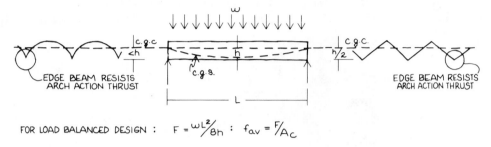

FOR LOAD BALANCED DESIGN : $F = \dfrac{\omega L^2}{8h}$: $f_{av} = \dfrac{F}{A_c}$

(Figure 11-16) has to be worked out to fit individual requirements, but in every case cylindrical shells will need end diaphragm action to provide overall stiffness and strength.

It is possible to use folded-plate and cylindrical shells to cantilever quite a distance. But concrete cantilevers of this type should be prestressed to avoid excessive deflections when the cantilever exceeds 30 or 40 ft. The basic calculation of a prestressed cantilever shell is shown in Figure 11-17, using the formula $F = wL^2/2h$. When prestressed to balance its dead weight, the cantilever should have zero or limited deflection; therefore, the secondary or local bending stresses produced by deflection are negligible. The deflections under live load should be estimated and controlled. But they are usually not too serious, unless the structure is too slender.

An interesting example of a 90-ft prestressed concrete cantilever shell is the grandstand roof for the Caracas National Racetrack (Photo 11-3). This shell is 3 in. thick and posttensioned in two directions: across the transverse catenary curve section of the shell and longitudinally in the edge beams that carry the cantilever loads together with the 3-in. shell acting in compression. The longitudinal prestress design is illustrated in Example 11-4.

FIGURE 11-17
OVERALL DESIGN OF A P.C. CANTILEVER SHELL.

(a) ELEVATION OF EDGE BEAM   (b) SECTION A-A THRU
                                                   EDGE BEAMS AND SHELL

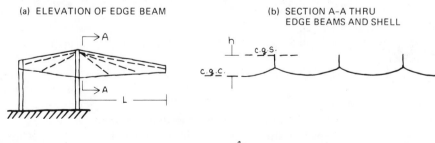

1. PRESTRESSING FORCE AT A–A ⇨ $F = \dfrac{wL^2}{2h}$ (FOR CANTILEVER)
2. NOTE THE ANALOGY TO UPSIDE DOWN T-BEAM ACTION

PHOTO 11-3
GRANDSTANDS FOR HIPPODROMO CARACAS, VENEZUELA. THREE-INCH SHELL CANTILEVERS 90 FT. (ARTHUR FROEHLICH, ARCHITECTS; T. Y. LIN INTERNATIONAL, STRUCTURAL ENGINEERS).

EXAMPLE 11-4: Design of The Hippodromo Cantilevers

This example will illustrate the basic calculation for a typical shell of the Caracas grandstand.

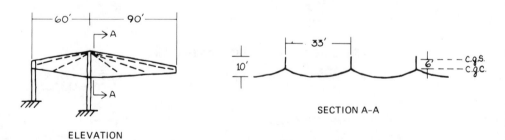

ELEVATION

SECTION A-A

Solution

As a trial section, try to prestress to balance *DL* only.

Weight of shell 3 in. at 33 ft and 150 pcf	= 1240 lbs./ft
Weight of beam, 12-in. thick, average 5 ft high =	750 lbs./ft
Total, say =	2 k/ft

$$-M_{max} = \frac{2 \times 90^2}{2} = 8100 \text{ kip-ft}$$

For arm = 6 ft, $T = \frac{8100}{6} = 1350 \text{ k}$

$A_s = \frac{1350}{150 \text{ ksi}} = 9 \text{ in.}^2$ of tendons

For $A_c = 2000 \text{ in.}^2$, $f_{avg} = \frac{1350 \text{ k}}{2000} = 675 \text{ psi}$, about right.

SECTION 5: Dome Shells

A dome can be visualized as an arch being revolved about its vertical axis to form a spheroid surface (Figure 11-18a). A dome can also be visualized as an arch translated over another arch (Figure 11-18b). In either case, the primary action in a dome is somewhat similar to that in an arch. Ideally, the main compressive force starts from the top of the arches and goes down along the radial or orthogonal ribs to the support system. But, as is the case for simple arch design, dome arches do not always coincide with the ideal line of pressure determined by the load distribution on each arch unit. Therefore, the arches will bend unless the dome surface is capable of providing *circumferential hooping action* to resist local bending (Figure 11-18c). When this horizontal hooping action is present, the surface behaves as a membrane and domes can be much thinner than simple arches.

Figure 11-19a shows a radial ribbed spherical dome that carries a vertical load. The line of pressure must be redirected by horizontal forces to make it follow the radial ribs of the dome. In this case, the necessary horizontal forces are supplied by hoop compression at various levels (rather than by local bending resistance, which is very limited). Finally, the edge forces are transmitted into a support subsystem along the bottom edge of the dome. Because these *edge forces will be tangential* to the shell surface, the support components can be of buttress type, as in Nervi's Olympic building (Photo 11-4). But the supports may also be vertical, as long as a *tension ring* is built to resist the outward thrust of the radial ribs. Note that either buttress or tension ring action will be required at the edge, whether the surface hoops above the edge are in tension or in compression.

Dome surfaces behave this way because they are not "developable." This means they cannot flatten out under load without tearing apart (Figure 11-19b). Thus, they are inherently efficient and can be employed to reduce dead load to a minimum. But note that a dome will have the least weight if it carries only the membrane compression along the arches (i.e., when the possibility of bending moment is minimized as far as possible). Shape is the important factor and, for a given dead-load condition, it is possible to have the radial system of ribs (or a grid of arches) of a shape such that the

FIGURE 11-18
DOMES GENERATED BY ROTATION OR TRANSLATION.

(a) SURFACES OF ROTATION

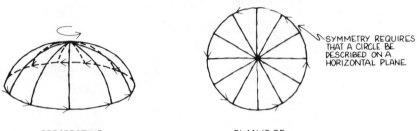

PERSPECTIVE
(ARCH IS ROTATED)

PLAN IS OF
RADIATING ARCHES

SYMMETRY REQUIRES
THAT A CIRCLE BE
DESCRIBED ON A
HORIZONTAL PLANE

(b) SURFACES OF TRANSLATION

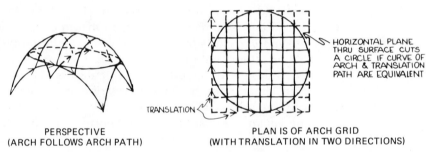

HORIZONTAL PLANE
THRU SURFACE CUTS
A CIRCLE IF CURVE OF
ARCH & TRANSLATION
PATH ARE EQUIVALENT

TRANSLATION

PERSPECTIVE
(ARCH FOLLOWS ARCH PATH)

PLAN IS OF ARCH GRID
(WITH TRANSLATION IN TWO DIRECTIONS)

(c) HOOPING ACTION ALLOWS A DOME SURFACE
TO ACT AS A THIN MEMBRANE

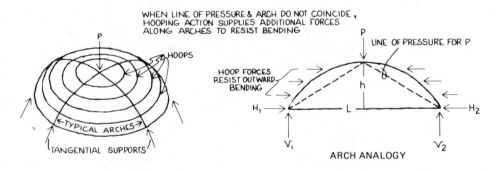

WHEN LINE OF PRESSURE & ARCH DO NOT COINCIDE,
HOOPING ACTION SUPPLIES ADDITIONAL FORCES
ALONG ARCHES TO RESIST BENDING

P
HOOPS

HOOP FORCES
RESIST OUTWARD
BENDING

TYPICAL ARCHES
TANGENTIAL SUPPORTS

P
LINE OF PRESSURE FOR P

h

H_1 L H_2

V_1 V_2

ARCH ANALOGY

resultant *DL* natural pressure line follows the surface curvature without producing any bending moment. If this is done, no circumferential hooping force is required around the dome (except possibly at the edge) to carry dead load. When live-load conditions are small compared to dead load, only small hooping resistance is required and the design is made simpler.

FIGURE 11-19
HOOPING ACTION IS ESSENTIAL FOR MOST DOMES.

(a) HOOPING ACTION COMPRESSIVE WHEN LINE
    OF PRESSURE ABOVE SURFACE

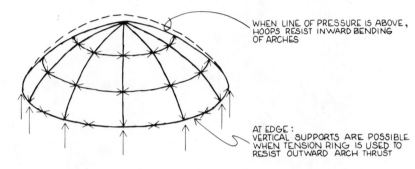

WHEN LINE OF PRESSURE IS ABOVE,
HOOPS RESIST INWARD BENDING
OF ARCHES

AT EDGE:
VERTICAL SUPPORTS ARE POSSIBLE
WHEN TENSION RING IS USED TO
RESIST OUTWARD ARCH THRUST

(b) HOOPING FORCES RESULT BECAUSE
    DOMES ARE NOT DEVELOPABLE

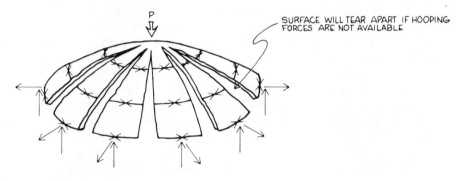

SURFACE WILL TEAR APART IF HOOPING
FORCES ARE NOT AVAILABLE

(c) SHAPE MAY BE FAR AWAY FROM NATURAL LINE OF
    PRESSURE IF HOOP FORCES ARE ADEQUATE

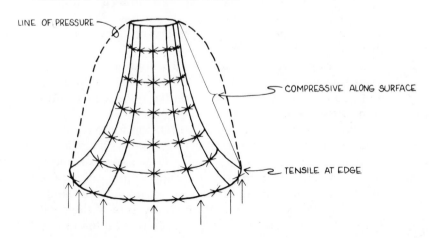

LINE OF PRESSURE

COMPRESSIVE ALONG SURFACE

TENSILE AT EDGE

PHOTO 11-4
OLYMPIC ARENA, ROME. (PIER LUIGI NERVI, DESIGNER).

Figure 11-19c shows a very unusual "dome," having the shape of a fisherman's hat. Actually, this shape is more suitable for a suspension design and does not carry loadings in the most efficient manner when supported from below. However, if base support is required for special functional or aesthetic reasons, it is possible to make it work structurally by adding heavy circumferential rings, most of which will carry compressive forces. Hence, aside from efficiency, the shape of the dome can be modified almost at will by using relatively heavy circumferential rings to control local bending tendancies.

We have now seen that a dome can be constructed of a system of arch and hoop components. Depending on the shape, the hoops above the edge may be either in tension or in compression. Further, the amount of hooping forces will depend on the degree to which the natural pressure line is approximated by the arches. If spaced fairly close together, each of these hoops may not have to take a lot of tension or compression force, and the local bending between rings is minimized. However, you should understand that, for partial or uneven loadings, larger forces can be produced in certain of these rings. Of course, the ideal would be a dome shell with a surface that can act as continuous radial ribs (or as a continuous "grid" of arches), and also as continuous circumferential rings, which can take both compression and tension.

A simple analysis will reveal the tension force (T) in a tension ring at the base of any given dome. Figure 11-20a shows a section of a dome with a circular base of diameter D. Figure 11-20b shows a half plane of the tension ring at the bottom of the dome. Using simple membrane theory, we can *assume* only membrane action, that is, no surface bending moment and radial edge forces tangent with the curvature of the shell surface. Hence, the edge forces can be resolved into a vertical load force (V) and a

FIGURE 11-20
FORCE IN TENSION RING.

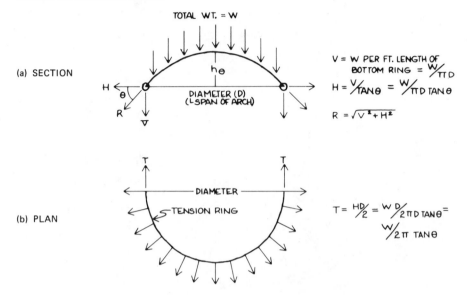

(a) SECTION

(b) PLAN

horizontal thrust (H). The angle of the slope at the edge of the dome is θ as shown. If the total weight of the dome is computed as W, the forces (V and H per foot of circumference) at the bottom of the dome can be computed:

$$V = \frac{W}{\pi D} \text{ (for vertical support design)},$$

$$H = \frac{V}{\tan \theta} = \frac{W}{\pi D \tan \theta} \text{ (for ring beam design), or}$$

$$R = \sqrt{H^2 + V^2} \text{ (for buttress design)}.$$

From Figure 11-20*b* it can be seen that the tension in the circular ring beam

$$T = \frac{HD}{2} = \frac{W}{2\pi \tan \theta}$$

The above formulae are basic for obtaining the tension force in the tension ring of a circular plan dome. And it should be noted that this tension force varies directly as the total weight W and inversely as $\tan \theta$. Therefore, the larger the angle θ, the smaller will be the tension force. In fact, if θ approaches 90° as a limit, $\tan \theta$ becomes infinity, and T becomes zero at the edge; *but for shallow angles, the force can become very large.*

Figure 11-21 shows a radial dome with its ribs built of trusses. If these trusses are reasonably deep, it is then possible to resist bending in the plane of the truss. Then it is not necessary to apply hoop components

FIGURE 11-21
CONCEPTUAL DESIGN OF A RADIAL RIB DOME.

(a) TRUSSED RIBS

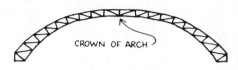

CROWN OF ARCH

(b) COMPRESSION RING AT CROWN OF DOME

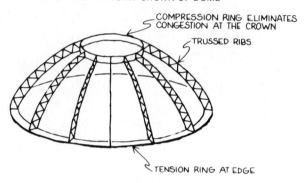

COMPRESSION RING ELIMINATES CONGESTION AT THE CROWN

TRUSSED RIBS

TENSION RING AT EDGE

(c) SECTION THRU DOME

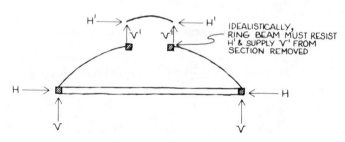

IDEALISTICALLY, RING BEAM MUST RESIST H' & SUPPLY V' FROM SECTION REMOVED

or to assume pure shell action in the membrane covering for such a dome. *This type of design is exactly analogous to that of an arch*, and the ribs will have to be as deep as for simple arch design.

Figure 11-21*b* shows another construction of a radially ribbed arch. When these ribs meet at the crown, they may become too congested, and therefore it may be better to provide a compression ring at the top of that dome to pick up the inward arch thrust (Figure 11-21*c*). The compression ring may carry a lot of force, especially if there are no other circumferential rings in that dome, such as in Figure 11-21*b*. On the other hand, if circumferential rings are provided as in Figure 11-19, then the force in the compression ring may be quite limited.

Using a truss-type construction, it is possible to build a dome of almost any curvature, including conical or pendentive domes, such as in Figures 11-22a and 11-22b. The vertical loads on such a dome may be transmitted in different ways. If the vertical loads will form pressure lines following the vertical curvature of that dome, there is no problem. But when the load pressure line deviates from the curvature of the dome, the presence of circumferential rings will produce tensile or compressive forces so as to redistribute the bending forces among the various ribs. However, if the ribs and the rings are reasonably thick by virtue of their truss design, some bending moments may be carried by them.

FIGURE 11-22
DOMES OF VARIOUS SHAPES.

(a) CONICAL DOME

(b) SQUARE DOME

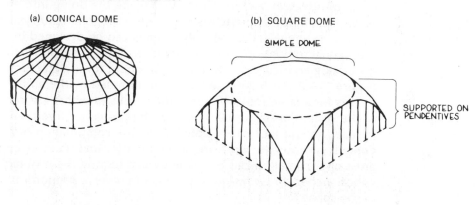

SIMPLE DOME

SUPPORTED ON
PENDENTIVES

(c) ELLIPTICAL OR SUPERELLIPTICAL DOME
(ORTHOGONAL GRID)

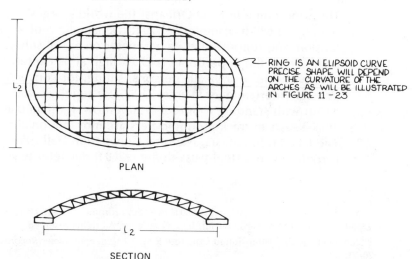

L_2

RING IS AN ELIPSOID CURVE
PRECISE SHAPE WILL DEPEND
ON THE CURVATURE OF THE
ARCHES AS WILL BE ILLUSTRATED
IN FIGURE 11 - 23

PLAN

L_2

SECTION

The precise distribution of the moments carried by radial ribs and the circumferential rings, either by axial force or by bending moments in the sections, cannot be easily determined by inspection. Although it is known, that *the stiffer element will carry the greater portion*, a complete computer analysis of domes is required for a final design. On the other hand, note that approximate designs must be made to ascertain preliminary dimensions as basic input for contemporary computer programs. Therefore, with heavy rings, the ribs will bend less; if there are no rings, the ribs will have to take all the moment via a combination of arch and bending action.

Another version of trussed dome design calls for the use of a skewed grid of arches, sometimes termed lamella domes (Photo 11-5a and b). Note that both skewed and radial designs were developed primarily for easy construction of domes with circular plan.

On the other hand, if the dome does not have a circular plan, as is common today, it may be difficult to divide the dome into equivalent radial sections. A better approach is to use an *orthogonal grid system* (Figure 11-22c). As suggested previously, domes can be designed by using an orthogonal translation approach by matching the sectional shape of the generating arch curves with the requirement for architectural plan shape (Figure 11-23). Such a dome has been designed by the authors, using steel trusses for a large football arena, the results of which are very interesting. At the scale of a 700 ft × 600 ft dome, the design required only 20 lb of steel for each square foot of projected area. The computation of forces in such a dome is obviously a very complicated matter that requires computers. But a gross approximation can be made by distributing the tributary load to the critical *arches on each axis* of the plan in inverse proportion to the square of the spans (L_1^2/L_2^2).

We refer you to two articles on radial truss rib steel domes, which are summarized here to illustrate the state of the art of circular plan dome design.

1. The Superdome in New Orleans, the world's largest steel trussed dome, is a 680-ft-diameter, ~120 ft rise lamella roof with a 9-ft deep tension ring supported at its perimeter on a wall subsystem ~150 ft high[1]. Thus, the Superdome is 273 ft high at its center. It will accommodate over 97,300 persons, sitting, for conventions and general entertainment events, 71,827 for regular-season football, and 80,101 with standing room for a Superbowl football play-off. (A total of 18,000 seats in track-mounted tiers are movable on steel tracks.)

2. The Jakarta, Indonesia, National Convention Hall, (Photo 11-6) is roofed with a steel-truss dome ~260 ft diameter having ~33 ft rise.[2]

---

[1]"World's Largest Steel Dome spans 680 ft over Superdrome," *Engineering News Record*, March 22, 1973, copyright ©, McGraw-Hill, Inc., New York. All rights reserved.

[2]"Steel Dome with Posttensioned Concrete Ring," *Engineering News Record*, April 25, 1974, pp. 20–21.

PHOTO 11-5
**SUPERDOME, NEW ORLEANS, LOUISIANA. (CURTIS AND DAVIS
CORPORATION, ARCHITECTS; S & P ASSOCIATES, INC.
ARCHITECT-ENGINEER) (*a*) COMPLETED BUILDING (*b*)
SUPERDOME LAMELLA RIBS UNDER CONSTRUCTION.**

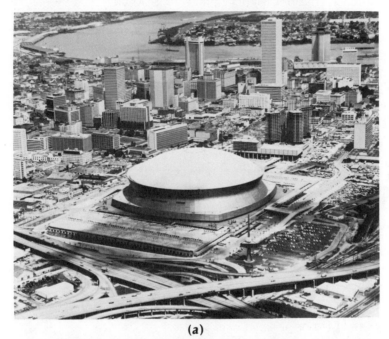

(a)

(b)

FIGURE 11-23
**DOMES OF VARYING SECTIONAL AND PLAN SHAPE CAN BE
GENERATED BY TRANSLATING AN (x–x) ARCH CURVE OVER A (y–y)
CURVE.**

(a) PERSPECTIVE OF ELLIPTICAL CURVE TRANSLATION

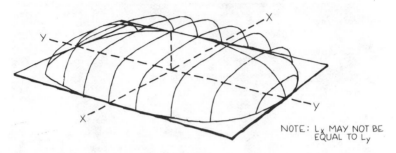

NOTE: L_x MAY NOT BE
EQUAL TO L_y

(b) THE TYPE OF ARCH CURVE DETERMINES
THE SECTIONS X–X AND Y–Y

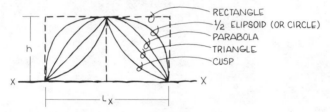

RECTANGLE
½ ELIPSOID (OR CIRCLE)
PARABOLA
TRIANGLE
CUSP

(c) DOME PLAN TYPES AS DETERMINED BY
TRANSLATION OF EQUAL ARCH SHAPES

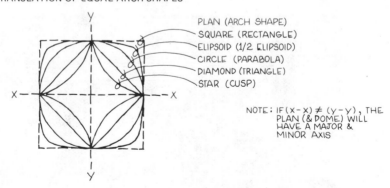

PLAN (ARCH SHAPE)
SQUARE (RECTANGLE)
ELIPSOID (1/2 ELIPSOID)
CIRCLE (PARABOLA)
DIAMOND (TRIANGLE)
STAR (CUSP)

NOTE: IF (x–x) ≠ (y–y), THE
PLAN (& DOME) WILL
HAVE A MAJOR &
MINOR AXIS

The tension ring at the lower end of the truss is prestressed concrete.
To avoid overstressing, the ring beam was stressed in two stages. The
first stage, amounting to 75 percent of the final prestress, was stressed
when the steel trusses were installed and before the central staging
(Figure 5) was removed. Note the minimal amount of falsework
required. The second stage was stressed after the ceiling system was
constructed and part of the roofing material was installed. Prestressing

PHOTO 11-6
**INDONESIA NATIONAL CONVENTION CENTER, JAKARTA. (T. Y. LIN
INTERNATIONAL, STRUCTURAL ENGINEERS.) (*a*) UNDER CONSTRUCTION. (*b*)
COMPLETED BUILDING.**

(a)

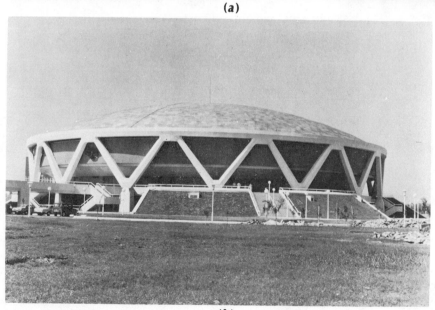

(b)

controls the deformation of the steel trusses and converts them from simple beams during erection into a trussed membrane dome. To span the ring beam supported on peripheral columns, the PC tendons were draped in a vertical plane, thus balancing the load on the beam (between columns) and at the same time applying a radial force. A total of only 610 kips of prestressing was required.

EXAMPLE 11-5: The Jakarta Dome Ring Beam

In the Jakarta National Convention Hall dome the tension ring is to be of posttensioned prestressed concrete. Estimate the amount of prestress and the size of the ring.

DOME SECTION

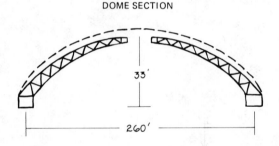

Solution

Assume: steel weight   10 psf
   roofing, mechanical   6 psf
   LL   <u>20 psf</u>
        Total   36 psf

W for dome $= 36 \times 130^2 \pi = 1920$ k

Tension in ring $T = \dfrac{W}{2\pi \tan \theta}$

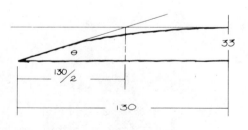

$\tan \theta$ approx $= \dfrac{2 \times 33}{130} = 0.5,$

$T = \dfrac{1920}{2\pi \times 0.5} = 610$ k

$A_s = \dfrac{610}{150 \text{ ksi}} = 4.1 \text{ in.}^2$

since the prestress for *DL* will be countered by T_{DL}, the concrete ring will have to absorb at least *LL* prestress, or *LL* + roofing = 26 psf out of 36 psf total

$$C = 610\frac{26}{36} = 400 \text{ k}.$$

To minimize creep, assume $f_c = 1$ ksi, $A_c = \frac{440}{1} = 440 \text{ in.}^2$

Try 20 in. × 24 in. ring section.

You can roughly approximate the rib design by treating it as a three-hinged arch loaded by a pie-shaped tributary area. The natural line of pressure for such a load will be roughly circular through the edges and the top, as shown in the figure.

SECTION 6: Dishes

A dish system is literally the structural reverse of a dome. It is a *suspension shell*, as a space enclosure, that has the advantage of being better acoustically than a dome, because it does not have a focal point inside the area covered by the dish. Such suspension structures are also more naturally efficient because buckling is not a problem. However, the concave shape does introduce a problem of drainage, which must be taken care of by providing pumps, pipes, and overflows to avoid overload. Figure 11-24a shows a circular dish system supported around its edges by a compression ring, which in turn is supported on a series of columns. The construction of such a dish commonly involves a number of radial tension cables (instead of compressive ribs as in a dome), which are connected to a tension ring near the center of this circle. But, as is true for domes, it may also be constructed of a grid of cables and the plan need not be circular.

In any case, it should be noted that the cables are very flexible and, unless stiffened, will assume the shape of the pressure curve created by a varying load applied along the length of the cable. Thus, if no circumferential cables or rings are supplied between the interior tension ring and the boundary compression ring, the position of these cables under a given load distribution can be determined by simple statics. However, to stiffen the system against wind or live load, additional intermediate rings or prestressed guy cables are necessary (Figure 11-24b). In prestressed concrete design, concrete decking or radial ribs are prestressed together to fill this need over the entire surface. The force in the circular *compression* ring is computed as it would be for (*T*) in a circular dome. In other words, the circular plan compression forces per foot of ring circumference

$$C = T = \frac{W}{2\pi \tan\theta}$$

FIGURE 11-24

WITH A CIRCULAR DISH THE EDGE FORCES ARE COMPRESSIVE:

$$C = \frac{W}{2\pi \tan \theta} \quad \text{(AS FOR } T \text{ IN FIGURE 11-20)}$$

(a) SIMPLE DISH

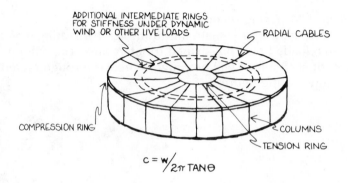

ADDITIONAL INTERMEDIATE RINGS
FOR STIFFNESS UNDER DYNAMIC
WIND OR OTHER LIVE LOADS

RADIAL CABLES

COMPRESSION RING

COLUMNS

TENSION RING

$$c = w/_{2\pi \text{ TAN}\theta}$$

(b) DISH WITH OVERHEAD CABLES

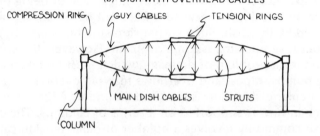

COMPRESSION RING

GUY CABLES

TENSION RINGS

MAIN DISH CABLES

STRUTS

COLUMN

where (*W*) *should include the effect of adding guy cables.* Each main cable is approximately circular in profile, since it is designed to carry a pie-shaped tributary load.

Figure 11-25 shows an *elliptical dish.* Such noncircular plans are often needed in order to fulfill arena requirements. Obviously, an elliptical dish would require radial cables of different length and curvature, which can make construction difficult. The forces would vary and the cables and the pie-segment panels between them would be different. One solution is to use an orthogonal approach whereby cables run in perpendicular directions to form a grid. The cable curvature can be held constant in each direction by translating a parabola of one curvature, along another parabola of less curvature, to generate an elliptical paraboloid dish, such as the one used for the PC Oklahoma State Fair Arena (Figure 11-26 and Photo 11-7*a* and *b*). To have the horizontal forces from these cables transmitted to the elliptical compression ring without creating any excessive bending moment in the

FIGURE 11-25
ELLIPTICAL DISH DESIGN.

(a) PLAN

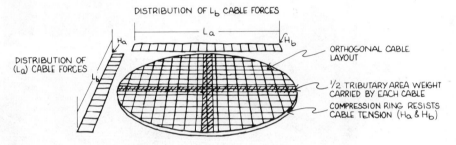

DISTRIBUTION OF L_b CABLE FORCES

L_a

H_a H_b

DISTRIBUTION OF
(L_a) CABLE FORCES

L_b

ORTHOGONAL CABLE
LAYOUT

½ TRIBUTARY AREA WEIGHT
CARRIED BY EACH CABLE

COMPRESSION RING RESISTS
CABLE TENSION (H_a & H_b)

(b) FORCES AT RING BEAM DUE TO ℄ (L_a) CABLE

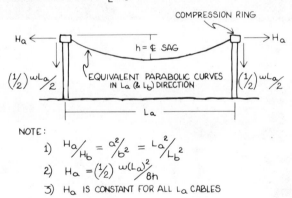

COMPRESSION RING

H_a ←   → H_a

h = ℄ SAG

$(\tfrac{1}{2})\, \omega L_a/2$ EQUIVALENT PARABOLIC CURVES
IN L_a (& L_b) DIRECTION $(\tfrac{1}{2})\, \omega L_a/2$

L_a

NOTE:

1) $H_a/H_b = a^2/b^2 = L_a^2/L_b^2$

2) $H_a = (\tfrac{1}{2}) \dfrac{\omega (L_a)^2}{8h}$

3) H_a IS CONSTANT FOR ALL L_a CABLES

PHOTO 11-7
**OKLAHOMA STATE FAIR ARENA (T. Y. LIN INTERNATIONAL, STRUCTURAL
ENGINEERS) (*a*) ROOF UNDER CONSTRUCTION. (*b*) COMPLETED BUILDING.**

(*a*)

(*b*)

FIGURE 11-26
GENERAL DIMENSIONS AND DETAILS FOR THE ELLIPTICAL PARABOLOID ROOF OF THE OKLAHOMA STATE FAIR ARENA.

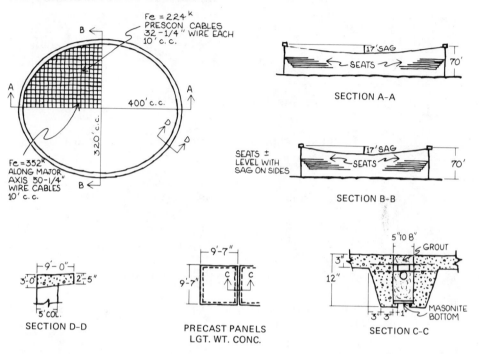

SECTION A-A

SECTION B-B

SECTION D-D

PRECAST PANELS
LGT. WT. CONC.

SECTION C-C

ring requires that

$$\frac{H_a}{H_b} = \frac{a^2}{b^2} = \frac{L_a^2}{L_b^2}$$

where a is the radius of the major axis of the ellipse and b is the radius of the minor axis of the ellipse. Applied to a circular plan, this means that $H_a/H_b = 1$; but in any case, it means that, if prestressed, *each cable carries one-half of its tributary load.* This is illustrated for the Oklahoma Shell in Example 11-6.

Example 11-6: Design of a Dish

For the Oklahoma shell (Figure 11-26), compute the prestressing force to balance the static design load in two directions and also the maximum compression in the elliptical ring.

Solution

$$DL = 48 \text{ psf}$$
$$LL = 12 \text{ psf}$$
$$\text{Total} = \overline{60 \text{ psf}}$$

In a prestressed elliptic paraboloid, 50 percent of the total load is carried in each of two directions, and force is constant for all cables in one direction. (If not prestressed, the load would be distributed according to the L_a/L_b ratio.)

Cables are at 10 ft c.c.

$$H_a \text{ along 400 ft direction} = \frac{wL_a^2}{8h} = \frac{30 \times 10 \times 400^2}{8 \times 17} = 352 \text{ k per cable.}$$

$$H_b \text{ along 320 ft direction} = \frac{wL_b^2}{8h} = \frac{30 \times 10 \times 320^2}{8 \times 17} = 224 \text{ k per cable.}$$

$$\text{Maximum compression in ring} = 352 \times \frac{160 \text{ ft}}{10} = 5650 \text{ k}$$

at extremities of minor axis, or

$$\text{Minimum compression} = 224 \times \frac{200}{10} = 4480 \text{ k}$$

at extremities of major axis.

For *controlling force* of 5650 k, ring area = 3500 in.^{2}

$$f_{avg} = \frac{5650}{3500} = 1.61 \text{ ksi which is high.}$$

But note that this maximum is reached only if roof is under full live load (*LL*), and looks reasonable.

SECTION 7:   Hyperbolic Paraboloid

Figure 11-27 illustrates that hyperbolic paraboloid (*HP*) shells, domes, and dishes involve surfaces of *double curvature* because they curve in both $\bar{x}$ and $\bar{y}$ directions. Similarly, cylindrical shells can be classified as those of single curvature. In contrast, folded plates are sometimes classified as shell systems of zero curvature. Folded and cylindrical shells are also termed *"developable,"* since their surface can be folded or rolled, from a flat surface. It also implies that their surfaces will tend to unfold, or unroll under loads, but not tear apart. On the other hand, domes and other shells of double curvature, such as the *HP* shells, cannot be formed by simply rolling or folding a flat surface. This means that they must be ruptured to be flattened out (Figure 11-19*b*).

Since the double curvatures of domes and dishes are either concave or convex, they are known as shells of *positive Gaussian curvature*. When the double curvatures are one concave and one convex, they are termed as having a *negative Gaussian curvature*, such as a hyperbolic paraboloid (saddle) surface.

FIGURE 11-27
**TYPES OF SHELL SURFACES ARE IDENTIFIED BY THE NATURE OF
THEIR CURVATURE.**

(a) SHELLS OF DOUBLE CURVATURE CANNOT FLATTEN
    WITHOUT TEARING APART (FIG. 2-19b)

DOMES (OR DISHES) ARE SHELLS
OF POSITIVE GAUSSIAN CURVATURE
(i. e. BOTH ARE EITHER CONVEX
OR CONCAVE)

HP (OR SADDLE) SHELLS ARE OF
NEGATIVE GAUSSIAN CURVATURE
(i. e. ONE CURVE CONVEX AND
THE OTHER CONCAVE)

(b) SHELLS OF 'O' OR SINGLE CURVATURE WILL
    TEND TO FLATTEN OUT UNDER LOAD

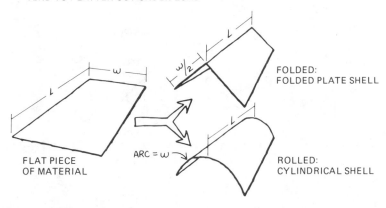

FOLDED:
FOLDED PLATE SHELL

FLAT PIECE
OF MATERIAL

ARC = ω

ROLLED:
CYLINDRICAL SHELL

The basic properties of a square plan hyperbolic paraboloid layout are
shown in Figure 11-28. Here we translate a concave parabola along another
convex parabola. In this manner, if we cut a vertical section along either x
or y axis, we get a *parabolic curve* (as with the elliptical dome). If we cut a
horizontal section, we get a pair of hyperbolas rather than the circle or
ellipsoid that we got with a dome or dish. Therefore, this surface is known
as a hyperbolic paraboloid. But if we draw a line along the surface at 45° (for
square plan) to the axes x and y, we will find it to be a *straight line*. In fact,
when a straight line is translated along two other straight lines *that are not
parallel to each other*, a hyperbolic paraboloid surface is formed. This is
important in that the hyperbolic paraboloid has the unusual quality that it
can be constructed of a series of straight lines. Thus a set of timber planks

FIGURE 11-28
**A HYPERBOLIC PARABOLOID (*HP*) SURFACE IS GENERATED BY TRANS-
LATING A PARABOLIC ARCH CURVE OVER A SUSPENSION CURVE OF THE
SAME TYPE.**

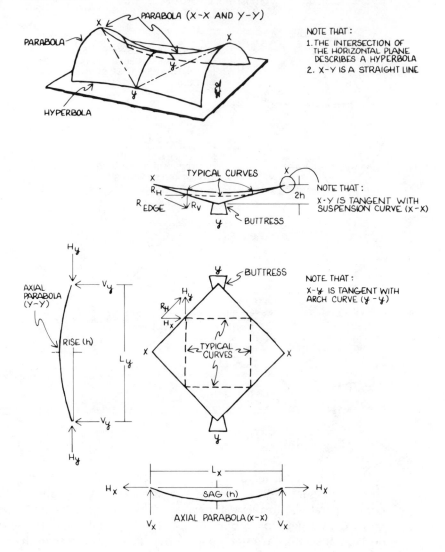

NOTE THAT:
1. THE INTERSECTION OF THE HORIZONTAL PLANE DESCRIBES A HYPERBOLA
2. X-Y IS A STRAIGHT LINE

NOTE THAT:
X-Y IS TANGENT WITH SUSPENSION CURVE (X-X)

NOTE THAT:
X-Y IS TANGENT WITH ARCH CURVE (Y-Y)

can be made to form an (*HP*) surface, and the plan will be square when
(*x-x*) = (*y-y*).

The edges of an *HP* membrane are usually connected to an *edge beam*.
The membrane force transmission in an *HP* shell can be discussed using the
axis parabolas shown in Figure 11-28. A suspension cable is, *for evenly*

distributed vertical loads, a parabola that concaves upward. Similarly, an ideal arch would be a convex parabola and each can be analyzed using the formula $H = wL^2/8h$. Applied to a symmetric *HP*, the axial sections represent arches that run at right angles to the suspension cables. And, importantly, all other sections through an *HP* in one direction (x, x or y, y) are made up of *portions of* the axis curves. Hence, the value H remains the same in all these individual parabolas, even though some are much longer than others.

When the plan is square, the H_x forces are equal to the H_y forces. This is so because in all cases, the values of L^2/h (in x, x or y, y) remain constant and therefore the force H also remains constant for an assumption of evenly distributed loading.

Similarly, both the cables and the arches would act on an edge beam at an angle of 45°, one producing a horizontal thrust and the other an equal pull. If we consider a plan of the edge beam acted on by these two sets of forces, we will see that their components add up along the length of an edge beam but cancel out in the horizontal direction perpendicular to it. Therefore, in a simple *HP* shell, using the membrane theory, the horizontal forces from the shell surface accumulate along the length of the edge beam starting from zero at the tip and adding up to a maximum near the support. It also happens that the vertical components of the cable and arch forces will always be such that the net effect of the horizontal and vertical edge forces is to coincide with the edge beam. Thus the edge beam can be considered as an inclined column acted on by axial forces transmitted to it from the two shell curves and accumulating along its length.

Theoretically, the edge beam would not bend at all, save for the fact that it has weight in itself. Of course, that weight acts on the beam to produce cantilever bending. Hence it is often said that the edge beam should be designed to carry its own weight as a cantilever in addition to its serving as a column carrying the loads from the shell. However, the situation is, in reality, more complicated because normally *the edge beam does not act by itself*, but is integrated into the shell, and they must move together under any bending as one unit. In other words, the beams together with certain parts of the shell around them act as a cantilever. This greatly strengthens the edge beam in comparison to the assumption of simple independent cantilever action.

Computer programs have been developed to analyze the interaction between the beam and the shell. For a final design, such advanced analysis must always be made unless the shell is of small span. But, for preliminary purposes, it is possible to approximate this effect as follows.

The four *HP* roof shells of the Ponce Coliseum (Photo 11-8 and Figure 11-29) were analyzed by the finite element method using a computer program. The sectored plan is slightly rectangular with each edge beam cantilevering 116 ft and 138 ft. But only one-sixth of its own weight was carried by each beam itself, whereas five-sixths of its weight is carried by composite action between the beam and the shell. This suggests that

PHOTO 11-8
**PONCE COLISEUM, PUERTO RICO. (V. MONSANTO, ARCHITECT; T. Y. LIN INTER-
NATIONAL AND RAYMOND WATSON, STRUCTURAL ENGINEERS)** (*a*) *HP* **CANTILEVER
ROOF SHELL** (*b*) **SKETCH OF 4-SECTOR PLAN.**

(a)

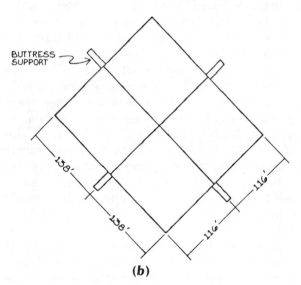

BUTTRESS
SUPPORT

138′

138′

116′

116′

(b)

preliminary design can be accomplished by means of an overall
approach.

An overall approach to *HP* design involves dealing with the shell as a
whole as in Figure 11-29*b*. The overall bending resistance of such a shell
may be estimated by assuming the entire shell to be a cantilever acting from
the supports. The overall depth at the supports is *h* (section *y-y*, Figure

FIGURE 11-29
THE BASIC PROPERTIES OF ONE SECTOR OF THE PONCE *HP*.

(a) PERSPECTIVE AND ELEVATIONS OF ONE SECTOR

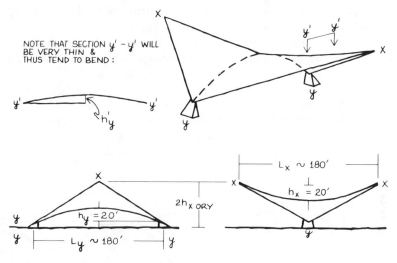

(b) APPROXIMATE ANALYSIS OF TENSILE AND COMPRESSIVE FORCES

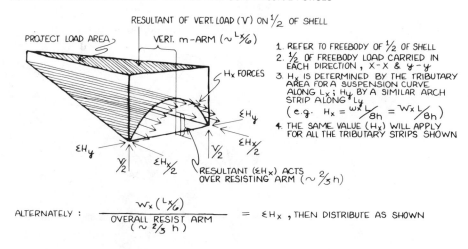

1. REFER TO FREEBODY OF $\frac{1}{2}$ OF SHELL
2. $\frac{1}{2}$ OF FREEBODY LOAD CARRIED IN EACH DIRECTION, X-X & y-y
3. H_x IS DETERMINED BY THE TRIBUTARY AREA FOR A SUSPENSION CURVE ALONG L_x; H_y BY A SIMILAR ARCH STRIP ALONG L_y
   $$\left(e.g.\quad H_x = \frac{w_x L}{8h} = \frac{W_x L}{8h}\right)$$
4. THE SAME VALUE (H_x) WILL APPLY FOR ALL THE TRIBUTARY STRIPS SHOWN

ALTERNATELY : $\dfrac{w_x\left(\frac{L_x}{6}\right)}{\text{OVERALL RESIST ARM} \ (\sim \frac{2}{3}h)} = \epsilon H_x$, THEN DISTRIBUTE AS SHOWN

11-29*a*). But it should be noted that section *y'-y'* near tip is flatter and thus more susceptible to bending deflections (Figure 11-29*a*).

There has not been enough theory or experience to set up rise- or sag-to-span ratios for *HP* shells. Suffice it to say that these shells should not be too flat, but the degree of flatness has yet to be defined. As a start, one can say that, with prestressed cables, $2h/L \geqslant 1/10$ (i.e., $h/L \geqslant 1/20$).

For reinforced design, h/L should be larger by 50 percent or more to limit deflections. Even so, when these shells cantilever in excess of 100 or 120 ft, bending stresses and deflections may become a problem. Then it will be desirable, or even necessary, to prestress both the shell and the edge beams, as was done for the Ponce Shell.

Being nearly square in plan, the Ponce Shell also represents a case of four-sector (or composite) shell design. A more common arrangement is a

FIGURE 11-30
SOME COMPOSITE SECTOR SHELLS.

COMPOSITE SECTOR SHELLS WITH
SECTORS SHAPED ACCORDING TO
A SURFACE OF SECOND ORDER

(a)   (b)   (c)

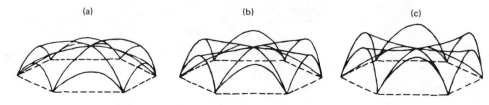

(d) COMPOSITE SECTOR SHELL, WITH SECTORS STRONGLY
CURVED IN DIRECTION OF THE POLYGON SIDES
AND SLIGHTLY IN PERPENDICULAR DIRECTION

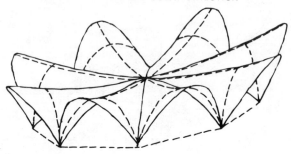

(e) CROSS VAULT–LIKE COMPOSITE SECTOR SHELL

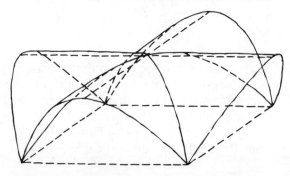

six-sector shell as in Figures 11-30a, 11-30b, 11-30c, whereas Figures 11-30d and 11-30e depict eight- or four-sector schemes. Sectored shells are useful because one can also add a central rooflight opening or rooflight strips *between sectors* as desired. They also minimize the height of the cantilevered portions relative to the low point of the interior space. Note in Figures 11-30d and 11-30e that one can design composite shells with the sectors shaped according to a surface of double or single curvature. Obviously the final design of these shells becomes a difficult problem. For final design, a finite element analysis using computer programs will be highly desirable for all shells, except for those of rather small spans. However, for preliminary purposes, they can be roughly approximated as shown in Figure 11-29 or by assuming simple arch action and dealing with basic statics. An overall analysis of the Ponce Shell follows in Example 11-7. For more information refer to Appendix G-1.

EXAMPLE 11-7: Design of an *HP* roof

Each quadrant of the Ponce Coliseum Shell (Figure 11-29) is made up of a rectangular plan *HP* shell 138 ft × 116 ft, with a rise of 20 ft between abutments, and a total rise of 40 ft from buttress level to the high points of the tips. It is desired to compute the membrane stresses as a guide to determining the shell thickness, then also to proportion the edge beams to carry its axial load, taking into account possible bending and prestressing effects. Refer to the freebody diagram (Figure 11-29b).

Diagonal distance between abutments: $L = \sqrt{138^2 + 116^2} = 180$ ft

4-in. concrete weighs 50 psf (no roof membranes);

$$DL = 50 \text{ psf}$$
$$LL = 56 \text{ psf}$$
$$\overline{DL + LL = 106 \text{ psf}}$$

If 50 percent is to be carried by arch action and 50 percent by suspension action, then $w = 53$ psf in each direction,

$$H = \frac{wL^2}{8h} = \frac{53 \times 180^2}{8 \times 20} = 10.7 \text{ k/ft, tension or compression.}$$

Thus, *excluding edge beams*, membrane stress = 10,700/4 × 12 = 223 psi tension or compression, which is quite low for concrete in compression but requires reinforcement of prestressing to resist the tension. Note that the edge beams add some 15% to *tension* only.

Alternately, one can treat the freebody as a cantilever with the total vertical load (V) acting over an arm of roughly ($L_x/6$) or 30 ft. This overall moment will be resisted by the resultant of tensile forces ($T = \Sigma H_x$) acting over an arm of roughly 2/3(h_x). T can be distributed evenly over the sectional arc of the freebody as illustrated in Figure 11-29b. Thus if edge beam adds approximately 15 percent to *DL* and *LL*;

Total Load = 1.15(106) = 122 psf,

$$V = \frac{138 \text{ ft} \times 116 \text{ ft}}{2}(.122 \text{ k}) = 978 \text{ k},$$

$$T = \frac{978 \text{ ft} \times 30 \text{ ft}}{2/3(20 \text{ ft})} \sim 2200 \text{ k}$$

For *projected* arc length assumed ~ 180 ft,

$$\frac{T}{\text{ft}} = \frac{1960 \text{ ft}}{180 \text{ ft}} \sim \frac{12.2 \text{ k}}{\text{ft}}$$

and the *tension* membrane stresses are shown to be slightly higher owing to the added weight of the edge beam (compression is still ~ 10.7 k/ft).

In either case, the thrust from the arch elements and the pull from the suspension elements accumulate an *axial load* into the edge beams. Using the first method:

Arch	$10.7 \times 0.707 =$	7.6 k/ft
Suspension	$10.7 \times 0.707 =$	7.6 k/ft
	Total $=$	15.2 k/ft

Since the interior beam (where sectors join) carries axial load from both sides,

$$15.2 \times 2 = 30.4 \text{ k/ft}$$

Consider a point on longest beam at 10 ft from support leaving 128 ft as the beam span, (approximately the average unsupported length for the two 116 and 138 ft beams); the axial load is (since the 1-ft strips act over a section distance of about 90 ft):

$$30.4 \times \sim 90 = 2740 \text{ k}$$

Note that one could also deal with the total vertical load and the average beam slope (i.e., 128 ft/40 ft).

Considering the shell load only, the interior beam at the 128 ft point is 60 in. wide and 47 in. deep, and

$$A_c = 60 \times 47 = 2820 \text{ sq in.}$$
$$f_{avg} = 2740/2820 = \sim (1.0) \text{ ksi}$$

This is a fairly high value, since the moment effect of the beam's own weight and prestressing have not yet been included. But fortunately, computer analysis indicates that only one-sixth of the beam's own weight is supported by the beam itself, with the rest carried by composite action between the shell and the beam. Hence, the design did work out satisfactorily.

An *HP* shell can be constructed on a circular plan, and it may be prestressed in two directions. Photo 11-9a and b and Figure 11-31 shows such a roof for the Arizona State Fair Coliseum. But one should note that

PHOTO 11-9
ARIZONA COLISEUM. (LESLIE MAHONEY, ARCHITECT; T. Y. LIN INTERNATIONAL, STRUCTURAL ENGINEERS (*a*) ROOF UNDER CONSTRUCTION. (*b*) COMPLETED BUILDING.

(*a*)

(*b*)

FIGURE 11-31
CIRCULAR *H-P* SHELL.

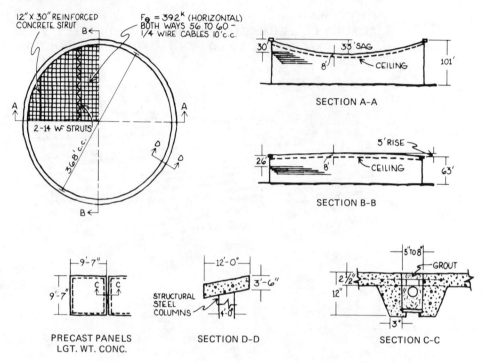

the sag (*h*) is *not* the same in each direction. It was also built of precast pieces of one shape (9 ft 7 in × 9 ft 7 in.) and posttensioned together to balance all the dead and live loads. This approach allowed constructive advantages and readers are referred to T. Y. Lin and Ben N. Young, "Two Large Shells of Post-tensioned Precast Concrete," *Civil Engineering*, July 1965.

SECTION 8: Lightweight Tension Systems

The design of lightweight tension structures is practically a field in itself. Here, pure membrane action is involved and the dynamics of wind and live load pose major problems since dead load is but a *small* part of the picture. We can only touch design basics, and interested readers are referred to *Tensile Structures* by Frei Otto of Germany, MIT Press, Cambridge Mass., 1973).

Perhaps the best-known lightweight tension structure is the Olympic Arena at Munich, Germany, and an excellent description appears in the July

PHOTO 11-10
**LAVERNE COLLEGE CAMPUS CENTER AND DRAMA LAB. (THE
SHAVER PARTNERSHIP, ARCHITECTS; BOB D. CAMPBELL & CO. AND
T. Y. LIN INTERNATIONAL, STRUCTURAL ENGINEERS).**

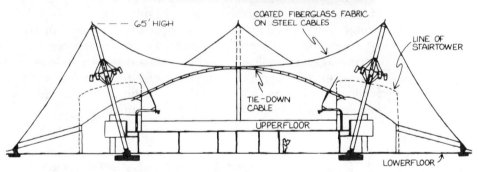

1972 issue of *Civil Engineering—ASCE*, pages 41–44. General descriptions of
some other fabric structures can be found in "Architectural Fabric
Structures" by Du Pont Co., Wilmington, Delaware, and "Fiberglas Fabric
Structures" by Owens-Corning Fiberglas Corp., Toledo, Ohio.

The following example summarizes the basic design issues and describes
a relatively simple but interesting tension roof constructed for LaVerne
College near Los Angeles in 1973 (Photo 11-10). Four columns are used to
form a 65-ft-high, 57,000 sq ft college student center that looks like an
intersecting cluster of circus tents.[3] Ten tons of fiberglass fabric are

---

[3]"10 Tons of Fabric Lifted, Joined to Form Tent-Like Student Center," *Engineering News Record*,
June 28, 1973, p. 44.

supported on radiating cables (i.e., total *DL* on the order of 1 psf). This unusual design was cost competitive at about $31 a square foot, compared with $38 to $40 for a more conventional structure, according to college officials. Cable sleeves kept the cables in position relative to the skin. The tents were then joined by bolting them between aluminum plates attached to tie-down cables in the intersections of the cones. With the skin in position, the columns, which have telescoping end sections, were jacked up, bringing the cables into position against the skin and stretching it into shape.

Basic design considerations for either air supported or tent structures are not different from those of thin shells. However, since the membranes have little weight and no practical stiffness in themselves, they must be *prestretched* to carry compressive stresses from *LL* without buckling. For tents a corollary problem is to *tie down critical points*, especially at edges, to avoid flutter under wind loading. Air supported designs should be *low in profile* to get lift, rather than downward pressure, from wind. Then pressurization of only about 0.04 psi (~6 psf) can support a stable roof.

As a gross approximation, one can design the cables to carry several (say 5 or 6) times the tributary area dead load of the membrane and cables (about 1 psf). In this way, some pretensioning of the membrane will be provided. The beginning designer can also use model construction and wind tunnel testing to appreciate the dynamic problems of profile or edge flutter. But, of course, the actual pretensioning needs are complex to ascertain, and all but the simplest of designs will require computer assistance for final design.

The more complicated applications are also expensive to construct. However, the future of lightweight construction is described well by the following excerpt about Fre; Otto's olympic tent design from the *Civil Engineering—ASCE* article mentioned above:

COMMENT ON CABLE-SUSPENDED ROOFS

A free-form roof over all olympic sports fields and intermediate areas was possible only by using a prestressed cable-suspended structure. Only with such a gravity-and-tension structure could the designers attain such harmony between shapes and forces.

Rope nets allow a greater variety of forms than any other type of structure, and no other system could have followed so faithfully the architect's ideas for the roof shape in relation to the landscape and buildings. Also, because prestressed rope nets cast almost no shadows, the requirement (made by television) for a translucent roof covering for the stadium could be met to an extent unequalled by other structures.

The design was chosen in an architectural competition, and construction budget seems not to have been the number one concern. The total cost of the roof is expected to be about 150 million DM ($48 million U.S.).

Leonhardt & Andra have designed a roof for another stadium of the same size where economics was given more weight. Low bid was about 1,000 DM per m² (about $39 U.S. per sq ft), about half the figure for the Olympic roof. A still-lower cost, about 60 percent as high as the revised olympic design, is possible with a conventional truss design, but esthetics are not as good and columns are in viewers way.

Another reason for the high cost of tentlike roofs is contractors' high bids on these first-of-a-kind structures. As soon as more are built, lower allowances for unanticipated problems can be expected.

Co-author Schlaich sees a healthy future for roofs of this type. He envisions "aircraft hangars, market halls and all other large-span halls," and even translucent covers for downtown shopping streets and plazas. One imaginative potential application: "We have designed large hyperbolic natural-draft cooling towers, 650 ft (200 m) high and 425 ft (130 m) in diameter, which cost about 25 percent less than the conventional concrete type, and which are structurally not limited in size as are the latter."

12
Foundation
Subsystems

SECTION 1: Introduction

A foundation is the interface between the buildings and earth. Foundation subsystems serve to transmit loads from the vertical subsystems of a building to the earth (Figure 12-1). As seen in plan, the vertical subsystems (columns, walls, and shafts) are distributed in some manner as points or lines of *load concentration*. But, the earth, which must ultimately support a structure, offers a more-or-less *distributed bearing capacity*.

If the bearing capacity of the earth is sufficient, a foundation can follow the distributed pattern of the vertical subsystems. But sometimes soil and/or loading conditions will require that foundations be like a mat resting on the earth rather than concentrated. Mat foundations have the advantage of generalizing the interface problem, but they often cost much more than isolated foundation components and are used only when needed. In either case, *the idea of a foundation is to distribute the concentrated load* to the earth without exceeding its long-term bearing capacity. Thus, the amount of distribution is inversely related to the bearing capacity.

Figure 12-1 illustrates some of the more common types of foundations. In some cases, a column can be supported by a square footing directly underneath it, a wall can be supported by a line footing along its length. When a building is only several stories high, a combination of line footings and square footings, spread throughout the entire area, is often sufficient to transmit the loads to the earth. These separate footings may even be connected with footing ties to link and strengthen their resistance under earthquake loadings.

A mat foundation throughout an entire building will not be required unless the earth supporting it is very weak or the building is relatively high (say, above 10 or 20 stories and it generates large overturn forces). In other words, in most buildings the vertical subsystems of the building, (walls, columns, and shafts), can be carried separately by individual footings onto the earth below. In intermediate conditions arch or prestressed beam-foundation ties may be required to spread loads more evenly than with independent bearing foundations, since they do not require bending deflection to distribute the loads.

As a first approximation for design, it is best to compute the approximate weight of the building structure and assume it to be uniformly spread over the entire plan area (i.e., as if for a mat design). Thus a generalized load

FIGURE 12-1
SOME FOUNDATION SUBSYSTEMS.

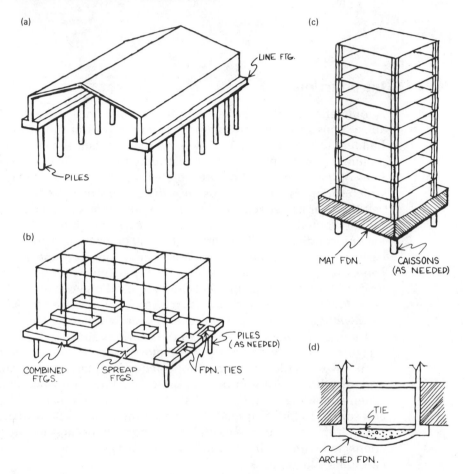

(a)

LINE FTG.

PILES

(b)

COMBINED FTGS.

SPREAD FTGS.

PILES (AS NEEDED)

FDN. TIES

(c)

MAT FDN.

CAISSONS (AS NEEDED)

(d)

TIE

ARCHED FDN.

intensity is obtained for comparison with the strength of the supporting earth itself. If the average load intensity over that area is much lower (say, $\leq$ one-fourth) the allowable supporting capacity of the earth, it is likely that individual subsystems would be more economical than a mat foundation. On the other hand, if that average load intensity is close to or greater than, let us say, one-half the capacity of the supporting soil, then it may be more economical to construct a mat foundation spread throughout the entire area than to use tied foundations.

Determination of the bearing capacity of the supporting earth is a complicated problem in itself because local conditions vary greatly, and oftentimes the earth is nonhomogeneous across the area of the structure. It also varies in layers as one goes deeper below the foundation surface. It is therefore standard practice to engage special consultants known as "foundation engineers" or "soil mechanics experts" to study the nature of the soil and foundation requirements at a site. This requires borings into the ground and testing of the samples taken in order to determine the characteristics of the supporting earth. Such on-site investigations are almost always carried out to some degree in order to determine the nature and the capacity of the foundation soil.

Nevertheless, as a start, one may simply look at the surface of the site or obtain boring data taken at or near the site from previous studies and explorations. Table 12-1 may serve as a guide to indicate the approximate bearing values allowed on different types of soils and rocks.

While bearing values reflect the strength of the soil, the main goal is actually to keep the settling of the structure as a whole within limits and to avoid any differential settling or lateral movement of components as well. If for some reason a building has to settle a large amount, say a few inches, such settling will not be detrimental if it takes place uniformly and provision is made for it. But differential settling between parts of the structure can cause serious problems. For example, one side of the structure may tend to settle more than the other, as with the Tower of Pisa. Another problem arises when an add-on structure may settle relative to the original structure whose settlings have already taken place.

For low buildings, it will often be sufficient to start design by computation

Table 12-1
Allowable Bearing Values on Soils, Tons per Sq Ft*

Massive crystalline bedrock: granite, gneiss, traprock— in sound condition	100
Foliated rock: schist and slate—in sound condition	40
Sedimentary rock: hard shales, siltstones, sandstones— in sound condition	15
Exceptionally compacted gravels or sands	10
Compact gravel or sand-gravel mixtures	6
Loose gravel; compact coarse sand	4
Loose coarses and or sand-gravel mixtures; compact fine sand, or wet, confined coarse sand	3
Loose fine sand or wet, confined fine sand	2
Stiff clay	4
Medium stiff clay	2
Soft clay	1

Note: Due to variations in the texture of the soil, or the amount of moisture in it, its confinement, and so on, this table is clearly insufficient to yield definite values. The values given here are intended only for schematic considerations at the early stage of the planning of a structure.
*Code Manual, New York State Building Construction Code.

of vertical load only. Usually, lateral loads are not an issue. But for tall and narrow buildings, the overturning effect of wind and earthquake forces (lateral loads) must always be evaluated. These forces can produce a significant change in the distribution of the vertical load on parts of the foundation system and can often concentrate large forces on certain areas.

When buildings get high, the vertical load intensity on the foundation can even reach a point such that the surface soil is no longer strong enough to support that weight. Then it becomes necessary to drive piles or drill caissons so as to transmit the load to better or stronger strata below the surface strata. This will enable high loads to be concentrated over relatively small areas at the top of the footings and transferred by piles or caissons to the earth.

In Sections 4 and 5 of this chapter we discuss pile and caisson design. In the following sections we briefly discuss the overall design requirements for each of the basic types of foundation identified above. For more information on the subject, refer to Section 6 (Foundation Engineering) of the *Civil Engineering Handbook* (McGraw-Hill, New York, 1968), edited by Frederick S. Merit, with T. Y. Lin as one of his consultants.

SECTION 2: Line Footings and Mat Foundations

Walls are often supported on a *line footing*, which, is placed directly beneath, but somewhat wider than, the wall in order to increase its bearing area on the earth. Figure 12.2a shows the total weight acting on a wall to be transmitted to the earth through such a footing.

A simple analysis of this footing is also shown in Figure 12.2b. Taking a

FIGURE 12-2
LINE FOOTINGS.

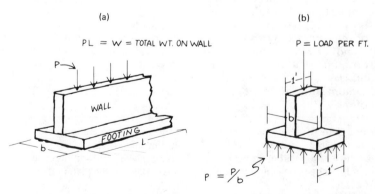

(a)

(b)

one-foot length of this wall together with its footing, if the load on this strip is (P/ft.) of wall, then P is assumed to be uniformly distributed across the width (b) of the footing. Therefore, the earth pressure $p = P/b$, in load per square foot.

If a moment exists in this wall due to lateral loads or eccentric loading, then the earth pressure down below will be somewhat different (Figure 12-3, parts a, b, and c). If the moment is relatively small so that the eccentricity (e) is less than the kern point ($k_b = b/6$ for rectangles), then the distribution is as shown in Figure 12-3a. Here the pressure p_1 is greater than p (i.e., $p < p_1 < 2p$), while p_2 is less than p. If the eccentricity equals $b/6$ (Figure 12-3b), then there will be zero pressure at one edge of the footing while the pressure at the other edge is doubled to $2p$.

Should it happen that the eccentricity is greater than $b/6$ (Figure 12-3c), then part of the footing will tend to be lifted from the earth and there will be no pressure on that part. Then, for a rectangular section, the active width of the footing is reduced to three times the maximum compressive edge to load centroid distance, $3(b/2 - e)$. The maximum edge pressure (p_1) will be twice the amount of the average pressure, thus $p_1 = 2P/3(b/2 - e)$.

This method of determining the combined vertical load and overturn pressure distribution for a line footing can also be applied to a *mat foundation* (Figure 12-4). If it is assumed that the weight of the entire building (W) coincides with the centroid of the mat foundation, then it is

FIGURE 12-3
FOOTINGS SUBJECTED TO OVERTURNING.

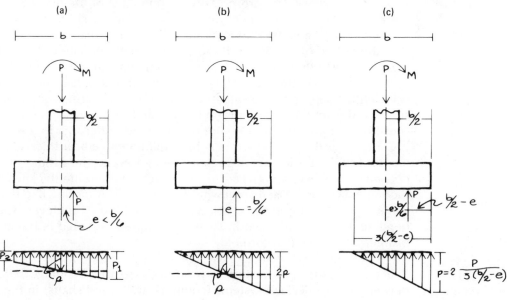

FIGURE 12-4
**A HEAVY MAT MAY SERVE
AS A FOUNDATION FOR
THE WHOLE BUILDING.**

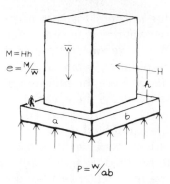

usually assumed that the pressure under the mat foundation equals W/ab, where ab is the area of the mat foundation. Should there be a lateral force H acting on the building, that force H will produce a moment $M = Hh$. The effect of that moment will be the same as that of an eccentricity of the building's weight; therefore, $e = M/W$. If one compares the amount of this eccentricity to the width of the mat foundation along the direction of the lateral force, one can approximate the pressure distribution across the mat foundation as with the line foundation. For example, if $e = a/6$ in Figure 12-4, then there will be zero pressure at one edge of the mat and twice the average pressure at the other edge, similar to Figure 12-2b. Of course, if the shape is not rectangular, the k_b point will be other than $a/6$.

In the design of either a wall footing or a mat foundation, it is usual practice to include the weight of the footing itself in the computation of the pressure on the foundation. However, when the foundation is excavated sufficiently deep, then the weight of the excavated earth is sometimes subtracted from the weight of the footing and only an excess of footing load is used in computing the pressure on the earth under that footing.

Concrete wall footings normally have minimum *vertical* thickness of about one foot. If the projection of such a footing beyond the width of the wall becomes excessive, say, over 2 or 3 ft each side, then the thickness of the footing may be increased somewhat (i.e., at 2 : 1). A footing like that, being loaded by the earth pressure below, acts as a double cantilever. The bending moment and shear in such a footing is computed in the normal manner and resisted accordingly.

A mat footing is usually at least 3 or 4 ft and up to 7 or 8 ft thick, depending upon its projection beyond the sides of a building, the distance between the vertical subsystems (columns, walls, and shafts) in the building, and the total amount and distribution of the building weight.

SECTION 3: **Spread and Combined Column Footings**

To support the concentrated load from a column, it is usual to use a *spread footing* that increases the area of support in two directions (Figure 12-5). If the load from the column is *P*, then the pressure on the earth is $p = P/ab$, where ab is the area of the footing. Such a footing often has a square shape, which is the most economical. But sometimes it has a rectangular shape, for reasons of space limitation. As might be expected, the design of such a footing is controlled by its resistance to bending as well as punching shear. To study the bending in rectangular footings, one would take two critical sections, just outside the faces of the column (Figure 12-5). Section 1-1 is a critical section for bending in the direction of the footing width *b*, and Section 2-2 represents a critical section for bending in the direction of width *a*.

The *punching shear* produced by the column on the footing is shown as a truncated core or pyramid surface with slopes about 1:2 (Figure 12-6). In this manner, one obtains a punching area larger than the perimeter of the column. Based on this enlarged perimeter, a punching shear is computed and compared to allowable values. For ordinary concrete construction, the allowable punching shear is on the order of $0.1 f'_c$, (e.g. 300 psi for $f'_c = 3000$ psi).

In the computation of the bending moment and the punching shear in a footing, the weight of the footing itself is always neglected. This is so because the footing's weight is directly counterbalanced by the earth

FIGURE 12-5
COLUMN FOOTINGS MUST BE SPREAD OUT TO TRANSFER A CONCENTRATED LOAD TO THE GROUND.

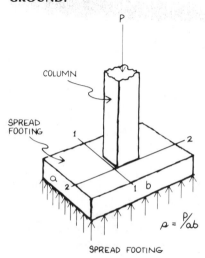

$$p = \frac{P}{ab}$$

SPREAD FOOTING

FIGURE 12-6
PUNCHING SHEAR IN A SPREAD FOOTING.

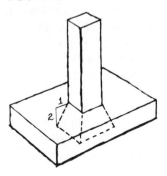

pressure so produced. Only the load from the column needs to be spread over the footing area when computing bending and shear in a spread footing. Example 12-1 illustrates how to make an overall design for a simple spread footing.

EXAMPLE 12-1: Design of a Spread Column Footing

For punching shear computation,

$p = 600/(8 \times 8) = 9.4 \text{ ksf}$

Base of truncated cone
$= 2 \text{ ft} + 2(18 \text{ in.}/2) = 3.5 \text{ ft}$

Force outside cone:

$9.4(8^2 - 3.5^2) = 490 \text{ k}$

Effective (perimeter) area
$= 4 \times 3.5 \times 1.5 = 21 \text{ ft}^2$

Punching shear
$= 490 \text{ k}/(21 \times 144) = 160 \text{ psi}$

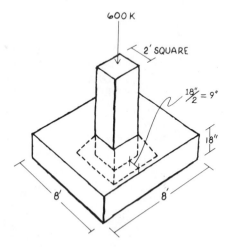

For concrete construction, say $f'_c = 3000$ psi, allowable punching shear is about $0.1f'_c = 300$ psi. Hence the above value of 160 psi is quite low. However, the total thickness of 18 in. is desired to minimize steel for bending reinforcement. Also, some 3 in. of concrete protection must be deducted to get the effective depth of footing, thus, 18 in. − 3 in. yields a

15 in. effective depth, which should be used in more accurate computation of punching area.

For moment computation, consider critical section at face of column. With a 3-ft cantilever, $w = 9.4 \times 8 = 75$ k/ft,

$$M = \frac{wL^2}{2} = 75 \times 3^2/2 = 350 \text{ k-ft.}$$

For $d = 15$ in.

$$a = \frac{7}{8}\left(\frac{15}{12}\right) = 1.1 \text{ ft,}$$

$$T = \frac{350}{1.1} = 320 \text{ k.}$$

$$As = \frac{320}{24} = 13.4 \text{ in.}^2;$$

Use 14 no. 9 with $A_s = 14$ in.^{2}, $p_s = 14/(15 \times 8 \times 12) = 1.0$ percent in each direction, which is o.k.

For various reasons, one footing that can support several columns may be designed. In this case it is called a *combined footing*, since it essentially involves *joining* two or more independent footings, each supporting one column (Figure 12-7). Figure 12-7a shows an edge column (A) that is limited by adjoining property line so that its footing has to be combined with another interior column (B) to avoid excessive eccentricity of load on the footing. Since this combined footing supports both columns A and B, the footing should be designed in such a way that the resulting load from the two columns P_1 and P_2 will more nearly coincide with the center of gravity (c.g.) of the combined footing. As a result, nearly uniform pressure on the supporting earth is produced. This of course would mean a more even settling of the combined footing as versus the possibility of a high eccentric load on the property line footing, producing perhaps a tilted settling if footings are independent.

Figure 12-7b shows a different arrangement for the same situation, with Column A tied to Column B through a strap. Again it is desirable to have the resultant of the two forces P_1 and P_2 coincide with the centroid (c.g.) of the footing.

Figure 12-7c shows a trapezoidal shape for a combined footing, carrying loads from two columns. Again, the centroid of this shape should best coincide with the centroid of the two loads.

The basic design of all these combined footings is essentially the same as that for a single-spread footing. Critical moment sections exist near the edge of the individual columns and about midway between them. Note that these critical sections run in two directions. Also, there may be punching shear limitations around the perimeter of each column.

Sometimes the bearing pressure under a footing is not controlled by the

FIGURE 12-7
COMBINED FOOTINGS.

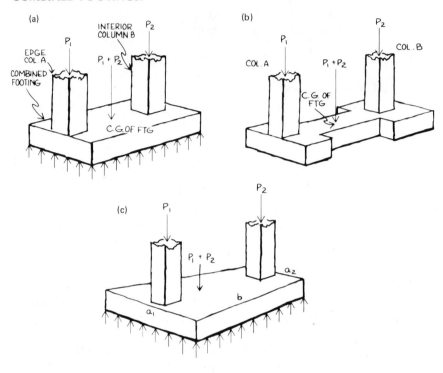

FIGURE 12-8
ASSUMED PRESSURE DISTRIBUTION UNDER ADJACENT FOOTINGS.

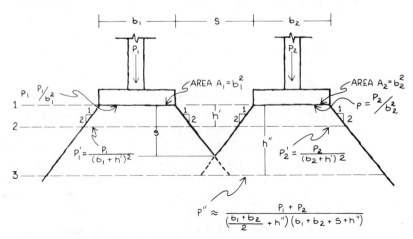

stratum immediately underneath the footing. Rather, another section further down below may be more critical because of a softer intermediate layer of soil, such as Layer no. 2 (Figure 12-8). In this case it may be assumed that the pressure immediately underneath the footing is distributed through the soil in a conical shape with a slope of one to two (~30°) so that the area at Layer 2 is increased. In case the soft layer is further down, say, at Layer 3, so that the two cones intersect each other, then the total available area to carry the two footings is somewhat reduced. This is shown by the intersecting area between the two cones at a deeper Layer 3. The lower-level bearing capacity should be checked and the footing size increased if necessary.

SECTION 4: Pile and Caisson Foundations

When the surface soil cannot adequately support a building, either because the soil is too weak or the loads too heavy, it is possible to transfer the load to strata below the top by means of piles or caissons. *Piles* are slender columns that support the loads through bearing at the top and then transfer these loads through friction along the length of the pile (due to its adhesion to the soil) or through direct bearing (at the bottom) to rock or some strong stratum.

Piles are usually grouped together with more than one or two underneath each footing. They can form a row underneath a line footing, or be arranged in any pattern underneath a square, rectangular, or combined footing, usually with their centroid concurrent with the resultant of the applied load so that the total load is assumed to be equally divided among the piles. Under certain loading conditions, a pile foundation may be subjected to an eccentric loading (Figure 12-9), in which case the applied moment would produce an eccentricity that would result in different loadings among the piles.

Generally speaking, pile or caisson foundations are more costly than spread foundations and therefore should be used only when spread foundations become uneconomical. Still, they are often less costly than mat foundations.

Piles are generally of two types. *End-bearing piles* transmit loads from the top to the bottom of the pile, which bears on a firmer stratum beneath. *Friction piles* transmit load from the pile gradually to the surrounding earth through friction or adhesion between the pile surface and its surrounding soil. Actually, most piles are a combination, transmitting loads through both bearing and friction. The percentage of loads transmitted by bearing or by friction will depend on the soil conditions.

To get an idea of pile capacities, one can refer to Table 12-2. This table gives maximum design loads allowed on different types of piles. But it

FIGURE 12-9
PILE FOUNDATION SUBJECTED TO ECCENTRIC LOADING.

(a) PLAN OF FOOTING

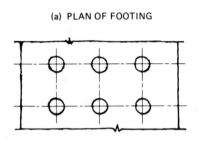

(b) VERTICAL SECTION

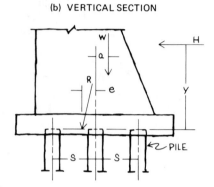

Table 12-2
Maximum Design Loads on Single Piles

TYPE OF PILE	ALLOWABLE STRESSES WHEN BEARING ON ROCK, PSI	MAXIMUM DESIGN LOAD, TONS		
		FOR PILES ON ROCK	FOR PILES ON SOFT ROCK, HARDPAN, OR GRAVEL OVER ROCK	FRICTION PILES
Timber piles		25†	25†	30
Closed-end pipe, cast-in-place concrete, and compacted concrete piles		120	80	60
concrete	$0.25f'_c$*			
Reinforcing steel	$0.40f_y$*			
Steel shell over $\frac{1}{4}$ in. thick	12,000			
Open-end concrete-filled pipe:				
18-in.-diameter or larger		250	80	60
Under 18 in. diameter		200	80	80
Concrete	$0.25f'_c$*			
Steel shell	12,000			
Steel bearing piles	12,000	150	80	60
Drilled-in caissons		No limit		
Concrete	$0.25f'_c$*			
Steel shell	12,000			
Steel core	18,000			

*f'_c = 28-day compressive strength of concrete test cylinder, psi
f_y = yield strength of steel reinforcing; limited to 40,000 psi
†Some building codes limit timber piles to as low as 12 tons; others place no limitation on design load when bearing on rock or other hard stratum other than allowable stresses on pile cross section at upper surface of soil supporting the pile.

Source: Civil Engineering Handbook, Standard Handbook For Civil Engineers, F. S. Merritt, editor; McGraw-Hill Book Company, copyright © 1968, New York.

should be understood that the capacity of each pile also depends greatly upon the actual soil properties surrounding the pile and hence cannot be predetermined for a given site without tests.

When a group of piles is under eccentric loading or overturning moment, some of the piles may be subjected to uplift forces. Resistance of piles to uplift is developed through friction or adhesion along the side of the piles. Again, the capacity to resist such uplift depends upon the local conditions. As a rule-or-thumb, it is often approximated that the uplift capacity of a friction pile is about 40 to 50 percent of its downward capacity.

EXAMPLE 12-2

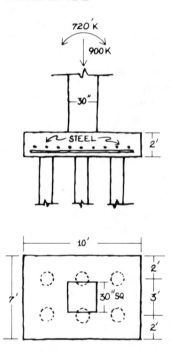

A 6-pile footing is to sustain an axial load $P = 900$ k and a moment of 720 kip-ft. Determine the load on the piles. $P = 900$ k is supported by 6 piles, $M = 720$ kip-ft. is resisted by only the end piles acting with an arm of 6 ft. Vertical load at $900/(6) = 150$ k per pile

$$\text{M-Load at } \frac{720}{6 \text{ ft}} = 120 \text{ k each end}$$

$$= 60 \text{ k per pile.}$$

Hence the end piles will be under $150 \pm 60 = 210$ k compression or 90 k compression. The two center piles are not affected by the moment, and carry 150 k each.

Since piles are under compression, they are *in fact* slender columns. However, the lateral support provided by the soil against the buckling of the piles is usually sufficient to prevent any actual buckling from taking place. Only when piles pass through water or saturated soft mud will they act like unsupported columns. Hence, in most cases, the rule-of-thumb strength of piles shown in Table 12-2 can be considered safe from buckling.

Piles are usually spaced *at least* (3) diameters apart in order to be more-or-less individually effective. When many piles are provided in a group, then there is a pile-group effect, which will reduce the average pile capacity. Such reduction can be minimized if the piles are spaced farther apart.

Although the bearing capacity of piles is best determined by a load test, the safe allowable load is about 50 percent of the ultimate load. Ultimate load may be defined as a load producing a permanent settling of a pile (on the order of $\frac{1}{4}$ in. or more) over a period of some two days. For buildings, the usual rule is that the allowable load is one-half that load causing a total settling not exceeding 1 in. When a load settling diagram shows a definite yield point, the safe load may be taken as 50 percent of the yield point load.

The capacity of a pile is often determined by empirical pile-driving formulas rather than soil analysis. These formulas, while not entirely reliable, usually yield fairly conservative, and thus safe, results. The most well-known formula is the *Engineering News* formula, which states

$$R = \frac{2WH}{(S+1)}$$

where R = safe load in a single pile, W = weight, in pounds, of striking parts of hammer, H = height of fall less twice the height of bounce in feet, S = average penetration in inches per blow for the last 5 to 10 blows. That is, R is a measure of the soil's resistance to being penetrated.

In regions where timber abounds, wood piles can be quite economical, especially when used as friction piles (when the pile's own strength does not govern). Cast-in-place concrete piles can carry heavier loads than either timber or driven piles. They are usually constructed by placing concrete in permanent steel pipes. These pipe forms can be driven with a removable mandrel placed inside to stiffen the shell. After driving, the mandrel is collapsed mechanically or pneumatically to allow withdrawal from the shell. In certain cases, it is also possible to drill holes and fill them with concrete (Figure 12-10). The holes may then be belled out at the bottom for load distribution.

Precast concrete piles are usually made in a factory, transported, and driven into position. Since the 1950s, these piles have been mostly prestressed and they can stand a great deal of abuse, during both transportation and driving (Table 12-3). They compete quite well economically with steel or reinforced concrete piles. For further information, one may refer to "Recommended Practice for Design, Manufacture and Installation of Prestressed Concrete Piling," *PCI Journal*, March–April, 1977.

FIGURE 12-10
DRILLED-IN CAISSONS. (*A*) TYPICAL SECTION OF CAISSON FOR A CONCRETE SUPERSTRUCTURE (*B*) TYPICAL SECTION OF CAISSON FOR A STEEL SUPERSTRUCTURE (*C*) TYPICAL CAISSON CROSS SECTION.

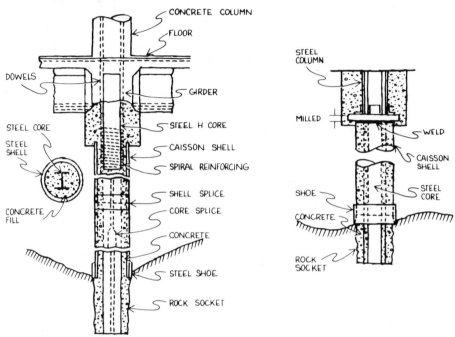

Steel-bearing piles are widely used because they can be subjected to hard driving and sustain a high compression load (Table 12-4). However, they may be more expensive than prestressed precast piles. In any case, these piles are usually installed by being driven with a hammer into the ground. In addition, preexcavating, jetting, vibrating, and hand-digging methods are sometimes used to put piles in place.

Drop hammers for pile-driving range from 500 to 3000 lb and drop a height of 1 to 4 ft. Single-acting and double-acting hammers weigh from 3000 to 40,000 lb. They are driven with a steam engine at a rate of 60 to 200 or more blows per minute.

Caissons are large hollow piles or large cylindrical enclosures that are sunk into the ground. They can sink under their own weight through soft soil or with the help of cutting edges. Their sinking can be assisted by jacking, jetting, excavating, and undercutting. Caissons for building foundations range from 3 to 10 ft or more in diameter. They can be belled out at the bottom to carry larger load. The hole in the caisson can be left open or filled with concrete. Large caissons, either of the open type without top and bottom, or of the pneumatic type using compressed air to equalize water pressure below ground, are used for heavy construction but not often applied to buildings.

Table 12-3
Prestressed Concrete Piles

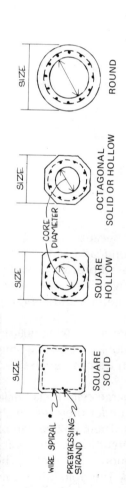

SQUARE SOLID   SQUARE HOLLOW   OCTAGONAL SOLID OR HOLLOW   ROUND

WIRE SPIRAL *
PRESTRESSING STRAND †
CORE DIAMETER

5 TURNS @ 1"
16 TURNS @ 3"
6" PITCH
16 TURNS @ 3"
5 TURNS @ 1"
1"

TYPICAL ELEVATION*

* WIRE SPIRAL VARIES WITH PILE SIZE
† STRAND PATTERN MAY BE CIRCULAR OR SQUARE

SIZE IN.	CORE DIA. IN.	AREA IN.^{2}	WEIGHT PLF	MOMENT OF INERTIA IN.^{4}	SECTION MODULUS IN.^{3}	RADIUS OF GYRATION IN.	PERIMETER (FT.)	ALLOWABLE CONCENTRIC SERVICE LOAD, TONS[2] f'_c			
				SECTION PROPERTIES[1]				5000	6000	7000	8000
				SQUARE PILES							
10	Solid	100	104	833	167	2.89	3.33	73	89	106	122
12	Solid	144	150	1728	288	3.46	4.00	105	129	152	176
14	Solid	196	204	3201	457	4.04	4.67	143	175	208	240
16	Solid	256	267	5461	683	4.62	5.33	187	229	271	314
18	Solid	324	338	8748	972	5.20	6.00	236	290	344	397
20	Solid	400	417	13,333	1333	5.77	6.67	292	358	424	490
20	11 in.	305	318	12,615	1262	6.43	6.67	222	273	323	373
24	Solid	576	600	27,648	2304	6.93	8.00	420	515	610	705
24	12 in.	463	482	26,630	2219	7.58	8.00	338	414	491	567
24	14 in.	422	439	25,762	2147	7.81	8.00	308	377	447	517
24	15 in.	399	415	25,163	2097	7.94	8.00	291	357	423	488

OCTAGONAL PILES

10	Solid	83	85	555	111	2.59	2.76	60	74	88	101
12	Solid	119	125	1134	189	3.09	3.31	86	106	126	145
14	Solid	162	169	2105	301	3.60	3.87	118	145	172	198
16	Solid	212	220	3592	449	4.12	4.42	154	189	224	259
18	Solid	268	280	5705	639	4.61	4.97	195	240	284	328
20	Solid	331	345	8770	877	5.15	5.52	241	296	351	405
20	11 in.	236	245	8050	805	5.84	5.52	172	211	250	289
22	Solid	401	420	12,837	1167	5.66	6.08	292	359	425	491
22	13 in.	268	280	11,440	1040	6.53	6.08	195	240	283	328
24	Solid	477	495	18,180	1515	6.17	6.63	348	427	506	584
24	15 in.	300	315	15,696	1308	7.23	6.63	219	268	318	368

ROUND PILES

36	26 in.	487	507	60,007	3334	11.10	9.43	355	436	516	596
48	38 in.	675	703	158,199	6592	15.31	12.57	493	604	715	827
54	44 in.	770	802	233,373	8643	17.41	14.14	562	689	816	943

Source. PCI Design Handbook.

(1) Form dimensions may vary with producers, with corresponding variations in section properties.

(2) Allowable loads based on N = $A_c(0.33 f'_c - 0.27 f_{pc})$; f_{pc} = 700 psi; Check local producer for available concrete strengths.

Table 12-4
Capacities of Standard Steel Bearing Piles*

NOMINAL DEPTH, IN.	WEIGHT, LB PER FT	AREA, SQ IN.	CAPACITY, TONS AT 12,000 PSI
8	36	10.60	63.60
10	42	12.35	74.10
10	57	16.76	100.56
12	53	15.58	93.48
12	74	21.76	130.56
14	73	21.46	128.76
14	89	26.19	157.14
14	102	30.01	180.06
14	117	34.44	206.64

*For maximum allowable capacity without load tests, see Table 7-6.
Source. PCI Design Handbook.

SECTION 5: Retaining Walls and Coffer Dams

When a building is constructed beneath the ground level, excavation results in a need to resist the *lateral earth pressure* below ground. Thus, during construction, it may be necessary to build temporary retaining walls or coffer dams to restrain the earth from collapsing inward around the building foundation. We will very briefly describe these systems.

When earth is retained by a vertical wall, it exerts a horizontal pressure against the wall. This pressure varies with soil type, but it can be conservatively approximated by considering such earth pressure in terms of equivalent fluid pressure, with the fluid having an assumed weight. If the weight of the fluid is w, then the horizontal pressure exerted by that fluid at any level is given by $p = wh$.

The pressure diagram is therefore a triangle, as shown in Figure 12-11. The total pressure for a 1-ft vertical strip of that wall is then given by the formula $P = wh^2/2$. And the moment produced by that pressure is then given by

$$M = \frac{wh^2}{2} \times \frac{h}{3} = \frac{wh^3}{6}$$

The difficulty in using this formula lies in the proper choice of the equivalent fluid weight w. For fairly stable soil, the value of w may be only 10 or 20 lb per cubic foot, whereas for loose soil, especially if saturated with water, the w can easily exceed 40 or 50 lb per cubic foot, going up to 100 lb per cubic foot or more. Thus, for the purpose of final design, one must consult with a soil foundation expert to determine the value of w. However, for usual conditions, an approximation value of $w = 30$ or 40 is often used.

FIGURE 12-11
PRESSURE DIAGRAM FOR WATER.

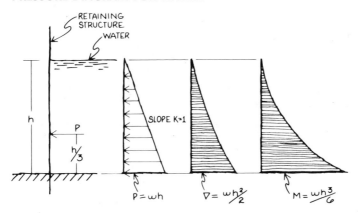

When additional weight is placed on top of the retained soil, a surcharge is applied to the back of this retaining wall and the active lateral pressure can be quite high. A usual rule is to put on top of the ground soil an additional depth of earth equal to the average weight of the building per square foot. Thus if the surcharge is 200 psf, it is about 2 ft of earth and the soil pressure on the top of the retaining wall is 2 ft of fluid with weight w.

One must also distinguish between the active (liquid) pressure and the passive pressure of soil, the latter being the capacity of the soil to resist force applied laterally to that soil. Generally speaking, the passive lateral pressure capacity is easily three to six times the value of the active pressure (say, on the order of 100–200 psf). However, to obtain accurate values, one must consult a soil mechanics expert.

The following example will illustrate these basic concepts.

Example 12-3: Retaining Wall Design

Preliminary design is to be made for an R.C. retaining wall 15 ft high.

Equivalent fluid pressure = 40 pcf. Use 2 ft surcharge.

Choose top wall $t = 7\frac{1}{2}$ in. (usual min. 6 in.–9 in.)

Bottom wall $t = 15$ in.

$L/t = 15 \text{ ft} \times 12/(15) = 12$, which is about right for this type of wall.

Foundation width = 60 percent × 15 ft = 9 ft with a two foot toe.

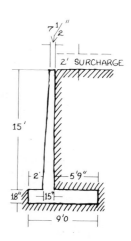

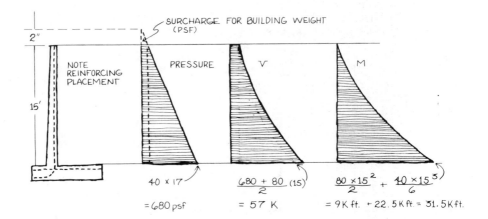

$M_{max} = 31.5$ k ft/ft width; $d = 15$ in. -2 in. $= 13$ in., $a = 7/8(13$ in./12$) \simeq 1$ ft
$T = (31.5/1) = 31$ k, Use 40 ksi steel, $A_s = (31/20) = 1.6$ in.^{2}/ft.
Use 2 no. 8 bars/ft (or one no. 8 at 6 in. c.c.)
$p_s = (1.6/13 \times 12) = 1$ percent which is just about right.

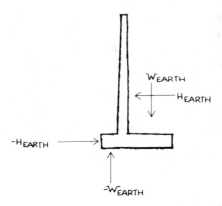

$M = 31.5$ k'/ft at base of wall must be resisted by the footing and stabilized by the weight of earth fill above the heel acting against the upward earth pressure on the toe of the footing. Reinforcement in the footing should be designed for these moments. In addition, the footing must be designed against horizontal sliding and not to produce excessive earth pressure below. But these matters are beyond the scope of this book and you are referred to any standard book on R.C. design for details.

The overturning moment on a retaining wall varies as the third power of the height of the wall. Therefore, it is important to keep retaining walls, particularly cantilevered walls (supported at the base only), to a minimum height. When walls must exceed 15 or 20 ft in height, it may no longer be economical to construct them as cantilevers.

When a building is completed, the floors of the building basement may be used to support the retaining walls, serving as horizontal diaphragms that transfer the earth pressure from one side of the building onto the other side (Figure 12-12). But, during the excavation of foundations for the basement of a building, it is often necessary to construct temporary walls of considerable height, or coffer dams, to protect the excavation process. Coffer dams may be constructed in a variety of ways.

A common method is to drive sheet piles around the periphery of the excavation. If the excavation is not too deep, say no more than 10 or 20 ft, cantilever sheet piles, generally of steel, can be used. When the depth exceeds 15 or 20 ft, it may be necessary to brace these sheet piles from the excavated area. Some common types of bracing are shown in Figure 12-13. When the foundation is small, it may be economical to use cross-lot bracing, holding one side against the other. But for large foundations, diagonal bracing may be more suitable.

Horizontal sheeting can be anchored to soldier piles driven at appropriate intervals (Table 12-5) and pile tops with concrete dead-man (Figure 12-13c). Or, when rock is close, the walls can be tied back with drilled and grouted rock anchors, as shown in Figure 12-14. These options have the advantage of providing an unobstructed excavation space for efficient construction. In fact, tie-back designs are becoming more and more popular, and in some cases application can be considered permanent.

Slurry trenches are sometimes used for constructing concrete walls (Figure 12-15). This method consists of excavating a trench 2 or 3 ft wide and filling that trench with a bentonite slurry that has a specific gravity of about 1.1, exerting a pressure against the side of the excavated trench, which

FIGURE 12-12
RETAINING WALLS MAY BE SUPPORTED BY FLOORS BELOW GRADE

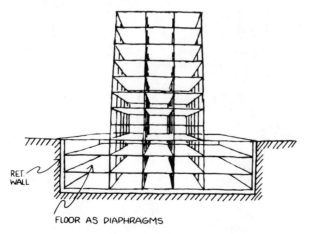

RET.
WALL

FLOOR AS DIAPHRAGMS

FIGURE 12-13
TYPES OF COFFERDAM BRACING.

(a) CIRCULAR COFFERDAM—
VERTICAL SECTION

(b) DIAGONALLY BRACED
COFFERDAM—VERTICAL
SECTION

(c) TOP-ANCHORED
COFFERDAM—
VERTICAL SECTION

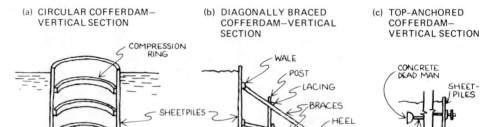

(d) ONE-WAY-BRACED
COFFERDAM

(e) TWO-WAY-BRACED
COFFERDAM

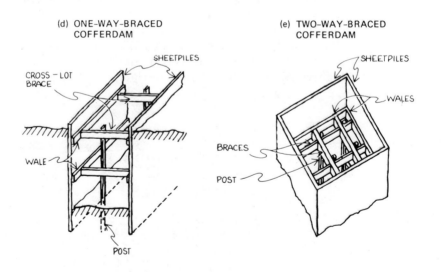

Table 12-5
Usual Maximum Spans of Horizontal Sheeting with Soldier Piles, Ft

NOMINAL THICKNESS OF SHEETING, IN.	IN WELL-DRAINED SOILS	IN COHESIVE SOILS WITH LOW SHEAR RESISTANCE
2	5	4.5
3	8.5	6
4	10	8

FIGURE 12-14
VERTICAL SECTION SHOWING PRESTRESSED TIE-BACKS FOR SOLDIER BEAMS.

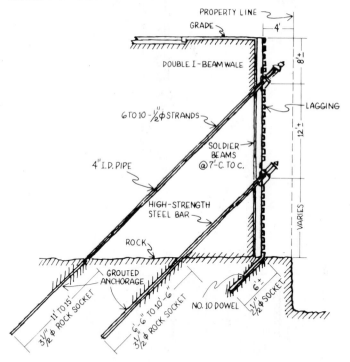

keeps it from collapsing. Excavation is carried out in sections, say, 20 ft long. When the bottom of the wall is reached, reinforcement is placed into that section and concrete is pumped into the trench, replacing the slurry, which may flow into the next section or may be pumped back for reuse. After the concrete has hardened, a wall is obtained.

FIGURE 12-15
**SLURRY-TRENCH METHOD FOR CONSTRUCTING A
CONTINUOUS CONCRETE WALL. (A) EXCAVATING ONE
SECTION (B) CONCRETING ONE SECTION WHILE ANOTHER IS
EXCAVATED.**

(a)

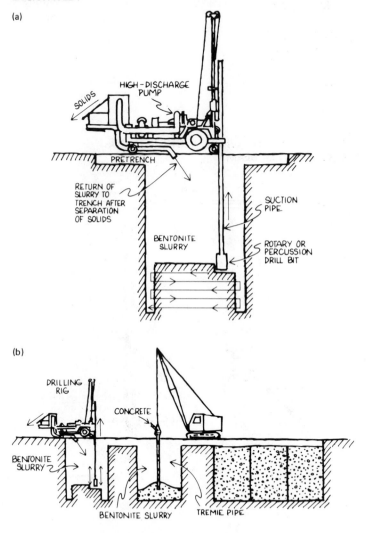

(b)

It is not possible to treat the issues of foundations design in any depth in
an introductory text. However, the above principles should be sufficient to
enable you to give schematic consideration to foundation design
requirements.

13
Construction

SECTION 1: Introduction

Construction of structures is a vital consideration in dealing with structural concepts and systems. In itself, it is an art heavily influenced by social and economic factors. It is, in fact, as much empirical as it is science and technology. Although construction consists of a field of activities *susceptible* to theoretical research, it must also be backed up by a great deal of experience. This makes it difficult to delve into the issues of construction in any great detail in this chapter. Our purpose is simply to introduce the relation between the more basic issues of construction and the overall design of structural systems.

This book has focused on conceptual and preliminary design of structural systems. Still, it is important to realize that construction feasibility and economics will depend on consideration of building methods, erection procedures, and management. It would be irresponsible to conceive of a scheme that looks beautiful on paper but ignores requirements for construction and availability of trained labor and material resources. Even during the overall stages of planning and design, one should be able to verify theoretical objectives in constructive terms that will fit particularized local conditions.

In broad terms, construction factors are not limited to work in the field or at the site. They embody the off-site issues of component fabrication, handling, storage, and transportation of building components as well as those of on-site erection and connection. To this list must be added the requirement of providing falsework when necessary to support the structure during construction and the availability of construction equipment required to accomplish the job.

One must first decide on a total scheme of construction that is consistent with the choice of materials or combination of materials. For example, a designer must make overall decisions as to whether the structure is to be built of steel, precast or prestressed concrete, or possibly of wood; whether the components are to be made in the shop or at the site; how the materials and/or components are to be transported and connected; the capacity, reach, and mobility of lifting equipment available. Then it is always desirable, even at the early stages of design, to consult a contractor who is familiar with the latest construction equipment and techniques are well as with the local conditions for availability of material, manpower, technical know-how, and supervision.

The total cost of a building's *structure* is made up of about 50 percent in cost of materials (which may include a certain amount of prefabrication cost in the shop) and another 50 percent in the cost of labor and equipment for erection at the site. In fact, the *basic* cost of materials is probably no more than one-third of the total cost of the structural system, with the rest going to the cost of labor and equipment. In turn, the total structural system cost will range from one-sixth to one-third of total construction costs (this is discussed in Chapter 14). Therefore, it would seem that to obtain economical construction, one must pay great attention not only to the saving of material, but even more to the saving of equipment and labor.

SECTION 2:   Steel Construction

Construction of steel structures in a building starts with shop fabrication and handling. Then the components are transported to the site, erected into position, and connected to form the structure. Since the in-place cost of steel in building structures ranges from two to four times the cost of steel as a raw material, ease of fabrication and erection is of prime significance to economy in construction.

In order to reduce fabrication costs, one must use standard and *available* shapes as much as possible. While the variety and availability varies with the local situation, angles, plates, and *WF* shapes are the most easily obtainable. Among the various *WF* and angle sections listed in steel construction handbooks, some sections are more available than others. One should always try to pick sections that are readily available in the area where construction is taking place.

Having chosen possible sections to be used in a given construction, one would try to combine these shapes and sections into various subsystems that can be easily erected and connected. Some sections and built-up shapes are shown in Figures 13-1 and 13-2.

The fabrication of these members and their connection is accomplished either by *bolting* or by *welding*. In earlier times, riveting was a common method. Riveting consists of a round steel bar forced in place to join several pieces of steel together. This method of connection and fabrication is now outdated by the use of high-tensile bolts (Figure 13-3). These high-tensile bolts depend on the clamping action produced when the nut or bolt is tightened to a predetermined tension. This tightening is done by calibrated torque wrenches to assure sufficient initial tension in these bolts. Where connection strength is not required, ordinary bolts without high initial tensioning are sometimes employed.

The second common method of connecting members is welding, which is

FIGURE 13-1
TYPICAL STEEL SHAPES.

(a) ROLLED SHAPES

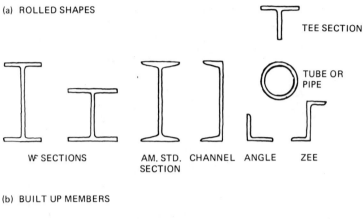

TEE SECTION

TUBE OR
PIPE

WF SECTIONS   AM. STD.   CHANNEL   ANGLE   ZEE
SECTION

(b) BUILT UP MEMBERS

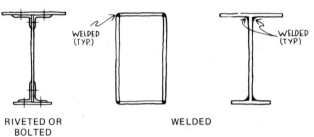

WELDED
(TYP.)

WELDED
(TYP.)

RIVETED OR   WELDED
BOLTED

a process applied by the use of heat to melt metals and to fuse components together (Figure 13-4). Briefly speaking, the heat is generated by an electric arc formed between a steel electrode and the steel part to be welded. The arc heat melts the base metal and the electrode simultaneously. The molten parts of the metal flow together, fuse, and are then allowed to cool off. Many techniques of shop and field welding have been developed, but discussion of them is beyond the scope of this book.

The fabrication of standard shapes and plates into components ready for connection at the site is again an art in itself. Fabrication is generally carried out at the shop, using well-developed methods, tools, and machinery, and employing mass-production techniques. To save labor and cost, one should minimize the types of shapes required for members and to mechanize the fabrication process as much as possible.

Figure 13-5a shows some typical connections for steel members using riveted, bolted, welded, or pinned connections. Figure 13-5b shows welded connections for trussed members using WF sections. Figure 13-5c illustrates some bolted connections for a heavy truss member in a bridge. Figure 13-6a illustrates some standard beam to column connections for buildings. Figure

FIGURE 13-2
NONSTANDARD SHAPES CAN BE BUILT UP.

(a) BUILT UP CHANNEL AND PLATE MEMBERS

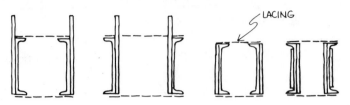

(b) BUILT UP ANGLE AND PLATE MEMBERS

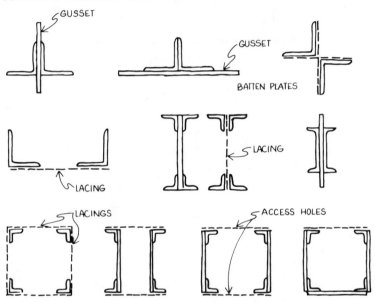

FIGURE 13-3
HIGH-TENSILE-STRENGTH BOLTS HAVE REPLACED RIVETING AS A MEANS OF CONNECTION.

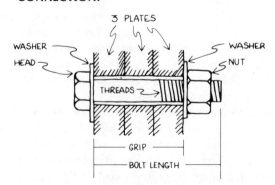

FIGURE 13-4
WELDING STEEL CONNECTIONS.

(a) PROCESS

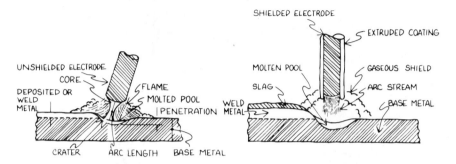

(b) TYPES OF WELDS:

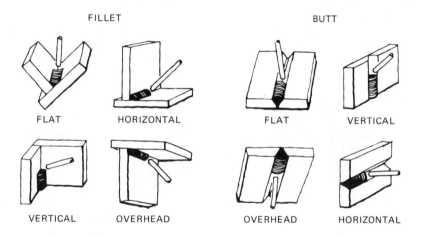

13-6*b* shows a typical column base connection. Figure 13-6*c* indicates some details for connecting roof purlins to supporting members.

After fabrication is completed in a factory, structural components are transported to the site by railroad flat cars, trucks, or barges. Then they are erected in a predetermined sequence, and they may require storage space. For multistory buildings, the components are usually erected in one-story, two-story, or even three-story tiers. For example, the World Trade Center (Photo 13-1*a* and *b*) was constructed of three-story prefabricated columns and spandrel subassemblies. Thus, after the foundations are completed, the columns are raised, set on the base plates, and bolted in place. During erection, the columns are braced laterally until the structure is completed.

After the columns are set, beams and girders are hoisted in place and temporarily bolted to the columns. Industrial buildings of one or two stories

FIGURE 13-5
EXAMPLES OF STEEL CONNECTIONS.

(a) RIGHT ANGLE AND SPLICE CONNECTIONS

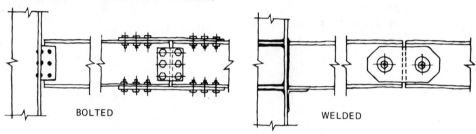

BOLTED

WELDED

(b) WELDED TRUSS CONNECTIONS

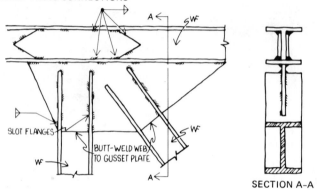

A ←— SWF

SLOT FLANGES

BUTT-WELD WEB
TO GUSSET PLATE

WF

S WF

WF

A ←

A WELDED SINGLE-GUSSET CONNECTION

SECTION A-A

(c) BOLTED TRUSS CONNECTIONS

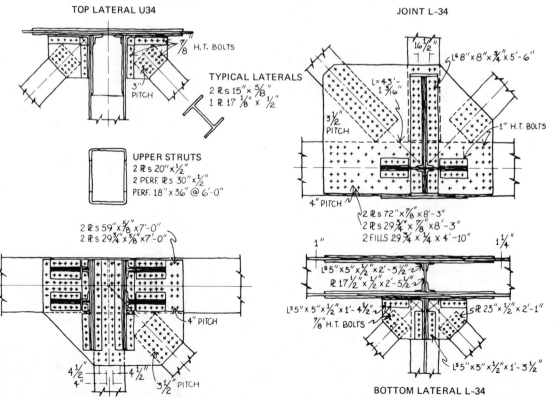

TOP LATERAL U34

JOINT L-34

$\frac{7}{8}$" H.T. BOLTS

3" PITCH

TYPICAL LATERALS
2 ℞s 15″ × $\frac{5}{8}$″
1 ℞ 17 $\frac{1}{8}$″ × $\frac{1}{2}$″

UPPER STRUTS
2 ℞s 20″ × $\frac{1}{2}$″
2 PERF. ℞s 30″ × $\frac{1}{2}$″
PERF. 18″ × 36″ @ 6′-0″

16 $\frac{1}{2}$″

L⁶ 8″ × 8″ × $\frac{3}{4}$″ × 5′-6″

L = 43′-1 $\frac{3}{16}$″

$3\frac{1}{2}$″ PITCH

1″ H.T. BOLTS

4″ PITCH

2 ℞s 72″ × $\frac{7}{8}$″ × 8′-3″
2 ℞s 29 $\frac{3}{4}$″ × $\frac{7}{8}$″ × 8′-3″
2 FILLS 29 $\frac{3}{4}$ × $\frac{1}{4}$ × 4′-10″

2 ℞s 59″ × $\frac{5}{8}$″ × 7′-0″
2 ℞s 29 $\frac{3}{4}$″ × $\frac{5}{8}$″ × 7′-0″

4″ PITCH

4 $\frac{1}{2}$″   4 $\frac{1}{2}$″
4″   3 $\frac{1}{2}$″ PITCH

JOINT U34

1″

1 $\frac{1}{4}$″

L⁹ 5″ × 5″ × $\frac{1}{2}$″ × 2′-5 $\frac{1}{2}$″
℞ 17 $\frac{1}{2}$″ × $\frac{1}{2}$″ × 2′-5 $\frac{1}{2}$″

L⁹ 5″ × 5″ × $\frac{1}{2}$″ × 1′-4 $\frac{1}{2}$″
$\frac{7}{8}$″ H.T. BOLTS

℞ 23″ × $\frac{1}{2}$″ × 2′-1″

L⁹ 5″ × 5″ × $\frac{1}{2}$″ × 1′-3 $\frac{1}{2}$″

BOTTOM LATERAL L-34

FIGURE 13-6
MORE EXAMPLES OF STEEL CONNECTIONS.

(a) BEAM-COLUMN CONNECTIONS

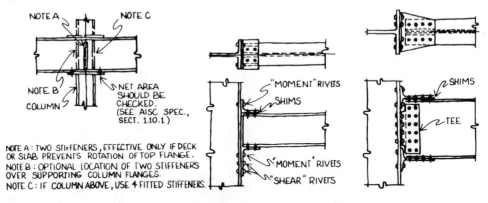

NOTE A : TWO STIFFENERS, EFFECTIVE ONLY IF DECK OR SLAB PREVENTS ROTATION OF TOP FLANGE.
NOTE B : OPTIONAL LOCATION OF TWO STIFFENERS OVER SUPPORTING COLUMN FLANGES.
NOTE C : IF COLUMN ABOVE, USE 4 FITTED STIFFENERS.

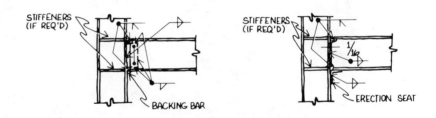

(b) COLUMN BASE CONNECTIONS   (c) ROOF PURLIN CONNECTIONS

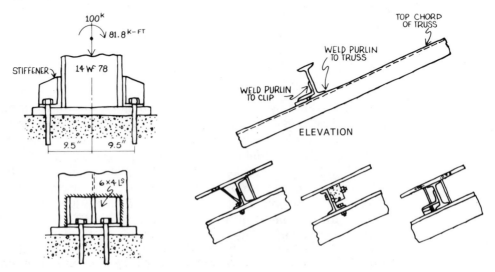

PHOTO 13-1
THE WORLD TRADE CENTER (a) COMPLETED BUILDINGS. (b) STEEL ERECTION OF WORLD TRADE CENTER.

(a)

can be erected by small truck cranes. Buildings up to 200 ft high are usually erected by larger truck cranes (Figure 13-7). Taller buildings require the use of special derricks, which are raised to the top of the completed portion of a structure as the building construction progresses upward.

Self-lifting guy derricks (Figure 13-8) are also used for multistory buildings. These derricks can jump themselves from one story to another. The boom serves as a gin pole to hoist the mast to a higher level.

The tower crane (Figure 13-9) is more expensive than other types, but has its own advantages. The control station can be located on the crane or at a distant position that enables the operator to see the load at all times. Also, the equipment can be used to place concrete and other materials directly in

(b)   Courtesy of the Port Authority of New York and New Jersey

the forms for floors and roofs, but eliminating chutes, hoppers, and barrows. Photo 13-2 and 13-3 show the erection of prefabricated concrete housing units and precast tee roof members.

For long-span steel construction, temporary staging may be required. The construction of a 200-ft dish, on the other hand, consisted of a tension ring near center and a compression ring around the periphery, Photo 13-4a. The roof was made up of 32 pieces of precast double tees erected with two cranes, Photo 13-4b. Falsework was set only for the inner ring, which was prestressed with a force of 3.2 million lb before removing the temporary supports.

464

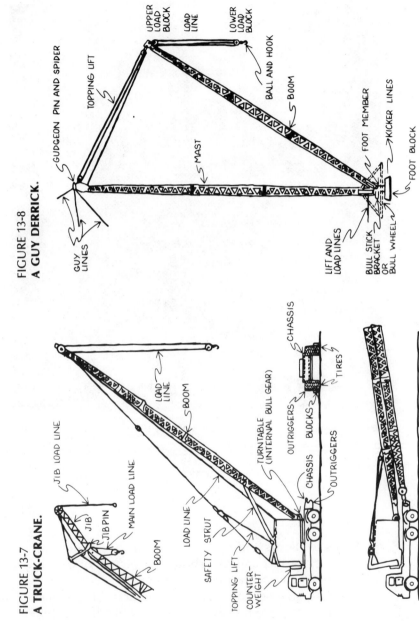

FIGURE 13-8
A GUY DERRICK.

GUDGEON PIN AND SPIDER

TOPPING LIFT

UPPER LOAD BLOCK

LOAD LINE

LOWER LOAD BLOCK

BALL AND HOOK

BOOM

MAST

FOOT MEMBER

KICKER LINES

FOOT BLOCK

GUY LINES

LIFT AND LOAD LINES

BULL STICK BRACKET OR BULL WHEEL

FIGURE 13-7
A TRUCK-CRANE.

JIB LOAD LINE

JIB

JIB PIN

MAIN LOAD LINE

BOOM

LOAD LINE

SAFETY STRUT

TOPPING LIFT

COUNTER-WEIGHT

LOAD LINE

BOOM

TURNTABLE (INTERNAL BULL GEAR)

CHASSIS

OUTRIGGERS

CHASSIS BLOCKS

OUTRIGGERS

CHASSIS

TIRES

FIGURE 13-9
A TOWER OR SLEWING CRANE.

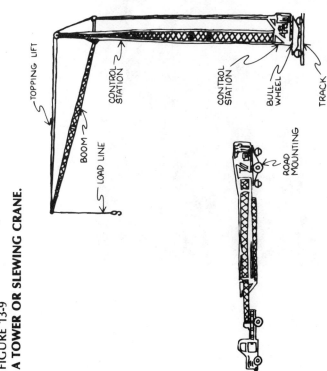

PHOTO 13-2
CRANE ERECTION OF PREFABRICATED CONCRETE HOUSING UNITS. THE ANDERSON SYSTEM, HONOLULU, HAWAII. (T. Y. LIN HAWAII, STRUCTURAL ENGINEERS)

PHOTO 13-3
**PRECAST TEE ROOF MEMBERS ERECTED WITH TWO CRANES FOR
9-STORY OFFICE BUILDING, UNIVERSITY OF CALIFORNIA, DAVIS
(T. Y. LIN INTERNATIONAL, STRUCTURAL ENGINEERS).**

PHOTO 13-4
**HIWAY HOUSE CONVENTION CENTER, PHOENIX, ARIZONA
(PERRY NEUSCHATZ, ARCHITECT; T. Y. LIN INTERNATIONAL,
STRUCTURAL ENGINEERS). (a) COMPLETED BUILDING. (b) UNDER
CONSTRUCTION.**

(a)

(b)

SECTION 3:   In-Place Concrete Construction

Concrete reinforced with steel is a more widely used structural material for construction. Concrete is a mixture of Portland cement, water, and aggregates (such as stone and sand) to form a more-or-less fluid mixture that can be placed in forms to produce almost any desired shape. Since concrete is strong in compression but weak in tension, it is reinforced with steel bars to resist the tensile forces. In modern prestressed concrete construction, steel tendons (bars, wires, or strands) are prestressed and anchored to the concrete at their ends, thus putting concrete into compression and enabling it to resist tension.

Of the total cost of concrete construction, concrete as a material is a relatively minor item (say, 15–20 percent), whereas steel reinforcement, in the form of either reinforcing bars or prestressing tendons, constitutes a larger portion of the cost (20–30 percent). But the most important part of the cost (50–60 percent) is *form work* and falsework.

Concrete as a material, its mix design, and so on is an entire technology in itself. This is evidenced by the enormous number of publications on concrete, such as those by the American Concrete Institute and the Prestressed Concrete Institute. But from the construction point of view, perhaps the most important item is the requirement of form-work for cast-in-place concrete. Form-work retains wet concrete until it sets and serves to

FIGURE 13-10
A SLIP FORM FOR A CONCRETE WALL.

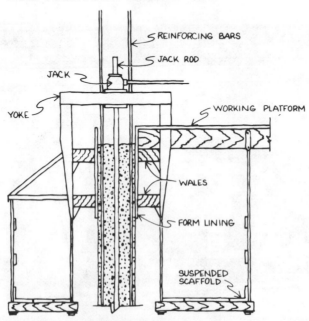

produce the desired shapes and surfaces for the concrete. Forms are supported on falsework of adequate strength and rigidity to hold the gravity and the lateral force of fluid concrete. Since the cost of forms is a major item in concrete construction, it must be minimized as much as possible. One approach is to make the forms reusable. This is accomplished in many ways. Figure 13-10 shows the essentials of a slip form for concrete walls, wherein the form (about 4 ft high) is jacked upward as the concrete hardens. These slip forms move at about 2 to 12 in. each hour as determined by the setting time and strength required of the concrete. This approach can cut costs and construction time for multistory buildings (Photo 13-5).

"Flying forms" afford another way to reuse forms. These forms do not really fly, but are taken from one floor to another by a crane so that one set of forms can be used to construct story upon story.

Sometimes lift slabs are used, whereby the slabs are cast at ground level

PHOTO 13-5
EIGHTEEN-STORY APARTMENT BUILDING USING SLIP-FORMED WALLS AND POSTTENSIONED SLABS, SOUTH SAN FRANCISCO, CALIFORNIA (CLAUDE OAKLAND, ARCHITECT; T. Y. LIN INTERNATIONAL, STRUCTURAL ENGINEERS).

FIGURE 13-11
COMPLEX FLOORS CAN BE DROP-FORMED.

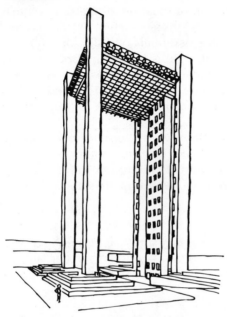

This drawing represents a concept proposed by Hanell and Hamilton, Architects, together with T. Y. Lin and Associates.

one on top of another with a separating agent between them. After all the slabs have been cast, they are lifted successively into position by synchronized jacks that ride up columns.

Figure 13-11 shows the process of drop forming, which thus far has been used only occasionally. When the bottom surface of the floor is of a complicated shape rather than flat, it may be worthwhile to make a special form and drop it from the top floor downward so as to serve as forming for all the floors.

There are, generally speaking, two types of plants for mixing concrete. One is known as *central mixing*. Central mixing is accomplished by a relatively small concreting plant that may be erected at the site rather easily. Then concrete raw materials would be transported to that plant dry, mixed with water, there, and immediately used for construction nearby.

As distinct from central mixing, *transit mixers* are often used. This means concrete is mixed at a large premixing concrete plant, from which fluid ready-mixed concrete will be transported to the site (generally within one hour's drive). This means concrete mixing will be under standardized control, but it does require special ready-mix trucks to carry the concrete to the site to be poured in place.

Table 13-1
Concrete Cover for Steel Reinforcement

Concrete deposited against the ground, 3 in.
Concrete exposed to seawater, 4 in., except precast-concrete piles, 3 in.
Concrete exposed to the weather or in contact with the ground after form
removal, 2 in. for bars larger than No. 5 and $1\frac{1}{2}$ in. for No. 5 or smaller.
Unexposed concrete, $\frac{3}{4}$ in. for slabs and walls, $1\frac{1}{2}$ in. for beams and girders, and
$\frac{3}{4}$ in. for joists not more than 30 in. apart.
Except for slabs and joists, cover should at least equal the bar diameter.

(ACI 318, Building Code Requirements for Reinforced Concrete, American Concrete Institute,
P.O. Box 4754, Redford Station, Detroit, Mich. 48219; Standard Specifications for Highway
Bridges, American Association of State Highway Officials, National Press Building, Washington,
D.C. 20004.)

After the concrete is discharged from the mixer, it is then distributed to
be poured in place using belt conveyers, chutes, steel buckets, and barrels.
Sometimes concrete is even pumped through a pipe that can be moved
around to place the concrete. This is quite convenient, but it has certain
problems, such as requiring too wet a mix, which can reduce strength
and/or exacerbate shrinkage and crack problems.

Steel *reinforcing bars* range in diameter from $\frac{1}{4}$ in. to $2\frac{1}{4}$ in. Sizes are
designed by numbers, as shown in Appendix B. Most of these steel bars
have yield points of 40, 50, or 60 ksi.

Wire meshes or wire fabric offer a convenient means for slab
reinforcement. They offer the advantage of easy placement. Both wires or
bars are shipped from a mill to a fabricator, who may cut, bend, and place
them at the site. Sometimes reinforcing steel subassemblies are
prefabricated in a shop, or even on the job.

Steel bars are cut, bent, and spliced according to certain standards, as
prescribed in various technical pamphlets. In a concrete component, bars
are spaced at least three diameters apart and they require a certain amount
of cover (from $\frac{3}{4}$ in. to about 2 in. or even 3 in.) for protecting the steel
against fire and/or corrosion. Table 13-1 shows the amount of concrete
cover for steel bars.

Prestressed concrete tendons require high-strength steel and can be of
wires, strands, or bars. The most commonly used one is the 270 ksi seven-
wire strand, $\frac{1}{2}$ in. in diameter. If used for pretensioned members, they are
simply placed in the molds and tensioned prior to placement of concrete.
For posttensioning, they may be threaded through conduits or, when
greased and wrapped in waterproof paper, placed in the forms before
casting. Table 13-2 shows the properties of some typical wires and strands
used for prestressing. When posttensioned, the tendons are anchored to
the concrete by various types of end anchorages. Some of these are shown
in Figure 13-12.

Table 13-2
Material Properties of Prestressing Steels

SMOOTH PRESTRESSING BARS, f_{pu} = 160 ksi

NOMINAL DIAMETER, in.	3/4	7/8	1	1 1/8	1 1/4	1 3/8
Area, sq in.	0.442	0.601	0.785	0.994	1.227	1.485
Weight, plf	1.50	2.04	2.67	3.38	4.17	5.05
0.7 f_{pu} A_{ps}, kips	49.5	67.3	87.9	111.3	137.4	166.3
0.8 f_{pu} A_{ps}, kips	56.6	77.0	100.5	127.2	157.0	190.1
f_{pu} A_{ps}, kips	70.7	96.2	125.6	159.0	196.3	237.6

SEVEN-WIRE STRAND, f_{pu} = 270 ksi

NOMINAL DIAMETER, in.	3/8	7/16	1/2	9/16	0.600
Area, sq in.	0.085	0.115	0.153	0.192	0.215
Weight, plf	0.29	0.40	0.53	0.65	0.74
0.7 f_{pu} A_{ps}, kips	16.1	21.7	28.9	36.3	40.7
0.8 f_{pu} A_{ps}, kips	18.4	24.8	33.0	41.4	46.5
f_{pu} A_{ps}, kips	23.0	31.0	41.3	51.8	58.1

PRESTRESSING WIRE

DIAMETER	0.105	0.120	0.135	0.148	0.162	0.177	0.192	0.196	0.250	0.276
Area, sq in.	0.0087	0.0114	0.0143	0.0173	0.0206	0.0246	0.0289	0.0302	0.0491	0.0598
Weight, plf	0.030	0.039	0.049	0.059	0.070	0.083	0.098	0.10	0.17	0.20
Ult. strength, f_{pu}, ksi	279	273	268	263	259	255	250	250	240	235
0.7 f_{pu} A_{ps}, kips	1.70	2.18	2.68	3.18	3.73	4.39	5.05	5.28	8.25	9.84
0.8 f_{pu} A_{ps}, kips	1.94	2.49	3.06	3.64	4.26	5.02	5.78	6.04	9.42	11.24
f_{pu} A_{ps}, kips	2.43	3.11	3.83	4.55	5.33	6.27	7.22	7.55	11.78	14.05

Source. PCI Design Handbook.

FIGURE 13-12
SOME POSTTENSIONING ANCHORAGES.

(a) BUTTON–HEADED ANCHORAGE STRESSING
HEAD THREADED FOR JACK ATTACHMENT

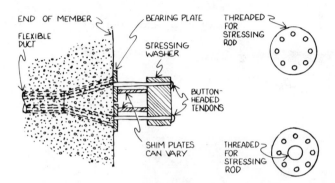

END OF MEMBER

FLEXIBLE DUCT

BEARING PLATE

STRESSING WASHER

THREADED FOR STRESSING ROD

BUTTON-HEADED TENDONS

SHIM PLATES CAN VARY

THREADED FOR STRESSING ROD

(b) ROEBLING SYSTEM HAS STRANDS SWAGED TO A
LONG THREADED STRESSING HEAD (NOTE LOCKNUT)

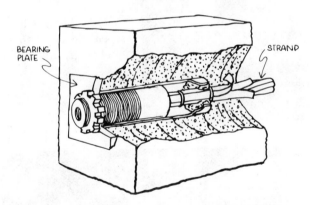

BEARING PLATE

STRAND

The usual strength of concrete is measured by standard cylinders, 6 in. in diameter and 12 in. high. The 28-day compressive strength of such cylinders is used as a measure of the strength of concrete. For foundation and concrete on the ground, a strength of 3000 psi is common. For above-ground construction, concrete strength is usually between 3000 and 4000 psi. For precast, prestressed concrete construction, a range of 5000 to 6000 psi early-strength concrete is often used.

To prestress tendons in prestressed concrete construction, hydraulic jacks, electrically operated, are commonly used. The capacity of these jacks may range from 25 tons up to 500 tons or more. Figure 13-13 shows one of these jacks in operation and a cone type anchorage.

Although in-place concrete construction is adaptable to special forms,

FIGURE 13-3
HYDRAULIC JACKS PULL STRANDS AGAINST ANCHORAGES CAST INTO CONCRETE.

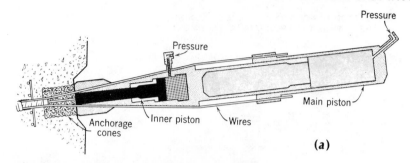

(a)

(b)

(a) Freysinnet double-acting jack pulls strand wires and then forces a gripping cone to anchor wires

(b, c) Anchorages (CCL type) for seven and single strand applications. Anchorages are cast in position.

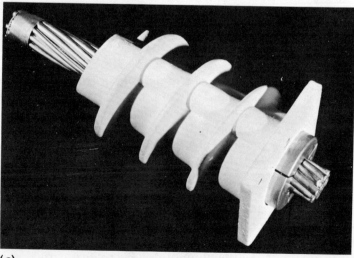

(c)

FIGURE 13-14
**UNUSUAL STRUCTURAL FORMS ARE POSSIBLE WITH
PRESTRESSED CONCRETE.**

(a) MODEL OF DETROIT METROPOLITAN AIRPORT ROOF MODULE

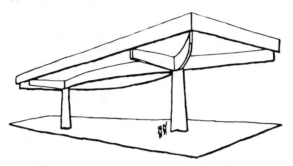

(b) SCHEMATIC OF PRESTRESSED CURVED BEAMS

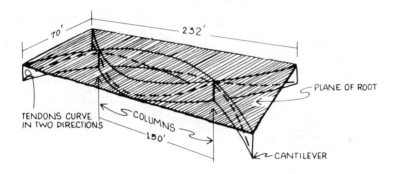

some of these forms have been standardized, such as joist forms with joists spaced on 3-ft centers, and waffle forms with the waffle ribs space on 3- to 10-ft centers. In addition to these standard forms, it is possible to use almost any form that the architect or engineer may wish, provided that form can be reused a number of times. These forms can be made of various materials—timber, plywood, plastic, or steel. The latter ones can be reused many times.

Some unusual structural forms can be designed into a concrete structure. An example of this is the curved beam roof of the Detroit Metropolitan Airport. Each panel of that roof covering an area 232 ft × 70 ft is supported by two curved girders resting on two columns (Figure 13-14). This structure is posttensioned in place with a center span of 150 ft and a cantilever of 40 ft at each end. Designed by the load-balancing method, this structure has no measurable deflection under its own weight. Extreme temperature variations produced deflections not exceeding $\frac{3}{4}$ in.

FIGURE 13-15
TYPICAL STANDARD TEE BEAM SECTIONS (LENGTH VARIES).

(a) SINGLE TEE

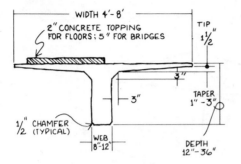

TABLE OF PROPERTIES

(WIDTH = 6'; WEB = 8"; TAPER = 2", NO TOPPING)

DEPTH, IN.	AREA, IN.²	I, IN.⁴	c_b, IN.
12	265	2360	9.2
16	297	6170	11.8
20	329	11,730	14.5
24	361	19,700	17.0
28	393	30,400	19.5
32	425	44,060	21.9
36	457	61,000	24.2

(b) DOUBLE TEE

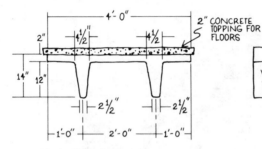

TABLE OF PROPERTIES

TOPPING	AREA, IN.²	I, IN.⁴	c_b, IN.
WITH 2" TOPPING	276	4456	11.74
WITHOUT TOPPING	180	2862	10.00

SECTION 4:   Precast Concrete Construction

To minimize forming and falsework at the job site, concrete members can be precast or produced in a factory. In a typical factory, the tendons are pretensioned from bulkheads at one end of a stressing bed to the other, with forms placed along the length of the bed (refer to Photo 8-3). Steel reinforcement is laid inside the forms and concrete is poured along the length. After the concrete has hardened, the tendons are cut from their bulkheads, thereby releasing the force to compress the concrete, producing pretensioned concrete.

Some popular forms for precast and pretensioned concrete are shown in Figures 13-15a and 13-15b. Figure 13-15a shows a single T section widely used for spans from 50 ft up to 80 ft or 100 ft. Figure 13-15b shows a shallower double T section, generally spanning from 30 to 60 ft. Figures 13-16a and 13-16b show some I sections and cored slabs used mostly for bridges. Figure 13-17a shows cored slabs used for building floors and roofs.

These figures show some standard sections, still, it is quite possible to

FIGURE 13-16
HEAVY LOADS REQUIRE HEAVY SECTIONS.

(a) STANDARD 'I' SECTIONS FOR BRIDGES

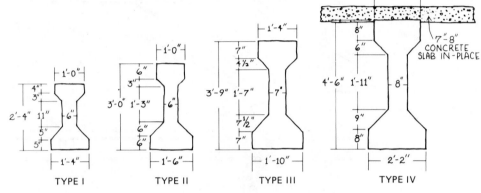

TYPE I TYPE II TYPE III TYPE IV

TABLE OF PROPERTIES
(WITHOUT IN-PLACE SLAB)

BEAM TYPE	AREA, IN.^{2}	I, IN.^{4}	c_b, IN.	RECOMMENDED SPAN LIMITS, FT.
I	276	22,750	12.59	30-45
II	369	50,980	15.83	40-60
III	560	125,390	20.27	55-80
IV	789	260,730	24.73	70-100

(b) STANDARD CORED SLAB SECTIONS FOR BRIDGES

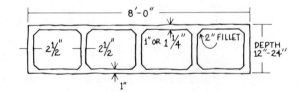

TABLE OF PROPERTIES
(TOP FLANGE = 1¼" THICK)

DEPTH, IN.	AREA, IN.^{2}	I, IN.^{4}	c_b, IN.	RECOMMENDED SPAN LIMITS, ft	
				ROOF	FLOOR
12"	370	7880	6.27	45	40
20"	470	27,100	10.39	70	60

FIGURE 13-17
PRECAST MEMBERS MAY BE OF VARIOUS SHAPES.

(a) STANDARD CORE SLABS FOR FLOORS AND ROOFS

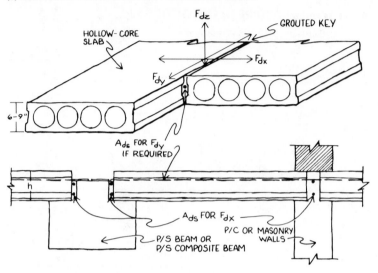

(b) SPECIAL SECTIONS OF ALMOST ANY SHAPE ARE POSSIBLE

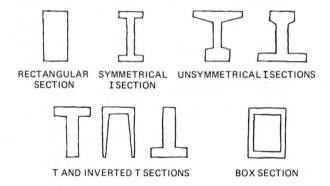

RECTANGULAR   SYMMETRICAL   UNSYMMETRICAL I SECTIONS
SECTION   I SECTION

T AND INVERTED T SECTIONS   BOX SECTION

devise almost any other sections (Figure 13-17b), including different shapes of wall panels, provided there is enough repetition to justify the cost of these special forms. Nominally this means repetition on the order of hundreds of special castings whereas standard sections can be purchased economically when there is much less repetition.

The detailed design of connections for precast members is an intricate problem in itself, still, it will be well to show here some of the typical types

FIGURE 13-18
CONNECTIONS OF CORED SLABS, DOUBLE AND SINGLE TEES TO BEAMS AND WALLS.

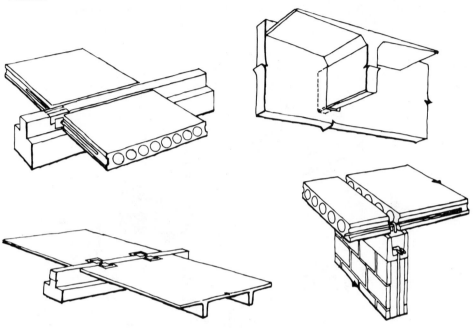

of connections now commonly used (Figures 13-18, 13-19, 13-20, and 13-21.)

Some special connections are possible to increase seismic resistance of precast construction. Precast wall panels can be integrated into a total wall system by using reinforced concrete columns and spandrel beams poured in place, employing the precast walls as the major part of their forming (Photo 13-6a). Precast cored slabs must be integrated by using an in-place concrete joint to integrate the slabs into a horizontal diaphragm for transferring seismic forces (Photo 13-6b). This allows the precast decking to first be set in place and then the supporting beams concreted, thus doing away with costly connections and bearings.

Large 90-ft-precast pieces of lightweight concrete were cast on site to form a 200-ft round roof for the Convention Center in Phoenix, Arizona (refer to Photo 13-4). The entire roof was made of 32 pieces of precast double Ts, weighing up to 45 tons each. Note that each roof piece is being handled by two cranes of 50- and 70-ton capacities. A main feature in this design is the minimizings of falsework, using only one circle of supports around the small inner tension ring, which was prestressed with a force of 3.2 million lb.

In the planning and design of precast components for construction, it is important that one consider the handling, transportation, erection, and

FIGURE 13-19
DOUBLE TEES MAY BE USED AS WALL COMPONENTS.
(a) DOUBLE TEES CONNECTED TO FOUNDATION WALLS

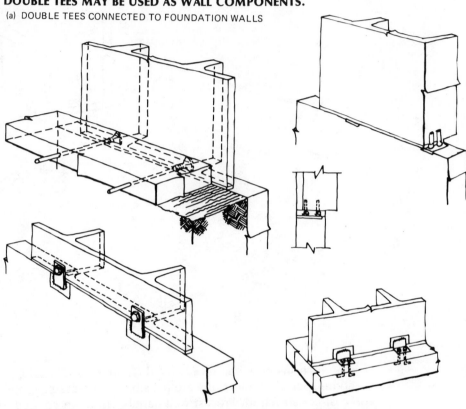

(b) DOUBLE TEES CONNECTED TO DOUBLE TEES

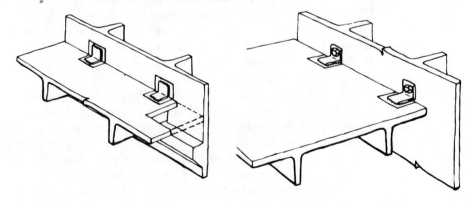

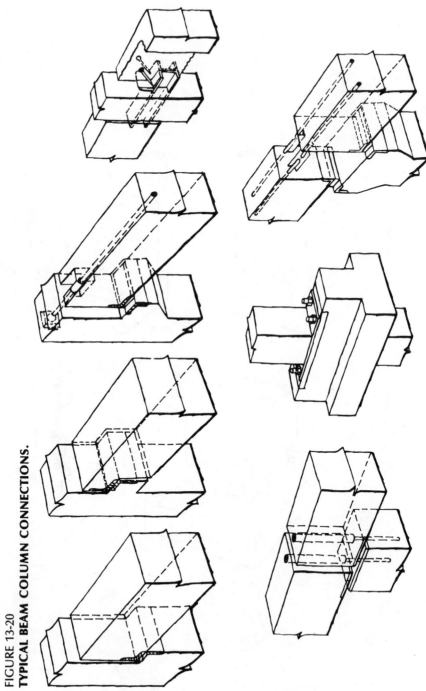

FIGURE 13-20
TYPICAL BEAM COLUMN CONNECTIONS.

FIGURE 13-21
TYPICAL COLUMN FOUNDATION CONNECTIONS.

CB-1   CB-2   CB-3

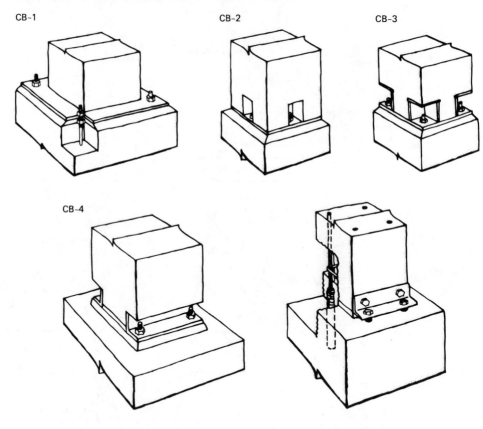

CB-4

PHOTO 13-6
**THE BUILDING AT 1625 ALAMEDA USED PRECAST WALL
COMPONENTS WHICH SERVED AS INTEGRATABLE COLUMN
FORMS AND TIED INTO FLOOR AND SPANDRELS AS WELL.**

(a)

Courtesy T. Y. Lin International

connection problems involved with these components. Generally speaking, it is advisable to use the maximum size that can be suitably transported and erected so as to minimize the number of connections. However, one must look into the capabilities of the local builders to determine the size and weight of these pieces. Furthermore, the accessibility of the entire structure to the lifting equipment must be considered.

On the usual highways and good roads in the United States, it is quite easy to transport pieces 60 ft, and often up to 80 ft long. Occasionally one can transport pieces of 80 to 100 ft; under especially favorable conditions, longer pieces, 120 or 140 ft, have been transported (Photo 13-7).

The allowable weight of each piece will primarily depend upon the crane capacity and generally should not exceed 10 tons if the crane has to reach out much to place the member into position (Figure 13-22). On the other hand, if the member is to be erected with the crane almost upright (i.e., at

PHOTO 13-7
120-FT LONG LIN TEES BEING HAULED OVER SECONDARY MOUNTAIN ROAD FOR BUENA VISTA SCHOOL GYMNASIUM, FT. HUACHUCA, ARIZONA.

FIGURE 13-22
PLACEMENT AND ERECTION ISSUES FOR PRECAST COMPONENTS.

(a) PRECAST MEMBERS MUST BE LIFTED AND LOADED PROPERLY

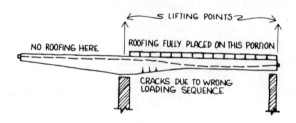

(b) CRANE CAPACITY IS RELATED TO REACH

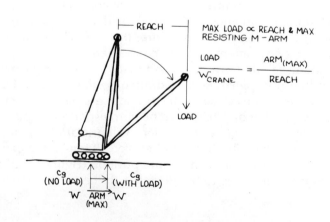

PHOTO 13-8
152-FT LIN TEES ARE ERECTED WITH TWO 60T CRANES (SKY HARBOR AIR TERMINAL, PHOENIX, ARIZONA).

maximum capacity), or with two cranes holding both ends of the member, then it can be much heavier (Photo 13-8).

The width of the members to be transported is usually limited to 8 ft for highway and road transportation. With special permit, 10 ft is commonly possible. Only under very unusual circumstances should one use more than 10 ft, with 12 ft being a maximum. In addition, one should remember that width can also limit length due to problems of clearance for turning corners on urban streets.

14
The Cost of Building Structures

SECTION 1: Introduction

In Chapter 1, the art of architectural design was characterized as one of dealing comprehensively with a complex set of physical and nonphysical design determinants. Structural considerations were cast as important physical determinants that should be dealt with in a hierarchical fashion if they are to have a significant impact on spatial organization and environmental control design thinking.

The economical aspect of building represents a nonphysical structural consideration that, in final analysis, must also be considered important. Cost considerations are in certain ways a constraint to creative design. But this need not be so. If something is known of the relationship between structural and constructive design options and their cost of implementation, it is reasonable to believe that creativity can be enhanced. This has been confirmed by the authors' observation that most creative design innovations succeed under competitive bidding and not because of unusual owner affluence as the few publicized cases of extravagance might lead one to believe. One could even say that a designer who is truly creative will produce architectural excellence within the constraints of economy. Especially today, we find that there is a need to recognize that elegance and economy can become synonymous concepts.

Therefore, in this chapter we will set forth a brief explanation of the parameters of cost analysis and the means by which designers may evaluate the overall economic implications of their structural and architectural design thinking.

The cost of structure alone can be measured relative to the total cost of building construction. Or, since the total *construction cost* is but a part of a total *project cost*, one could include additional consideration for land (10–20 percent), finance and interest (100–200 percent), taxes and maintenance costs (on the order of 20 percent). But a discussion of these so-called *nonconstruction costs* is beyond the scope of this book, and we will focus on the cost of construction only.

On the average, purely *structural costs* account for about 25 percent of total construction costs. This is so because it has been traditional to discriminate between purely structural and other so-called *architectural* costs of construction. Thus, in tradition we find that architectural costs have been taken to be those that are not necessary for the structural strength and physical integrity of a building design.

"Essential services" forms a third construction cost category and refers to the provision of mechanical and electrical equipment and other service systems. On the average, these *service costs* account for some 15 to 30 percent of the total construction cost, depending on the type of building. Mechanical and electrical refers to the cost of providing for air-conditioning equipment and the means of air distribution as well as other services, such as plumbing, communications, and electrical light and power.

The salient point is that this breakdown of costs suggests that, up to now, an average of about 45 to 60 percent of the total cost of constructing a typical design solution could be considered as architectural. But this picture is rapidly changing. With high interest costs and a scarcity of capital, client groups are demanding leaner designs. Therefore, one may conclude that there are two approaches the designer may take towards influencing the construction cost of building.

The first approach to cost efficiency is to consider that wherever architectural and structural solutions can be achieved *simultaneously*, a potential for economy is evident.[1] Since current trends indicate a reluctance to allocate large portions of a construction budget to purely architectural costs, this approach seems a logical necessity. But, even where money is available, any use of structure to play a basic architectural role will allow the nonstructural budget to be applied to fulfill other architectural needs that might normally have to be cut back. The second approach achieves economy through an integration of service and structural subsystems to round out one's effort to produce a total architectural solution to a building design problem.

The final pricing of a project by the constructor or contractor usually takes a different form. The costs are broken down into (1) cost of materials brought to the site, (2) cost of labor involved in every phase of the construction process, (3) cost of equipment purchased or rented for the project, (4) cost of management and overhead, and (5) profit. The architect or engineer seldom follows such an accurate path but should perhaps keep in mind how the actual cost of a structure is finally priced and made up.

Thus, the percent averages stated above are obviously crude, but they can suffice to introduce the nature of the cost picture. The following sections will discuss the range of these averages and then proceed to a discussion of square footage costs and volume-based estimates for use in rough approximation of the cost of building a structural system. In addition, the

---

[1] This is what F. L. Wright meant when he said "Form and Function are one," and Mies Van Der Rohe meant when he said, "Less is More," Alvar Aalto and Louis Kahn call for maximizing the number of jobs done by each design component. Mother Nature has always been doing this.

variables affecting the growth or diminution of such costs will be discussed. We will have to leave it to the reader's imagination to continue to explore how to integrate structure with other subsystems to fulfill one's total architectural needs.

SECTION 2: Percentage Estimates

The type of building project may indicate the range of percentages that can be allocated to structural and other costs. As might be expected, highly decorative or symbolic buildings would normally demand the lowest percentage of structural costs as compared to total construction cost. In this case the structural costs might drop to 10–15 percent of the total building cost because more money is allocated to the so-called architectural costs. Once again this implies that the symbolic components are conceived independent of basic structural requirements. However, where structure and symbolism are more-or-less synthesized, as with a church or Cathedral, the structural system cost can be expected to be somewhat higher, say, 15 to 20 percent (or more).

At the other end of the cost scale are the very simple and nonsymbolic industrial buildings, such as warehouses and garages. In these cases, the nonstructural systems, such as interior partition walls and ceilings, as well as mechanical systems, are normally minimal, as is decoration, and therefore the structural costs can account for 60 to 70 percent, even 80 percent of the total cost of construction.

Buildings such as medium-rise office and apartment buildings (5–10 stories) occupy the median position on a cost scale at about 25 percent for structure. Low and short-span buildings for commerce and housing, say, of three or four stories and with spans of some 20 or 30 ft and simple erection requirements, will yield structural costs of 15–20 percent of total building costs.

Special-performance buildings, such as laboratories and hospitals, represent another category. They can require long spans and a more than normal emphasis on mechanical equipment. As a result, a larger than average portion of the total costs will be allocated to services (i.e., 30–50 percent), with about 20 percent going for the purely structural costs. Tall office buildings (15 stories or more) and/or long-span buildings (say, 50 to 60 ft) can require a higher percentage for structural costs (about 30 to 35 percent of the total construction costs), with about 30 to 40 percent allocated to services.

In any case, these percentages are typical and can be considered as a measure of average efficiency in design of buildings. For example, if a low, short-span and nonmonumental building were to be bid at 30 percent for the structure alone, one could assume that the structural design may be comparatively uneconomical. On the other hand, the architect should be aware of the confusing fact that economical bids depend on the practical ability of *both* the designer and the contractor to interpret the design and construction requirements so that a low bid will ensue. Progress in structural design is often limited more by the designer's or contractor's lack of experience, imagination, and/or absence of communication than by the idea of the design. If a contractor is uncertain, he will add costs to hedge the risk he will be taking. It is for this reason that both the architect and the engineer should be well-versed in the area of construction potentials if innovative designs are to be competitively bid. At the least the architect must be capable of working closely with imaginative structural engineers, contractors and even fabricators wherever possible. Conversely, one cannot count on efficiency via handbook structural design even if the architecture is very ordinary. *Efficiency always requires knowledge and above all imagination, and these are essential when designs are unfamiliar.*

The foregoing percentages can be helpful in approximating total construction costs if the assumption is made that structural design is at least of average (or typical) efficiency. For example, if a total office building construction cost budget is $5,000,000, and 25 percent is the "standard" to be used for structure, a projected structural system should cost no more than $1,250,000. If a very efficient design were realized, say, at 80 percent of what would be given by the "average" efficient design estimate stated above the savings, (20 percent), would then be $250,000 or 5 percent of total construction costs $5,000,000. If the $5,000,000 figure is committed, then the savings of $250,000 could be applied to expand the budget for "other" costs.

All this suggests that creative integration of structural (and mechanical and electrical) design with the total architectural design concept can result in either a reduction in purely construction costs or more architecture for the same cost. Thus, the degree of success possible depends on knowledge, cleverness, and insightful collaboration of the designers and contractors.

The above discussion is only meant to give the reader an overall perspective on total construction costs. The following sections will now furnish the means for estimating the cost of structure alone. Two alternative means will be provided for making an *approximate* structural cost estimate: one on a square foot of building basis, and another on volumes of structural materials used. Such costs can then be used to get a rough idea of total cost by referring to the "standards" for efficient design given above. At best, this will be a crude measure, but it is hoped that the reader will find that it makes him somewhat familiar with the type of real economic problems that responsible designers must deal with. At the least, this capability will be useful in comparing alternative systems for the purpose of determining their relative cost efficiency.

SECTION 3: Square-Foot Estimating

As before, it is possible to empirically determine a "standard" per-square-foot cost factor based on the average of costs for similar construction at a given place and time. More-or-less efficient designs are possible, depending on the ability of the designer and/or contractor to use materials and labor efficiently, and vary from the average. Some figures are given here for 1979 average U.S. conditions, and they can be updated for future times by referring to the ENR cost index curve (Figure 14-1). Note that an index is required because of cost variability (e.g., $1913 = 100 \rightarrow 1979 \sim 3000$).

The range of square-foot costs for "normal" structural systems is $10 to $16 psf. For example, typical office buildings average between $12 and $16 psf, and apartment-type structures range from $10 to $14. In each case, the lower part of the range refers to short spans and low buildings, whereas the upper portion refers to longer spans and moderately tall buildings.

Ordinary industrial structures are simple and normally produce square-foot costs ranging from $10 to $14, as with the more typical apartment buildings. Although the spans for industrial structures are generally longer than those for apartment buildings, and the loads heavier, they commonly have fewer complexities as well as fewer interior walls, partitions, ceiling requirements, and they are not tall. In other words, simplicity of design and erection can offset the additional cost for longer span lengths and heavier loads in industrial buildings.

Of course there are exceptions to these averages. The limits of variation depend on a system's complexity, span length over "normal" and special loading or foundation conditions. For example, as shown in Figure 5-1, the Crown Zellerbach high-rise bank and office building in San Francisco is an exception, since its structural costs were unusually high. However, in this case, the use of 60 ft steel spans and free-standing columns at the bottom, which carry the considerable earthquake loading, as well as the special foundation associated with the poor San Francisco soil conditions, contributed to the exceptionally high costs. The design was also unusual for its time and a decision had been made to allow higher than normal costs for all aspects of the building to achieve open spaces and for both functional and symbolic reasons. Hence the proportion of structural to total costs probably remained similar to ordinary buildings.

The effect of spans longer than normal can be further illustrated. The "usual" floor span range is as follows: for apartment buildings, 16 to 25 ft; for office buildings, 20 to 30 ft; for industrial buildings, 25 to 30 ft loaded heavily at 200 to 300 psf (i.e., about three or four times that of apartments or offices); and garage-type structures span, 50 to 60 ft, carrying relatively light (50–75 psf) loads (i.e., similar to those for apartment and office structures). Where these spans are doubled, the structural costs can be expected to rise about 20 to 30 percent.

The increased loading in the case of industrial buildings offers another

492

FIGURE 14-1
COST INDEX CURVE TO MARCH 1979 (TAKEN FROM ENR COST INDEX CURVE FOR MARCH 1974).

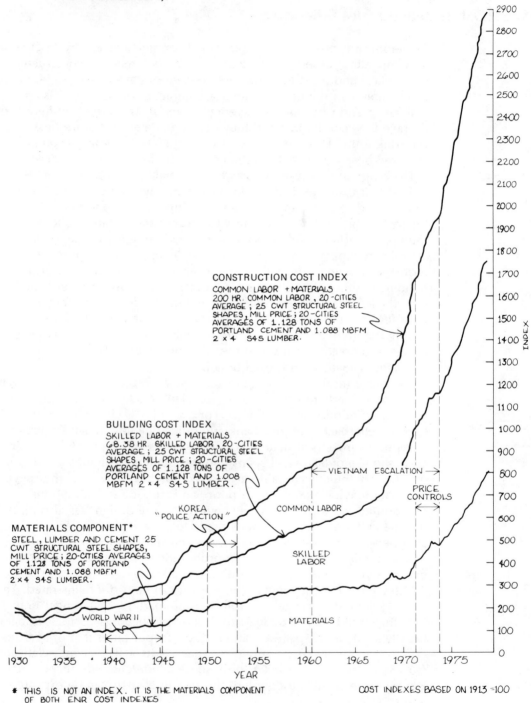

* THIS IS NOT AN INDEX. IT IS THE MATERIALS COMPONENT OF BOTH ENR COST INDEXES

COST INDEXES BASED ON 1913 =100

insight into the dependency of cost estimates on "usual" standards. If the loading in an industrial building were to be increased to 500 psf (i.e., two or three times), the additional structural cost would be on the order of another 20 to 30 percent.

The reference in the above cases is for floor systems. *For roofs* using efficient orthotropic (flat) systems, contemporary limits for economical design appear to be on the order of 150 ft, whether of steel or prestressed concrete. Although space-frames are often used for steel spans over 150 ft, the fabrication costs begin to rise considerably.

For roof spans of over 150 ft and up to 400 ft and beyond, the cost of structure is a major part of total costs (i.e., 50 percent or more). Suspension, arch, shell, or combination-membrane, and space-frame systems can be constructed at a structural cost range of $15 to $25 psf. But this will depend heavily on the abilities of the designer to optimize the use of materials and allow for the local erection conditions. For example, the Oklahoma Arena, (see Photo 11-7a and b), an elliptical dish roof spanning 320 ft × 400 ft, was built in 1966 at $4 psf (which would be some $12 psf today), using prestressed concrete shell design. The original steel space-frame design for this same structure was to have cost $8 psf (which would be some $24 psf today). In the case of the Arizona State Fair Building (see Photos 11-9a and b), a circular hyperbolic paraboloid roof of 386 ft diameter (the first time such a large shell of this type was built) was constructed at a 1967 cost of just over $5 psf. In contrast, the 400-ft circular dish roof of the Oakland Coliseum, also built in 1967, cost about $12 psf.

At any rate, it should be recognized that very long-span subsystems are special cases and can in themselves have a great or small effect on total costs. When the cost of special vertical and/or foundation subsystems is added, structural costs for special buildings can vary greatly from design to design. The more special the form, the more that design *knowledge and creativity, as well as construction skill*, will determine the potential for achieving cost efficiency.

SECTION 4: Volume-Based Estimates

When more accuracy is desired, estimates of costs can be based on the volume of materials used to do a job. At first glance it might seem that the architect would be ill equipped to estimate the volume of material required in construction with any accuracy, and much less speed. But it is possible, with a moderate learning effort, to achieve some capability for making such estimates.

Volume-based estimates are given by assigning in-place value to the pounds or tons of steel, or the cubic yards of reinforced or prestressed concrete required to build a structural system. For such a preliminary estimate, one does not need to itemize detailed costs. For example, in-place concrete costs include the cost of forming, falsework, reinforcing steel, labor, and overhead. Steel includes fabrication and erection of components.

Costs of *structural steel* as measured by weight range from $0.50 to $0.70 per pound *in place* for building construction. For low-rise buildings, one can use stock wide-flange structural members that require minimum fabrication, and the cost could be as low as $0.50 per pound. More complicated systems requiring much cutting and welding (such as a complicated steel truss or space-frame design) can go to $0.70 per pound and beyond. For standard tall building designs (say, exceeding 20 stories), there would typically be about 20 to 30 pounds of steel/psf, which one should wish not to exceed. A design calling for under 20 psf would require a great deal of ingenuity and the careful integration of structural and architectural components and would be a real accomplishment. You may wish to refer to Figure 5-1 in Chapter 5 to compare per-square-foot weights of some contemporary steel structure buildings.

Concrete costs are volumetric and should range from an *in-place low* of $150 per cu yd for very simple reinforced concrete work to $300 per cu yd for expensive small quantity precast and prestressed work. This large range is due to the fact that the contributing variables are more complicated, depending upon the shape of the precast components, the erection problems, and the total quantity produced.

Form work is generally the controlling factor for any cast-in-place concrete work. Therefore, to achieve a cost of $150 per cu yd, only the simplest of systems can be used, such as flat slabs that require little cutting and much reuse of forms. Where any beams are introduced that require special forms and difficulty in placement of concrete and steel bars, the range begins at $180 per cu yd and goes up to $300. Since, in this country, high labor costs account for high forming costs, this results in pressure to use the simplest and most repetitive of systems to keep costs down. When cast-in-place work begins to approach $240 per cu yd, it becomes rewarding to consider the possibility of mass-produced precast and prestressed components, which may bring a savings in costs and/or construction *completion time*. The latter results in savings due to lower construction financing costs for the contractor plus quicker earnings for the owner.

One important exception to the above cost picture is that of concrete work in *foundations*. Here the cost of forming and casting simple foundations (i.e., for spread foundations with very little steel, such as subgrade bearing walls and mat foundations) should be considered at about $90 per cu yd. But in case pile foundations are required, costs can rapidly climb, since one 16 in. diameter pile can cost $12 per ft or more in place, of course depending on soil conditions.

It is enlightening to pay some attention to the makeup of these in-place concrete estimates. The cost of concrete alone for ordinary reinforced concrete work is about $40 per cu yd delivered. For special concrete, such as lightweight and/or high-strength quick-setting concrete, the cost can go to $50 or even $60 per cu yd. Mild reinforcing steel, depending on the cutting and fabrication complexity of the required reinforcing design, can range from 30¢ to 46¢ per lb in place. For an average of about 150 lb of steel per cubic yard of ordinary reinforced concrete, the steel cost would range from about $45 to $60 per sq yd. Labor, including placing of reinforcing and concrete as well as finishing of concrete, costs about $20 to $40 per cu yd depending on the complexity of placing and working the concrete.

Form work represents the largest single cost factor for most concrete work. The cost can be stated as per square feet of contact area, with slabs requiring single-side and walls double-side forming. In either case, efficiency depends on reusability and the simplicity of form design. For the simplest reusable plywood forms, such as for a flat slab, the costs will run a minimum of $1 psf of contact area. Since walls require forming on two sides, their forming cost will easily be $2 psf. This amounts to some $80 of forming cost per cu yd of concrete for an ordinary 8-in. wall. When beams are introduced, cutting and erection costs are much affected by high labor costs, and the forming costs can easily go to $2.50 or $3.00 psf of contact area. Special designs for very complicated forming, such as for nonstandard waffle systems, or for shell and suspension design, will often contribute a large portion to cast-in-place concrete cost, unless the forms are reused.

The *mass of concrete per square foot of plan* area affects the form/cost ratio. This is pronounced in the case of, say, a simple 3-in. shell as compared with an 8-in. flat slab. At $1 psf form cost, one cubic yard of concrete placed for a 3-in. shell will require 108 sq ft of form, at a cost of $108. Thus, the thinner the system, the greater the influence of form costs on total costs.

Prestressing costs can now be compared with nonprestressed concrete work. The material and labor for prestressing steel cost about $40 to $60 per cu yd for pretensioned precast concrete and $60 to $80 per cu yd for posttensioned in-place concrete. But with competent design, prestressed structural members are designed thinner in comparison with reinforced concrete design, and the overall cost of prestressed concrete construction could often be cheaper than ordinary reinforced concrete work. The other advantages of weight reduction and minimum deflection are additional.

Often where prestressing is not found to be less expensive in terms of immediate construction cost, the ability to design for longer spans and lighter elements with less wall, column and foundation loading, as well as the increased architectural freedom, determine the desirability of going to prestressed elements. The point for the designer to remember is that good design in either material will be competitive and frequently one's decision is in a context of many important building design determinants, only one of which is the structural system.

To summarize, the range of cost per cubic yard of standard types of poured-in-place concrete work will average from $150 to $250, the minimum being for simple reinforced work and the maximum for moderately complicated posttensioned work. This range is large and any estimate that ignores the effect of variables discussed above will be commensurately inaccurate.

SECTION 5: Some Generalizations about the Variables of Cost

Understanding the variables inherent in the design of a structural system will help the designer to create more efficient designs.

Since the proportioning of most structural subsystems and components is normally controlled more by moment than by shear, the relation of span, depth and loads to moment resistance will offer insight into the cost determinants derived therefore. For a simple beam, the maximum evenly distributed load moment is given by $wL^2/8$. Since this moment is an expression of a load (w) per unit of span length and the length (L), all that is needed to express the relation of the three controling factors is the effect of depth.

From the fact that the external moment on a subsystem or component must be resisted by an internal moment, the relation is $wL^2/8 = Fa$, where (a) may be termed the *effective* depth, or overall moment resisting arm, of the member and F is the total resisting force required, both in tension (T) or compression (C). For a typical wide-flange type of steel beam where the greater part of the mass is concentrated in the flanges, the moment-resisting arm a is only slightly less than the total depth d, since the centroid of the compression and tension forces is near the flange area centroid. In contrast, for a rectangular beam, the arm a is only ($2/3d$) in the elastic range and $d/2$ in the ultimate range, as discussed in Chapter 8. But for the purpose of discussion, d is a measure of a and can be used as a representative of the efficiency issue. We can say

$$F = \frac{wL^2/8}{a} \propto \frac{wL^2/8}{d}$$

Looking at the relation, it can be seen that the (L) factor is squared and the inverse d factor is not. Assuming the section depth as constant, if the length of any bending member is increased by two, the force in the flanges will increase by four times! This would be an extravagant way to design, since four times the material must be supplied to resist compressive and tensile forces in order to span twice the distance. However, by keeping the span-to-depth ratio constant (i.e., doubling the depth as well as the span), we see that only two times the force is required to span twice the distance. Hence, the issue is to maintain an efficient span-to-depth ratio as the span is manipulated in building design.

Generally, the effect of load increase on costs is less than that of a span increase for two reasons. First, F varies with the first power of w, not the second power. Then, many designs call for proportions larger than that described by load considerations alone. Therefore, it is often possible to increase the load while the dimensions will remain practically the same. Thus, construction costs do not always vary directly with the increase in load.

Although the L^2 relation is a deterrent to the designer's freedom to go to longer spans, the ability to do so economically depends on the ability of a designer to achieve *total-system* rather than elemental design competence. For example, at one time, 30-ft spans were common for traditional reinforced concrete parking garage designs. However, a skillful designer can go to 60 ft with little or no extra cost by combining the floor surface with a prestressed girder to form a T-section subsystem, using the roof or floor surface as a compression flange and high-strength prestressing steel for the tension force at the bottom of the T stem (Photo 14-1). The structural depth

PHOTO 14-1
PARKING STRUCTURE, UNIVERSITY OF CALIFORNIA, BERKELEY. THE ROOF SUPPORTED 18 IN. OF EARTH AND A NATURAL GRASS PLAYING FIELD UNTIL RECENTLY. (ANSHEN & ALLEN, ARCHITECTS; T. Y. LIN INTERNATIONAL STRUCTURAL ENGINEERS).

would remain nearly the same; the structural cost would increase by only about 50¢ psf, and the free area thus obtained would more than offset (in convenience and rentable stalls) this cost increase (Photo 14-2). This is a good example of the value of using a total-system design approach, and the opportunities for rewards to be gained by such an approach with any building type are many.

The present economical span limit of such an integrated floor or roof design, as in the above example, appears to be about 120 ft carrying a design *live load* of about 50 lb per sq ft (Photo 14-3). The girders would be 4 ft deep with stems 14 in. thick, placed at 20 ft centers and integrated with a 5-in. slab. This has already been done several times in the Los Angeles area of California, with the cost being only about 10 to 20 percent more than for a 60-ft design.

Let us examine the effect of loading on the cost of such spans. Assuming the use of 120 psf lightweight concrete, the 5-in. slab would weigh about 50 psf, which, together with the design live load of 50 psf, brings the total design load to 100 psf. Over a tributary width of 20 ft, this yields a total design load of 2000 plf on the girders (i.e., not including girder web *DL* at ~240 plf). If the live load were to be increased to 100 psf, the total load on the T-designed subsystem, still using a 5-in. slab, will be increased by 50 percent to 3000 plf. The moment so produced is easily resisted by the slab under compression, and the only increase would be additional steel for the tendons. Hence, the structural cost would not rise in direct proportion to the loads. The really crucial factor here is in maximizing the effectiveness of

PHOTO 14-2
LONG SPANS ARE CONVENIENT FOR PARKING. HOLIDAY INN, MEMPHIS, TENNESSEE. (T. Y. LIN INTERNATIONAL, STRUCTURAL ENGINEERS).

PHOTO 14-3
ERECTION OF 120-FT T-BEAM, TELECOMPUTING BUILDING. (T. Y. LIN INTERNATIONAL, STRUCTURAL ENGINEERS).

the structural depth by means of an optimizing approach to the design of total systems.

The property of *span continuity* in a design has a significant effect on the effectiveness of members and thus on the span-to-depth ratio possible. The general relationship for maximum moment in evenly distributed load-carrying systems is wL^2/C, where C is 8 for a simple beam and 2 for a cantilever. Thus the cantilever, *where not part of an integrated system*, is the least effective means of carrying a load. On the other hand, if a series of simple spans are made continuous, the total system interaction is such that an opportunity for savings is possible. Providing foundation conditions are stable, continuity by cantilever at both ends of a span approximates fixed-end conditions. The maximum moment in a *fully* fixed-end design would be found by assigning $C = 12$, for $wL^2/12 = \frac{2}{3}(wL^2/8)$ at the ends and only $wL^2/24 = \frac{1}{3}(wL^2/8)$ at midspan. For partially fixed conditions, the reduction would be somewhat less as discussed in chapter 8.

SECTION 6: Economics of Long-Span Roof Systems

For floor systems using standard steel *WF* sections, the *economic* span limit is probably around 60 to 80 ft, with the depth of the web (4–6 ft) limiting the architectural acceptibility and cost effectiveness of the system. Where depth is not so limited, a one-way steel truss design can often be efficiently utilized up to about 100 ft for floors and 150 ft or more for roofs. The truss is more efficient because it attains greater depth with a minimum of web material and it offers space through which mechanical equipment can pass. However, fabrication can be expensive.

Still, it is interesting to contrast the 120 ft prestressed concrete spans referred to in the previous section. With careful total-system design, mechanical systems can be integrated by means of *properly placed* holes in the *T* stem. Holes of 2 ft × 4 ft or more have been attained economically within a structural depth of only four feet, whereas a steel truss would need at least 8 ft depth to span 120 ft. However, if depth is not a problem, the truss offers the opportunity for design of service systems *after* the design of structural systems.

For moderately long spans of 120 ft to, say, 150 ft, where level surfaces are desired, the steel space-frame becomes attractive (Photo 14-4). Steel space frames spanning 120 ft and 150 ft cost some $15 psf and offer the freedom to distribute the mechanical and air-conditioning systems within the frame space. Larger flat spans to, say, 300 ft might also be more efficiently spanned with steel space frames. However, at that scale, a depth of 20 ft or more is required.

PHOTO 14-4
FLAT ROOF STEEL SPACE FRAME, BROOME COUNTY CONVENTION CENTER, NEW YORK. (ELSF, ARCHITECTS; T. Y. LIN INTERNATIONAL, STRUCTURAL ENGINEERS).

For (very) long spans where level surfaces are not required, suspension systems may prevail. Really good *one way* suspension designs could yield costs of about $12 psf, for overall span-to-depth ratios of 6 to 10. When a prestressed suspension *grid* is used, as in the case of dish shells, the overall span-to-depth ratio may be as high as 20. In contrast, an arch system will require a ratio as low as 8 or 6. But in either case, we have a system capable of utilizing very little material to span very large distances.

An example of simple suspension is the Golden Gate Bridge (Photo 14-5), with a main span of 4200 ft and a relatively shallow sag (depth) of about 400 ft ($\sim \frac{1}{10}$). However, suspension requires that means be provided for resisting the horizontal pull of cables by means of a massive (and costly) buttress at each end. Similarly, a simple arch produces thrust at the springing that may either be buttressed or tied to take the horizontal force. Where the suspension dish or arch-dome systems are used, these horizontal forces can be carried by tension or compression rings and the whole structure can be elevated on columns if desired.

PHOTO 14-5
GOLDEN GATE SUSPENSION BRIDGE. (COURTESY OF REDWOOD EMPIRE ASSOCIATION).

502

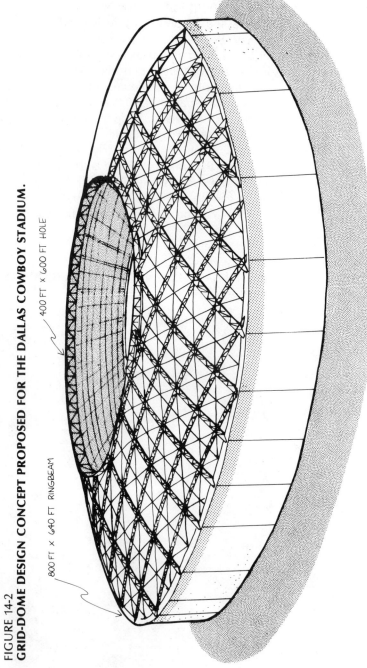

FIGURE 14-2
GRID-DOME DESIGN CONCEPT PROPOSED FOR THE DALLAS COWBOY STADIUM.

400 FT × 600 FT HOLE

800 FT × 640 FT RINGBEAM

There is also a problem with the simple suspension system because the cable itself is very flexible. For example, the main cable of the Golden Gate Bridge is only 3 ft in diameter over 4200 ft of span, and has no moment-resisting capacity. Therefore, to render the structure stable under varying live loads and wind, the *overall* system must somehow be stiffened, with trusses, for example. For design of simple suspension roof structures, the use of tie-down cable systems may also be used.

For dome and dish construction, the problem of stability is inherent to the three-dimensional membrane action of the total system. But as very large domes of, say, 800 to 1000-ft span become feasible, the issue of *localized stiffening* will be more of a problem. Local flexibility of thin membranes under wind loading will prove to be a major problem requiring careful study of the force systems potentially operating on the system. A membrane design for a trussed grid dome spanning 800 ft × 640 ft with a 400 ft × 300 ft hole in the center was prepared in 1970 that called for about 25 lb of steel per square foot (Fig. 14-2). But a very sophisticated computer simulation study of performance under uneven horizontal loading conditions was an important factor in the evolution of the design. In any case, the larger the span, the more useful becomes the potential for prefabrication and economical erection to arrive at optimum solutions. All this seems to indicate that, as the height, span, and magnitude of a building increase, technology in all of its aspects plays a more important part in building economics.

SELECTED REFERENCES

A. Philosophy of Structures

Candela: The Shell Builder, Colin Faber, Reinhold, New York, 1963.
Concepts of Structures, William Zuk, Reinhold, New York, 1963.
Eliel Saarinen, Alan Temko, George Braziller, New York, 1962.
Developments in Structural Form, Rowland Mainstone, M. I. T. Press, Cambridge, Mass., 1975.
On Growth and Form, D'Arcy Thompson (abridged edition by John T. Bonner), Cambridge University Press, New York, 1961.
Philosophy of Structures, Edward Torroja, McGraw-Hill, New York, 1958.
Structures, Pier Luigi Nervi, McGraw-Hill, New York, 1956.
Structure in Architecture, Mario Salvadori, Prentice-Hall, Englewood Cliffs, N. J., 1963.
Structure in Art and Science, Gyorgy Kepes, George Braziller, New York, 1965.

B. Statics and Mechanics of Materials

Architectural Structures, an Introduction to Structural Mechanics, Henry J. Cowan, American Elsevier, New York, 1976.
Mechanics of Materials, E. P. Popov, 2nd Ed., Prentice-Hall, Englewood Cliffs, N.J., 1976.
Mechanics of Materials, Archie Higdon, Edward H. Ohlsen, William B. Stiles, John A. Weese, and William F. Riley, Wiley, New York, 1976.
Statics, Engineering Mechanics Vol. 1, J. L. Meriam, Wiley, New York, 1978.

C. Structural Analysis and Design

Structural Analysis for Engineering Technology, Jack D. Bakos, Jr., Merrill, 1973.
Structural Analysis, Jack C. McCormac, Intext Educational, 1975.
Elementary Structural Analysis, Charles H. Norris, John B. Wilbur, and Senol Utku, McGraw-Hill, New York, 1976.
Structural Design in Architecture, Mario Salvadori and Matthys Levy, Prentice-Hall, Englewood Cliffs, N.J., 1967.
Structures for Architects, Bezaleel S. Benjamin, Ashnorjen Bezaleel Publishing Company, 1974.
Fundamentals of Structural Design: Steel, Concrete, and Timber, Louis A. Hill, Jr., Intext Educational, 1975.
High-Rise Building Structures, Wolfgang Schueller, Wiley, New York, 1977.
Tensile Structures, Frei Otto, M. I. T. Press, Cambridge, Mass., 1973.

D. Building Codes and Engineering Handbooks

Uniform Building Code, International Conference of Building Officials, Whittier, Cal., 1979.

Structural Engineering Handbook, E. H. Gaylord and C. N. Gaylord, 2nd Ed., McGraw-Hill, New York, 1979.

Building Construction Handbook, edited by Frederick S. Merritt, McGraw-Hill, New York, 1975.

Standard Handbook for Civil Engineers, edited by Frederick S. Merritt, McGraw-Hill, New York, 1968.

E. Steel Structures

Manual of Steel Construction, American Institute of Steel Construction, New York, 1970 (new Ed. expected 1979).

Standard Specifications and Load Tables, Steel Joist Institute, Arlington, Va, 1977.

Design of Steel Structures, B. Bresler, T. Y. Lin, and J. Scalzi, Wiley, 1968.

Steel Buildings: Analysis and Design, Stanley M. Crawley and Robert M. Dillon, Wiley, New York, 1970.

Structural Steel Design, Jack C. McCormac, Intext Educational, 1971.

Basics of Structural Steel Design, Samual H. Marcus, Reston, 1977.

F. Concrete Structures

Building Code Requirements for Reinforced Concrete American Concrete Institute, Detroit, Mich., 1977.

PCI Design Handbook, Prestressed Concrete Institute, Chicago, 2nd Ed., 1978.

Post-tensioning Manual, Post-tensioning Institute, Phoenix, Ariz., 1978.

CRSI Handbook, Concrete Reinforcing Steel Institute, Chicago, 1975.

Handbook of Concrete Engineering, edited by Mark Fintel, Van Nostrand Reinhold, New York, 1974.

Reinforced Concrete Engineering., edited by B. Bresler, Wiley, 1972.

Reinforced Concrete Fundamentals, Phil M. Ferguson, Wiley, New York, 1965.

Design of Reinforced Concrete Structures, George Winter and Arthur Nilson, McGraw-Hill, New York, 1975.

Design of Prestressed Concrete Structures, T. Y. Lin, 2nd Ed., Wiley, 1963 (expected new edition, 1980, with N. Burns).

G. Timber and Plastics

Wood Engineering, German Gurfinkel, Southern Forest Products Association, New Orleans.

Manual of Timber Construction, American Institute of Timber Construction, New York.

National Design Specification for Stress-Grade Lumber and Its Fastenings, National Forest Products Association, Washington, DC. 1977.

Wood Structural Design Data, National Forest Products Association, Washington, DC, 1970.

Structural Design with Plastics, Bezaleel S. Benjamin, Van Nostrand Reinhold, New York 1969.

Appendix

Appendix A

SOME USEFUL PRINCIPLES OF STRUCTURAL STATICS AND MECHANICS OF MATERIALS

Students using this book should have prior understanding of the following basic concepts:

A-1 STATICS
A-2 MECHANICS OF MATERIALS

A-1　STATICS

a. Resolution of a force (*F*) into components (F$_x$ and F$_y$) by parallelogram.
   1. Components at right angles:

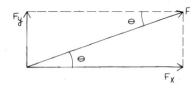

$$F_y = F \sin \theta$$

$$F_x = F \cos \theta$$

   2. Components *not* at right angles:

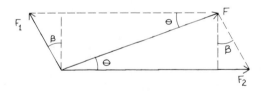

$$F_1 = \frac{F \sin \theta}{\cos \beta}$$

$$F_2 = F \cos \theta + F_1 \sin \beta$$

b. Resultant of 2 forces by parallelogram.
   1. Symmetric case: (F$_1$ = F$_2$)

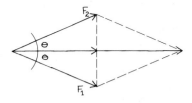

$$F = 2F_1 \cos \theta$$

   2. Asymmetric case: (F$_1$ ≠ F$_2$, θ ≠ β)

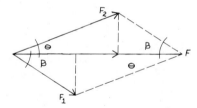

$$F = F_1 \cos \beta + F_2 \cos \theta$$

c. Freebody diagrams and requirement for equilibrium.
Equilibrium of a body in space requires fulfillment of the following two conditions:

(1) *Force Equilibrium*
$$\sum F_x = 0$$
$$\sum F_y = 0$$
$$\sum F_z = 0$$

(2) *Moment Equilibrium*
$$\sum M_x = 0$$
$$\sum M_y = 0$$
$$\sum M_z = 0$$

1. All forces and reactions in plane:

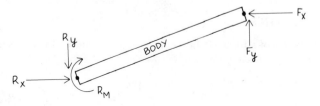

2. More general case:

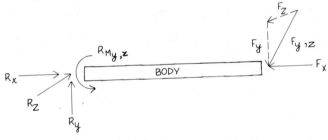

3. Moments and force reactions:

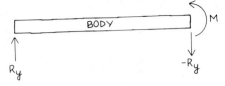

4. Moments and moment reactions:

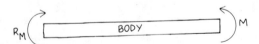

A-2   MECHANICS OF MATERIALS

a. Unit stresses and unit strains.

   1. Axial stress:

$$6 = \frac{F}{A} \, , \quad 6 = \text{unit normal stress}$$

$$F = \text{force acting normal to sectional area}$$

$$A = \text{corresponding sectional area}$$

   2. Shearing stress:

$$\text{rule } v = \frac{V}{A} \, , \quad v = \text{unit shearing stress}$$

$$V = \text{force acting parallel to the sectional area}$$

$$A = \text{corresponding area}$$

   3. Unit strain:

$$\epsilon = \frac{\Delta}{L} , \quad \epsilon = \text{unit strain (shortening or elongation per unit of length)}$$

$$\Delta = \text{total shortening or elongation of member}$$

$$L = \text{original length of member}$$

b. Modulous of elasticity.

   1. Axial modulous—Hooke's law states that, *within the elastic limit,* the relationship between unit stress and unit strain is linear for all materials. Thus,

$$E = \frac{6}{\epsilon} , \quad E = \text{elastic (Young's) modulous}$$

   2. Poisson's ratio:

$$\nu = \frac{\text{lateral strain}}{\text{axial strain}}$$

   3. Shearing modulous:

$$G = \frac{v}{r} , \quad v = \text{unit shearing stress}$$

$$r = \text{shearing strain}$$

$$G = \text{shearing modulous of elasticity}$$

It is known that, $G = \dfrac{E}{2\,(1 + \nu)}$

c. Stresses in beams.

   1. The flexure formula by the elastic theory.
      Tensile or compressive stress (6_y) required at level (y) of a section:

$$6_y = \frac{M(y)}{I} , \quad M = \text{moment } on \text{ section}$$

$$y = \text{distance of fiber in question from the neutral axis } (N\text{-}A) \text{ of the section.}$$

$$I = \text{moment of inertia for the section}$$

2. Shearing stress and shear flow in beams:

$$q = vb = \frac{VQ}{I}, \quad q = \text{shear flow or total shear force per unit of length at a certain fiber}$$
distance (y) from $N\text{-}A$

b = section width at (y)

Q = statical moment about $N\text{-}A$ of the sectional area *above* the shearing plane in question (at y)

Appendix B

TABLES TO ASSIST IN DESIGN OF STRUCTURAL SYSTEMS & COMPONENTS

B-1 APPROXIMATE STRUCTURAL SPAN-DEPTH RATIOS FOR HORIZONTAL SUBSYSTEMS AND COMPONENTS

	AVERAGE	MAXIMUM	TYPICAL[a] SPAN FT.
I. Timber:			
Plywood	36	40	3-5
Planks	28	32	2-6
Joists	22	26	10-25
Beams	16	20	15-30
Girders	12	16	20-35
Gable bents	26	30	30-50
Trusses	4	8	30-100
II. I-beams and joists	18	Beams 24, Joists 25	15-60
Plate and I-girders	14	20	40-100
Trusses	12	18	40-80
Gable bents	30	40	50-120
Arches (span to rise)	8	16	80-200
(span to thickness)	40	50	—
Simple suspension (span to rise)	10	15	150-300
Cable stayed	6	10	150-300
III. Reinforced concrete:			
Solid slabs	28	32	10-25
Slabs with drops & capitals (or two-way slab on beams)	30	36	20-35
Waffle slabs	20	24	30-40
Joists	22	26	25-45
Beams	16	20	15-40
Girders	12	16	20-60
Gable bents	24	30	40-80
Arches (span to rise)	8	12	60-150
(span to thickness)	30	40	—
Cylindrical thin shell roofs (min. thickness may govern)			
longitudinal span to structural depth	12	15	50-70
transverse span to thickness	50	60	12-30
IV. Prestressed Concrete:			
Solid slabs	40	44	20-35
Slabs with drops or two-way slab on beams	44	48	35-45
Waffle slabs	28	32	35-70
Cored slabs	36	40	30-60
Joists	32	36	40-60
Beams	24	28	30-80
Girders	20	24	40-120
Cylindrical thin shell roofs (Min. thickness may govern)			
longitudinal span to structural depth	15	20	(60-120)
transverse span to thickness	60	70	15-35

[a] For cantilever spans, use about $\frac{1}{3}$ of those shown (span-depth at about 40%).

B-2   APPROXIMATE AVERAGE DESIGN STRESSES, KSI

This table is prepared for use with approximate preliminary designs, based on WSD (Working stress design) or USD (Ultimate strength design).

MATERIAL	DESCRIPTION		FLEXURAL MEMBERS						COLUMNS (short)	
			Tension		Compression		Shear		Compression	
			WSD	USD	WSD	USD	WSD	USD	WSD	USD
Steel	A36		24	36	24	36	15	25	20	30
	Suspension cable		60	150						
	Reinforcing (hard) bars (interm.)		24 / 20	60 / 40	24 / 20	60 / 40	—	—	20 / 16	60 / 40
	Tendons	Wires	150	250						
		Strands	160	270						
		Bars	90	150						
Concrete	Reinforced concrete 3 ksi				1.3	2.5	0.1* / 0.2*	0.2* / 0.4*	0.7	2
	Prestressed concrete 5 ksi				2.2	4.2	0.2* / 0.4*	0.3* / 0.7*	1.2	4
Timber			1.2		1.2		0.1		1.2	
Aluminum			22		22		10		18	

Concrete % A_{net}, RC ; A_{gross} PC

Percent tension steel in cross section Area

Low	Avg.	High

Note that High value may require compressive reinforcement.

		1. BEAMS (% Area)			2. SLABS (% Area)			3. T-BEAMS (% Area)		
RC		.7	1.2	2.0	.3	.5	1.0	.2	.4	.8
PC		.3	.4	.85	.15	.2	.3	.1	.15	.4

1. Moment usually controls
2. Deflection usually controls
3. % is lower because flange area contribution is large:
   a) For RC common to use 16 times flange thickness
   b) For PC conservative to assume $\frac{1}{2}$ span = max-flange

*Lower values with little or no reinforcement; higher values with heavy reinforcement.

B-3   PROPERTIES OF RE-BARS

B-3a   DESIGNATIONS, AREAS, PERIMETERS, AND WEIGHTS OF STANDARD BARS

BAR DESIGNATION[a]	DIAMETER, in.	CROSS-SECTIONAL AREA, sq. in.	PERIMETER, in.	UNIT WEIGHT PER ft, lb
No. 2	$\frac{1}{4} = 0.250$	0.05	0.79	0.167
No. 3	$\frac{3}{8} = 0.375$	0.11	1.18	0.376
No. 4	$\frac{1}{2} = 0.500$	0.20	1.57	0.668
No. 5	$\frac{5}{8} = 0.625$	0.31	1.96	1.043
No. 6	$\frac{3}{4} = 0.750$	0.44	2.36	1.502
No. 7	$\frac{7}{8} = 0.875$	0.60	2.75	2.044
No. 8	$1 = 1.000$	0.79	3.14	2.670
No. 9	$1\frac{1}{8}\dagger = 1.128$	1.00	3.54	3.400
No. 10	$1\frac{1}{4}\dagger = 1.270$	1.27	3.99	4.303
No. 11	$1\frac{3}{8}\dagger = 1.410$	1.56	4.43	5.313

[a] Based on the number of eighths of an inch included in the nominal diameter of the bars. The nominal diameter of a deformed bar is equivalent to the diameter of a plain bar having the same weight per foot as the deformed bar. Bar No. 2 in plain rounds only. All others in deformed rounds.

B-3b   AREAS OF BARS IN SLABS, SQUARE INCHES PER FOOT

SPACING, INCHES	BAR DESIGNATION								
	No. 3	No. 4	No. 5	No. 6	No. 7	No. 8	No. 9	No. 10	No. 11
3	0.44	0.78	1.23	1.77	2.40	3.14	4.00	5.06	6.25
$3\frac{1}{2}$	0.38	0.67	1.05	1.51	2.06	2.69	3.43	4.34	5.36
4	0.33	0.59	0.92	1.32	1.80	2.36	3.00	3.80	4.68
$4\frac{1}{2}$	0.29	0.52	0.82	1.18	1.60	2.09	2.67	3.37	4.17
5	0.26	0.47	0.74	1.06	1.44	1.88	2.40	3.04	3.75
$5\frac{1}{2}$	0.24	0.43	0.67	0.96	1.31	1.71	2.18	2.76	3.41
6	0.22	0.39	0.61	0.88	1.20	1.57	2.00	2.53	3.12
$6\frac{1}{2}$	0.20	0.36	0.57	0.82	1.11	1.45	1.85	2.34	2.89
7	0.19	0.34	0.53	0.76	1.03	1.35	1.71	2.17	2.68
$7\frac{1}{2}$	0.18	0.31	0.49	0.71	0.96	1.26	1.60	2.02	2.50
8	0.17	0.29	0.46	0.66	0.90	1.18	1.50	1.89	2.34
9	0.15	0.26	0.41	0.59	0.80	1.05	1.33	1.69	2.08
10	0.13	0.24	0.37	0.53	0.72	0.94	1.20	1.52	1.87
12	0.11	0.20	0.31	0.44	0.60	0.78	1.00	1.27	1.56

B-4   PROPERTIES OF TIMBER FASTENERS

B-4a   LOADS ON NAILS. NORMAL LOAD DURATION

	SIZE OF NAIL							
d	8	10	12	16	20	30	60	Spikes
L (in.)	$2\frac{1}{2}$	3	$3\frac{1}{4}$	$3\frac{1}{2}$	4	$4\frac{1}{2}$	6	8–12
SPECIES	Withdrawal Load (per inch penetration of side-driven nails)							
Douglas fir	32	36	36	40	47	53	68	92
Southern pine	48	55	55	60	71	76	97	138
Red oak	64	72	72	75	94	101	128	183

d = pennyweight

For spikes of equal length, add about 10% to loads for 60d nails and about 20% to loads for 30d and smaller nails. When driven in unseasoned wood, which will season under load, or in wood pressure-impregnated with fire-retardant chemicals, the allowable load in withdrawal from side grain shall be one fourth of the tabulated value, except that for hardened deformed-shank nails, the full load may be used. (*National Design Specification*, 1962.)

B-4b   LATERAL OR SHEAR LOAD PER NAIL

SPECIES	LATERAL OR SHEAR LOAD PER NAIL							
Hemlock, white pine	51	62	62	70	91	101	146	248
Cedar, cypress	64	77	77	88	113	126	182	310
Birch, elm, soft maple	71	85	85	97	126	140	202	345
Douglas fir, southern pine	78	94	94	107	139	154	223	380
Oak, hard maple, southern locust	97	116	116	132	171	191	276	469

For spikes of equal length, add about 30% to loads for nails. Above loads require penetration in main member of 2/3L in softwoods and 1/2L in hardwoods. Minimum penetration allowed is L/2 in softwoods and 2/5L in hardwoods. When penetration is between minimum and standard required to develop full load, take load value as proportional to penetration from 0 load at 0 penetration to full load at standard penetration.

Based on Appendix J of *National Design Specification*, 1960.

B-4c LOADS ON BOLTED JOINTS

Bolt Diameter in.	$\frac{1}{2}$		$\frac{3}{4}$		1	
EFFECTIVE LENGTH	*P*	*Q*	*P*	*Q*	*P*	*Q*
$1\frac{5}{8}$	1010	480	1550	600	2070	730
$2\frac{5}{8}$	1280	780	2430	980	3340	1170
$3\frac{5}{8}$	1290	1020	2860	1350	4430	1620
$5\frac{1}{2}$			2890	1940	5120	2460

P = load parallel to grain.
Q = load perpendicular to grain.

   From Table 12 of the *National Design Specification for Stress Grade Lumber and its Fastenings,* 1962 Edition, giving values for bolts to $1\frac{1}{4}$-in. diameter with effective length of $11\frac{1}{2}$ in. for 18 species of wood. This table gives allowable loads in pounds on one bolt loaded at both ends for Douglas fir and southern pine.

B-4d GLUE LAMINATE STRESSES

GLUE LAMINATE

Use appropriate wood shearing stress since modern glues are equal to or stronger than the wood. Range is considerable:	psi
1. Douglas fir, southern pine, oak, hard maple, southern locust	~100–150
2. Hemlock, white pine cedar, cypress, birch, elm, soft maple	~60–100
3. Pre-fabricated (selected)	~150

Appendix C

SOME SECTION PROPERTIES FOR STRUCTURAL COMPONENTS

C-1   SECTION MODULOUS FOR SOME SHAPES USED AS BEAMS (S_x)

S_x In.³	SHAPE	S_x In.³	SHAPE	S_x In.³	SHAPE
1110	W 36 × 300	300	W 30 × 108	130	W 24 × 61
		300	W 27 × 114	129	W 18 × 70
1030	W 36 × 280	300	W 24 × 120	128	S 20 × 75
		284	W 21 × 127	128	W 16 × 78
952	W 36 × 260	276	W 24 × 110	127	W 21 × 62
				121	W 14 × 78
894	W 36 × 245	270	W 30 × 99	118	S 20 × 65.4
		267	W 27 × 102	118	W 18 × 64
837	W 36 × 230	252	S 24 × 120	116	W 16 × 71
813	W 33 × 240	250	W 24 × 100	116	W 12 × 85
		250	W 21 × 112		
742	W 33 × 220			114	W 24 × 55
		243	W 27 × 94	112	W 14 × 74
671	W 33 × 200	236	S 24 × 105.9		
				110	W 21 × 55
665	W 36 × 194	221	W 24 × 94	108	W 18 × 60
651	W 30 × 210	220	W 18 × 114	107	W 12 × 79
				104	W 16 × 64
622	W 36 × 182	212	W 27 × 84	103	S 18 × 70
587	W 30 × 190	202	W 18 × 105	103	W 14 × 68
		199	S 24 × 100		
580	W 36 × 170	198	W 21 × 96	98.4	W 18 × 55
				97.5	W 12 × 72
542	W 36 × 160	197	W 24 × 84	94.4	W 16 × 58
530	W 30 × 172	189	W 14 × 119		
		187	S 24 × 90	93.3	W 21 × 49
504	W 36 × 150	185	W 18 × 96	92.2	W 14 × 61
494	W 27 × 177			89.4	S 18 × 54.7
487	W 33 × 152	176	W 24 × 76		
		176	W 14 × 111	89.1	W 18 × 50
448	W 33 × 141	175	S 24 × 79.9	88.0	W 12 × 65
446	W 27 × 160	169	W 21 × 82		
		166	W 16 × 96	81.6	W 21 × 44
440	W 36 × 135	164	W 14 × 103	80.8	W 16 × 50

(continued)

C-1 (continued)

S_x In.³	SHAPE	S_x In.³	SHAPE	S_x In.³	SHAPE
414	W 24 × 160	161	S 20 × 95	79.0	W 18 × 45
		157	W 18 × 85	78.1	W 12 × 58
406	W 33 × 130			77.8	W 14 × 53
404	W 27 × 145	153	W 24 × 68	75.1	MC 18 × 58
380	W 30 × 132	152	S 20 × 85	73.7	W 10 × 66
373	W 24 × 145	151	W 21 × 73	72.5	W 16 × 45
		151	W 16 × 88	70.7	W 12 × 53
359	W 33 × 118	151	W 14 × 95	70.2	W 14 × 48
355	W 30 × 124	142	W 18 × 77	69.7	MC 18 × 51.9
332	W 24 × 130				
		140	W 21 × 68		
329	W 30 × 116	138	W 14 × 87		
317	W 21 × 142	131	W 14 × 84		
63.4	W 18 × 40	28.9	W 14 × 22	10.5	W 10 × 11.5
67.1	W 10 × 60	27.4	W 8 × 31	10.5	S 7 × 15.3
64.8	S 15 × 50	27.0	C 12 × 30	10.2	W 6 × 16
64.7	W 12 × 50	26.6	M 10 × 29.1	10.0	W 6 × 15.5
		26.5	W 10 × 25	9.90	W 8 × 13
64.6	W 16 × 40				
64.3	MC 18 × 45.8	25.3	W 12 × 22	9.23	MC 12 × 10.6
62.7	W 14 × 43	24.7	S 10 × 25.4	9.09	C 8 × 13.75
61.6	MC 18 × 42.7	24.3	W 8 × 28	8.77	S 6 × 17.25
60.4	W 10 × 54	24.1	C 12 × 25	8.14	C 8 × 11.5
59.6	S 15 × 42.9	23.6	M 10 × 22.9		
58.2	W 12 × 45			7.76	M 10 × 9
		21.5	C 12 × 20.7	7.80	W 8 × 10
57.9	W 18 × 35	21.5	W 10 × 21	7.78	C 7 × 14.75
56.5	W 16 × 36			7.37	S 6 × 12.5
54.7	W 14 × 38	21.3	W 12 × 19	7.25	W 6 × 12
54.6	W 10 × 49			6.93	C 7 × 12.25
53.8	C 15 × 50	21.1	M 14 × 17.2		
51.9	W 12 × 40	20.8	W 8 × 24	6.40	MC 10 × 8.4
50.8	S 12 × 50	20.7	C 10 × 30	6.08	C 7 × 9.8
49.1	W 10 × 45			6.09	S 5 × 14.75
		18.8	W 10 × 19	5.80	C 6 × 13
48.6	W 14 × 34	18.2	C 10 × 25	5.08	W 6 × 8.5
				5.06	C 6 × 10.5
47.2	W 16 × 31	17.6	W 12 × 16.5	4.92	S 5 × 10
46.5	C 15 × 40	17.0	W 8 × 20		
46.0	W 12 × 36	16.7	W 6 × 25	4.62	M 8 × 6.5
45.4	S 12 × 40.8	16.2	W 10 × 17	4.42	MC 10 × 6.5
42.2	W 10 × 39	16.2	S 8 × 23	4.38	C 6 × 8.2
42.0	C 15 × 33.9	15.8	C 10 × 20	3.56	C 5 × 9
41.9	W 14 × 30	14.8	W 12 × 14	3.44	M 7 × 5.5

S_x	SHAPE		S_x	SHAPE		S_x	SHAPE	
In.³			In.³			In.³		
39.5	W	12 × 31	14.4	S	8 × 18.4	3.39	S	4 × 9.5
			14.1	W	8 × 17	3.00	C	5 × 6.7
38.3	W	16 × 26	13.8	W	10 × 15	3.04	S	4 × 7.7
38.2	S	12 × 35	13.5	C	10 × 15.3			
36.4	S	12 × 31.8	13.5	C	9 × 20	2.40	M	6 × 4.4
			13.4	W	6 × 20	2.29	C	4 × 7.25
35.1	W	14 × 26	12.1	S	7 × 20	1.95	S	3 × 7.5
35.0	W	10 × 33				1.93	C	4 × 5.4
34.2	W	12 × 27	12.0	M	12 × 11.8	1.68	S	3 × 5.7
31.1	W	8 × 35	11.8	W	8 × 15	1.38	C	3 × 6
30.8	W	10 × 29	11.3	C	9 × 15	1.12	C	3 × 5
29.4	W	10 × 35	11.0	C	8 × 18.75			
			10.6	C	9 × 13.4	1.17	C	3 × 4.1

C-2   GENERAL STEEL SECTION PROPERTIES

C-2a   PROPERTIES FOR DESIGNING STEEL W SHAPE

DESIGNATION	AREA A	DEPTH d	FLANGE		WEB THICK-NESS t_w	ELASTIC PROPERTIES					
			WIDTH b_f	THICK-NESS t_f		AXIS X-X			AXIS Y-Y		
						I	S	r	I	S	r
	In.²	In.	In.	In.	In.	In.⁴	In.³	In.	In.⁴	In.³	In.
W 36 × 300	88.3	36.72	16.655	1.680	0.945	20300	1110	15.2	1300	156	3.83
× 280	82.4	36.50	16.595	1.570	0.885	18900	1030	15.1	1200	144	3.81
× 260	76.5	36.24	16.551	1.440	0.841	17300	952	15.0	1090	132	3.77
× 245	72.1	36.06	16.512	1.350	0.802	16100	894	15.0	1010	123	3.75
× 230	67.7	35.88	16.471	1.260	0.761	15000	837	14.9	940	114	3.73
W 36 × 194	57.2	36.48	12.117	1.260	0.770	12100	665	14.6	375	61.9	2.56
× 182	53.6	36.32	12.072	1.180	0.725	11300	622	14.5	347	57.5	2.55
× 170	50.0	36.16	12.027	1.100	0.680	10500	580	14.5	320	53.2	2.53
× 160	47.1	36.00	12.000	1.020	0.653	9760	542	14.4	295	49.1	2.50
× 150	44.2	35.84	11.972	0.940	0.625	9030	504	14.3	270	45.0	2.47
× 135	39.8	35.55	11.945	0.794	0.598	7820	440	14.0	226	37.9	2.39
W 33 × 240	70.6	33.50	15.865	1.400	0.830	13600	813	13.9	933	118	3.64
× 220	64.8	33.25	15.810	1.275	0.775	12300	742	13.8	841	106	3.60
× 200	58.9	33.00	15.750	1.150	0.715	11100	671	13.7	750	95.2	3.57

(continued)

C-2a (continued)

DESIGNATION	AREA A	DEPTH d	FLANGE WIDTH b_f	FLANGE THICK-NESS t_f	WEB THICK-NESS t_w	AXIS X-X I	AXIS X-X S	AXIS X-X r	AXIS Y-Y I	AXIS Y-Y S	AXIS Y-Y r
	In.²	In.	In.	In.	In.	In.⁴	In.³	In.	In.⁴	In.³	In.
W 33 × 152	44.8	33.50	11.565	1.055	0.635	8160	487	13.5	273	47.2	2.47
× 141	41.6	33.31	11.535	0.980	0.605	7460	448	13.4	246	42.7	2.43
× 130	38.3	33.10	11.510	0.855	0.580	6710	406	13.2	218	37.9	2.38
× 118	34.8	32.86	11.484	0.738	0.554	5900	359	13.0	187	32.5	2.32
W 30 × 210	61.9	30.38	15.105	1.315	0.775	9890	651	12.6	757	100	3.50
× 190	56.0	30.12	15.040	1.185	0.710	8850	587	12.6	673	89.5	3.47
× 172	50.7	29.88	14.985	1.065	0.655	7910	530	12.5	598	79.8	3.43
W 30 × 132	38.9	30.30	10.551	1.000	0.615	5760	380	12.2	196	37.2	2.25
× 124	36.5	30.16	10.521	0.930	0.585	5360	355	12.1	181	34.4	2.23
× 116	34.2	30.00	10.500	0.850	0.564	4930	329	12.0	164	31.3	2.19
× 108	31.8	29.82	10.484	0.760	0.548	4470	300	11.9	146	27.9	2.15
× 99	29.1	29.64	10.458	0.670	0.522	4000	270	11.7	128	24.5	2.10
W 27 × 177	52.2	27.31	14.090	1.190	0.725	6740	494	11.4	556	78.9	3.26
× 160	47.1	27.08	14.023	1.075	0.658	6030	446	11.3	495	70.6	3.24
× 145	42.7	26.88	13.965	0.975	0.600	5430	404	11.3	443	63.5	3.22
W 27 × 114	33.6	27.28	10.070	0.932	0.570	4090	300	11.0	159	31.6	2.18
× 102	30.0	27.07	10.018	0.827	0.518	3610	267	11.0	139	27.7	2.15
× 94	27.7	26.91	9.990	0.747	0.490	3270	243	10.9	124	24.9	2.12
× 84	24.8	26.69	9.963	0.636	0.463	2830	212	10.7	105	21.1	2.06
W 24 × 160	47.1	24.72	14.091	1.135	0.656	5120	414	10.4	530	75.2	3.35
× 145	42.7	24.49	14.043	1.020	0.508	4570	373	10.3	471	67.1	3.32
× 130	38.3	24.25	14.000	0.900	0.565	4020	332	10.2	412	58.9	3.28
W 24 × 120	35.4	24.31	12.088	0.930	0.556	3650	300	10.2	274	45.4	2.78
× 110	32.5	24.16	12.042	0.855	0.510	3330	276	10.1	249	41.4	2.77
× 100	29.5	24.00	12.000	0.775	0.468	3000	250	10.1	223	37.2	2.75
W 24 × 94	27.7	24.29	9.061	0.872	0.516	2690	221	9.86	108	23.9	1.98
× 84	24.7	24.09	9.015	0.772	0.470	2370	197	9.79	94.5	21.0	1.95
× 76	22.4	23.91	8.985	0.682	0.440	2100	176	9.69	82.6	18.4	1.92
× 68	20.0	23.71	8.961	0.582	0.416	1820	153	9.53	70.0	15.6	1.87
W 24 × 61	18.0	23.72	7.023	0.591	0.419	1540	130	9.25	34.3	9.76	1.38
× 55	16.2	23.55	7.000	0.503	0.396	1340	114	9.10	28.9	8.25	1.34
W 21 × 142	41.8	21.46	13.132	1.095	0.659	3410	317	9.03	414	63.0	3.15
× 127	37.4	21.24	13.061	0.985	0.588	3020	284	8.99	365	56.1	3.13
× 112	33.0	21.00	13.000	0.865	0.527	2620	250	8.92	317	48.8	3.10
W 21 × 96	28.3	21.14	9.033	0.935	0.575	2100	198	8.61	115	25.5	2.02
× 82	24.2	20.86	8.962	0.795	0.499	1760	169	8.53	95.6	21.3	1.99

DESIGNATION	AREA A	DEPTH d	FLANGE WIDTH b_f	FLANGE THICKNESS t_f	WEB THICKNESS t_w	ELASTIC PROPERTIES AXIS X-X I	S	r	AXIS Y-Y I	S	r
	In.²	In.	In.	In.	In.	In.⁴	In.³	In.	In.⁴	In.³	In.
W 21 × 73	21.5	21.24	8.295	0.740	0.455	1600	151	8.64	70.6	17.0	1.81
× 68	20.0	21.13	8.270	0.685	0.430	1480	140	8.60	64.7	15.7	1.80
× 62	18.3	20.99	8.240	0.615	0.400	1330	127	8.54	57.5	13.9	1.77
× 55	16.2	20.80	8.215	0.522	0.375	1140	110	8.40	48.3	11.8	1.73
W 21 × 49	14.4	20.82	6.520	0.532	0.368	971	93.3	8.21	24.7	7.57	1.31
× 44	13.0	20.66	6.500	0.451	0.348	843	81.6	8.07	20.7	6.38	1.27
W 18 × 114	33.5	18.48	11.833	0.991	0.595	2040	220	7.79	274	46.3	2.86
× 105	30.9	18.32	11.792	0.911	0.554	1850	202	7.75	249	42.3	2.84
× 96	28.2	18.16	11.750	0.831	0.512	1680	185	7.70	225	38.3	2.82
W 18 × 85	25.0	18.32	8.838	0.911	0.526	1440	157	7.57	105	23.8	2.05
× 77	22.7	18.16	8.787	0.831	0.475	1290	142	7.54	94.1	21.4	2.04
× 70	20.6	18.00	8.750	0.751	0.438	1160	129	7.50	84.0	19.2	2.02
× 64	18.9	17.87	8.715	0.686	0.403	1050	118	7.46	75.8	17.4	2.00
W 18 × 60	17.7	18.25	7.558	0.695	0.416	986	108	7.47	50.1	13.3	1.68
× 55	16.2	18.12	7.532	0.630	0.390	891	98.4	7.42	45.0	11.9	1.67
× 50	14.7	18.00	7.500	0.570	0.358	802	89.1	7.38	40.2	10.7	1.65
× 45	13.2	17.86	7.477	0.499	0.335	706	79.0	7.30	34.8	9.32	1.62
W 18 × 40	11.8	17.90	6.018	0.524	0.316	612	68.4	7.21	19.1	6.34	1.27
× 35	10.3	17.71	6.000	0.429	0.298	513	57.9	7.05	15.5	5.16	1.23
W 16 × 96	28.2	16.32	11.533	0.875	0.535	1360	166	6.93	224	38.8	2.82
× 88	25.9	16.16	11.502	0.795	0.504	1220	151	6.87	202	35.1	2.79
W 16 × 78	23.0	16.32	8.586	0.875	0.529	1050	128	6.75	92.5	21.6	2.01
× 71	20.9	16.16	8.543	0.795	0.486	941	116	6.71	82.8	19.4	1.99
× 64	18.8	16.00	8.500	0.715	0.443	836	104	6.66	73.3	17.3	1.97
× 58	17.1	15.86	8.464	0.645	0.407	748	94.4	6.62	65.3	15.4	1.96
W 16 × 50	14.7	16.25	7.073	0.628	0.380	657	80.8	6.68	37.1	10.5	1.59
× 45	13.3	16.12	7.039	0.563	0.346	584	72.5	6.64	32.8	9.32	1.57
× 40	11.8	16.00	7.000	0.503	0.307	517	64.6	6.62	28.8	8.23	1.56
× 36	10.6	15.85	6.992	0.428	0.299	447	55.5	6.50	24.4	6.99	1.52
W 16 × 31	9.13	15.84	5.525	0.442	0.275	374	47.2	6.40	12.5	4.51	1.17
× 26	7.67	15.65	5.500	0.345	0.250	300	38.3	6.25	9.59	3.49	1.12
W 14 × 136	40.0	14.75	14.740	1.063	0.660	1590	216	6.31	568	77.0	3.77
× 127	37.3	14.62	14.690	0.998	0.610	1480	202	6.29	528	71.8	3.76
× 119	35.0	14.50	14.650	0.938	0.570	1370	189	6.26	492	67.1	3.75
× 111	32.7	14.37	14.620	0.873	0.540	1270	176	6.23	455	62.2	3.73
× 103	30.3	14.25	14.575	0.813	0.495	1170	164	6.21	420	57.6	3.72
× 95	27.9	14.12	14.545	0.748	0.465	1060	151	6.17	384	52.8	3.71
× 87	25.6	14.00	14.500	0.688	0.420	967	138	6.15	350	48.2	3.70

C-2a (continued)

DESIGNATION	AREA A In.²	DEPTH d In.	FLANGE WIDTH b_f In.	FLANGE THICK-NESS t_f In.	WEB THICK-NESS t_w In.	AXIS X-X I In.⁴	AXIS X-X S In.³	AXIS X-X r In.	AXIS Y-Y I In.⁴	AXIS Y-Y S In.³	AXIS Y-Y r In.
W 14 × 84	24.7	14.18	12.023	0.778	0.451	928	131	6.13	225	37.5	3.02
× 78	22.9	14.06	12.000	0.718	0.428	851	121	6.09	207	34.5	3.00
W 14 × 74	21.8	14.19	10.072	0.783	0.450	797	112	6.05	133	26.5	2.48
× 68	20.0	14.06	10.040	0.718	0.418	724	103	6.02	121	24.1	2.46
× 61	17.9	13.91	10.000	0.643	0.378	641	92.2	5.98	107	21.5	2.45
W 14 × 53	15.6	13.94	8.062	0.658	0.370	542	77.8	5.90	57.5	14.3	1.92
× 48	14.1	13.81	8.031	0.593	0.339	485	70.2	5.86	51.3	12.8	1.91
× 43	12.6	13.68	8.000	0.528	0.308	429	62.7	5.82	45.1	11.3	1.89
W 14 × 38	11.2	14.12	6.776	0.513	0.313	386	54.7	5.88	26.6	7.86	1.54
× 34	10.0	14.00	6.750	0.453	0.287	340	48.6	5.83	23.3	6.89	1.52
× 30	8.83	13.86	6.733	0.383	0.270	290	41.9	5.74	19.5	5.80	1.49
W 14 × 26	7.67	13.89	5.025	0.418	0.255	244	35.1	5.64	8.86	3.53	1.08
× 22	6.49	13.72	5.000	0.335	0.230	198	28.9	5.53	7.00	2.80	1.04
W 14 × 730	215	22.44	17.889	4.910	3.069	14400	1280	8.18	4720	527	4.69
× 665	196	21.67	17.646	4.522	2.826	12500	1150	7.99	4170	472	4.62
× 605	178	20.94	17.418	4.157	2.598	10900	1040	7.81	3680	423	4.55
× 550	162	20.26	17.206	3.818	2.386	9450	933	7.64	3260	378	4.49
× 500	147	19.63	17.008	3.501	2.188	8250	840	7.49	2880	339	4.43
× 455	134	19.05	16.828	3.213	2.008	7220	758	7.35	2560	304	4.37
W 14 × 426	125	18.69	16.695	3.033	1.875	6610	707	7.26	2360	283	4.34
× 398	117	18.31	16.590	2.843	1.770	6010	657	7.17	2170	262	4.31
× 370	109	17.94	16.475	2.658	1.655	5450	608	7.08	1990	241	4.27
× 342	101	17.56	16.365	2.468	1.545	4910	559	6.99	1810	221	4.24
× 314	92.3	17.19	16.235	2.283	1.415	4400	512	6.90	1630	201	4.20
× 287	84.4	16.81	16.130	2.093	1.310	3910	465	6.81	1470	182	4.17
× 264	77.6	16.50	16.025	1.938	1.205	3530	427	6.74	1330	166	4.14
× 246	72.3	16.25	15.945	1.813	1.125	3230	397	6.68	1230	154	4.12
W 14 × 237	69.7	16.12	15.910	1.748	1.090	3080	382	6.65	1170	148	4.11
× 228	67.1	16.00	15.865	1.688	1.045	2940	368	6.62	1120	142	4.10
× 219	64.4	15.87	15.825	1.623	1.005	2800	353	6.59	1070	136	4.08
× 211	62.1	15.75	15.800	1.563	0.980	2670	339	6.56	1030	130	4.07
× 202	59.4	15.63	15.750	1.503	0.930	2540	325	6.54	930	124	4.06
× 193	56.7	15.50	15.710	1.438	0.890	2400	310	6.51	930	118	4.05
× 184	54.1	15.38	15.660	1.378	0.840	2270	296	6.49	883	113	4.04
× 176	51.7	15.25	15.640	1.313	0.820	2150	282	6.45	838	107	4.02
× 167	49.1	15.12	15.600	1.248	0.780	2020	267	6.42	790	101	4.01
× 158	46.5	15.00	15.550	1.188	0.730	1900	253	6.40	745	95.8	4.00
× 150	44.1	14.88	15.515	1.128	0.695	1790	240	6.37	703	90.6	3.99
× 142	41.8	14.75	15.500	1.063	0.680	1670	227	6.32	660	85.2	3.97

DESIGNATION	AREA A	DEPTH d	FLANGE WIDTH b_f	FLANGE THICK-NESS t_f	WEB THICK-NESS t_w	ELASTIC PROPERTIES AXIS X-X I	S	r	AXIS Y-Y I	S	r
	In.²	In.	In.	In.	In.	In.⁴	In.³	In.	In.⁴	In.³	In.
W 14 × 320	94.1	16.81	16.710	2.093	1.890	4140	493	6.63	1640	196	4.17
W 12 × 190	55.9	14.38	12.670	1.735	1.060	1890	263	5.82	590	93.1	3.25
× 161	47.4	13.88	12.515	1.486	0.905	1540	222	5.70	486	77.7	3.20
× 133	39.1	13.38	12.365	1.236	0.755	1220	183	5.59	390	63.1	3.16
× 120	35.3	13.12	12.320	1.106	0.710	1070	163	5.51	345	56.0	3.13
× 106	31.2	12.88	12.230	0.986	0.620	981	145	5.46	301	49.2	3.11
× 99	29.1	12.75	12.192	0.921	0.582	859	135	5.43	278	45.7	3.09
× 92	27.1	12.62	12.155	0.856	0.545	789	125	5.40	256	42.2	3.08
× 85	25.0	12.50	12.105	0.796	0.495	723	116	5.38	235	38.9	3.07
× 79	23.2	12.38	12.080	0.736	0.470	663	107	5.34	216	35.8	3.05
× 72	21.2	12.25	12.040	0.671	0.430	597	97.5	5.31	195	32.4	3.04
× 65	19.1	12.12	12.000	0.606	0.390	533	88.0	5.28	175	29.1	3.02
W 12 × 58	17.1	12.19	10.014	0.641	0.359	476	78.1	5.28	107	21.4	2.51
× 53	15.6	12.06	10.000	0.576	0.345	426	70.7	5.23	96.1	19.2	2.48
W 12 × 50	14.7	12.19	8.077	0.641	0.371	395	64.7	5.18	56.4	14.0	1.96
× 45	13.2	12.06	8.042	0.576	0.336	351	58.2	5.15	50.0	12.4	1.94
× 40	11.8	11.94	8.000	0.516	0.294	310	51.9	5.13	44.1	11.0	1.94
W 12 × 36	10.6	12.24	6.565	0.540	0.305	281	46.0	5.15	25.5	7.77	1.55
× 31	9.13	12.09	6.525	0.465	0.265	239	39.5	5.12	21.6	6.61	1.54
× 27	7.95	11.96	6.497	0.400	0.237	204	34.2	5.07	18.3	5.63	1.52
W 12 × 22	6.47	12.31	4.030	0.424	0.260	156	25.3	4.91	4.64	2.31	0.847
× 19	5.59	12.16	4.007	0.349	0.237	130	21.3	4.82	3.76	1.88	0.820
× 16.5	4.87	12.00	4.000	0.269	0.230	105	17.6	4.65	2.88	1.44	0.770
× 14	4.12	11.91	3.968	0.224	0.198	88.0	14.8	4.62	2.34	1.18	0.754
W 10 × 112	32.9	11.38	10.415	1.248	0.755	719	126	4.67	235	45.2	2.67
× 100	29.4	11.12	10.345	1.118	0.685	625	112	4.61	207	39.9	2.65
× 89	26.2	10.88	10.275	0.998	0.615	542	99.7	4.55	181	35.2	2.63
× 77	22.7	10.62	10.195	0.868	0.535	457	86.1	4.49	153	30.1	2.60
× 72	21.2	10.50	10.170	0.808	0.510	421	80.1	4.46	142	27.9	2.59
× 66	19.4	10.38	10.117	0.743	0.457	382	73.7	4.44	129	25.5	2.58
× 60	17.7	10.25	10.075	0.683	0.415	344	67.1	4.41	116	23.1	2.57
× 54	15.9	10.12	10.028	0.618	0.358	306	60.4	4.39	104	20.7	2.56
× 49	14.4	10.00	10.000	0.558	0.340	273	54.6	4.35	93.0	18.6	2.54
W 10 × 45	13.2	10.12	8.022	0.618	0.350	249	49.1	4.33	53.2	13.2	2.00
× 39	11.5	9.94	7.990	0.528	0.318	210	42.2	4.27	44.9	11.2	1.98
× 33	9.71	9.75	7.964	0.433	0.292	171	35.0	4.20	36.5	9.16	1.94
W 10 × 29	8.54	10.22	5.799	0.500	0.289	158	30.8	4.30	16.3	5.61	1.38
× 25	7.36	10.08	5.762	0.430	0.252	133	26.5	4.26	13.7	4.76	1.37
× 21	6.20	9.90	5.750	0.340	0.240	107	21.5	4.15	10.8	3.75	1.32

C-2a (continued)

DESIGNATION	AREA A	DEPTH d	FLANGE WIDTH b_f	FLANGE THICK-NESS t_f	WEB THICK-NESS t_w	ELASTIC PROPERTIES AXIS X-X I	S	r	AXIS Y-Y I	S	r
	In.²	In.	In.	In.	In.	In.⁴	In.³	In.	In.⁴	In.³	In.
W 10 × 19	5.61	10.25	4.020	0.394	0.250	96.3	18.8	4.14	4.28	2.13	0.874
× 17	4.99	10.12	4.010	0.329	0.240	81.9	16.2	4.05	3.55	1.77	0.844
× 15	4.41	10.00	4.000	0.269	0.230	68.9	13.8	3.95	2.88	1.44	0.809
× 11.5	3.39	9.87	3.950	0.204	0.180	52.0	10.5	3.92	2.10	1.06	0.757
W 8 × 67	19.7	9.00	8.287	0.933	0.575	272	60.4	3.71	88.6	21.4	2.12
× 58	17.1	8.75	8.222	0.808	0.510	227	52.0	3.65	74.9	18.2	2.10
× 48	14.1	8.50	8.117	0.683	0.405	184	43.2	3.61	60.9	15.0	2.08
× 40	11.8	8.25	8.077	0.558	0.365	146	35.5	3.53	49.0	12.1	2.04
× 35	10.3	8.12	8.027	0.493	0.315	126	31.1	3.50	42.5	10.6	2.03
× 31	9.12	8.00	8.000	0.433	0.288	110	27.4	3.47	37.0	9.24	2.01
W 8 × 28	8.23	8.06	6.540	0.463	0.285	97.8	24.3	3.45	21.6	6.61	1.62
× 24	7.06	7.93	6.500	0.398	0.245	82.5	20.8	3.42	18.2	5.61	1.61
W 8 × 20	5.89	8.14	5.268	0.378	0.248	69.4	17.0	3.43	9.22	3.50	1.25
× 17	5.01	8.00	5.250	0.308	0.230	56.6	14.1	3.36	7.44	2.83	1.22
W 8 × 15	4.43	8.12	4.015	0.314	0.245	48.1	11.8	3.29	3.40	1.69	0.876
× 13	3.83	8.00	4.000	0.254	0.230	39.6	9.90	3.21	2.72	1.36	0.842
× 10	2.96	7.90	3.940	0.204	0.170	30.8	7.80	3.23	2.08	1.06	0.839
W 6 × 25	7.35	6.37	6.080	0.456	0.320	53.3	16.7	2.69	17.1	5.62	1.53
× 20	5.88	6.20	6.018	0.367	0.258	41.5	13.4	2.66	13.3	4.43	1.51
× 15.5	4.56	6.00	5.995	0.269	0.235	30.1	10.0	2.57	9.67	3.23	1.46
W 6 × 16	4.72	6.25	4.030	0.404	0.260	31.7	10.2	2.59	4.42	2.19	0.967
× 12	3.54	6.00	4.000	0.279	0.230	21.7	7.25	2.48	2.98	1.49	0.918
× 8.5	2.51	5.83	3.940	0.194	0.170	14.8	5.08	2.43	1.98	1.01	0.889
W 5 × 18.5	5.43	5.12	5.025	0.420	0.265	25.4	9.94	2.16	8.89	3.54	1.28
× 16	4.70	5.00	5.000	0.360	0.240	21.3	8.53	2.13	7.51	3.00	1.26
W 4 × 13	3.82	4.16	4.060	0.345	0.280	11.3	5.45	1.72	3.76	1.85	0.991

C-2b STEEL SECTION PROPERTIES FOR Q DECKING (NON-COMPOSITE)

	SECTION AND GAUGE	ACTUAL WT./SQ. FT. LBS.	I FOR DEFL. IN.⁴	TOP SM IN³ +MOMENT	BOTTOM SM IN³ −MOMENT
	3-22	1.8	.180	.203	.219
	3-20	2.2	.230	.265	.273
	3-18	2.9	.337	.398	.380
	3-16	3.5	.442	.506	.480
	3-14	4.4	.562	.633	.592
	3-12	5.9	.756	.880	.880
	UKX 20-20	3.8	.381	.310	.439
	UKX 20-18	4.3	.411	.317	.487
	UKX 18-20	4.4	.520	.462	.542
	UKX 18-18	4.8	.566	.472	.591
	UKX 18-16	5.3	.603	.481	.641
	UKX 16-16	5.8	.763	.654	.745
	UKX 16-14	6.5	.820	.667	.810
	UKX 14-14	7.3	1.011	.893	.939
	UKX 12-12	9.9	1.373	1.353	1.330
	21-22	2.1	.675	.386	.467
	21-20	2.6	.855	.500	.575
	21-18	3.5	1.258	.755	.787
	21-16	4.2	1.703	.982	.991
	21-14	5.2	2.264	1.261	1.230
	21-12	6.9	3.381	1.823	1.699
	NKX 20-20	4.2	1.431	.600	1.030
	NKX 20-18	4.7	1.543	.613	1.185
	NKX 18-20	5.0	1.951	.884	1.079
	NKX 18-18	5.4	2.125	.909	1.391
	NKX 18-16	5.8	2.226	.923	1.623
	NKX 16-16	6.5	2.888	1.260	1.760
	NKX 16-14	7.2	3.084	1.285	2.092
	NKX 14-14	8.1	3.903	1.746	2.226
	NKX 12-12	11.1	6.049	2.833	3.178
	NKC 18-18	5.5	2.765	1.405	1.096
	NKC 16-16	6.8	3.572	1.881	1.458
	NKC 14-14	8.5	4.514	2.460	1.947
	12-20	3.6	2.933	1.126	1.062
	12-18	4.9	4.078	1.610	1.534
	12-16	5.9	5.195	2.107	2.038
	12-14	7.3	6.180	2.694	2.570
	12-12	10.0	8.587	3.433	3.604
	FKX 18-18	6.5	5.93	1.90	1.834
	FKX 18-16	7.0	6.30	1.94	2.392
	FKX 16-16	7.9	7.57	2.49	2.491
	FKX 16-14	8.5	8.06	2.54	3.068
	FKX 14-14	9.8	9.02	3.18	3.140
	FKX 12-12	13.4	12.59	4.10	4.375

TABLE OF PROPERTIES NOTES:

1. Section properties for all sections have been computed in accordance with the A.I.S.I. "Specification For The Design of Cold-Formed Steel Structural Members" (Latest Edition).
2. All values given in the table are for one foot widths of units.

C-2c STANDARD LOAD TABLE FOR OPEN WEB JOISTS, J-SERIES†. BASED ON ALLOWABLE STRESS OF 22,000 PSI AND ALLOWABLE TOTAL END REACTIONS

Joist designation	8J3	10J3	10J4	12J3	12J4	12J5	12J6	14J3	14J4	14J5	14J6	14J7	16J4	16J5	16J6	16J7	16J8
*Depth in inches	8	10	10	12	12	12	12	14	14	14	14	14	16	16	16	16	16
Resisting moment in inch kips	70	89	111	108	135	161	196	127	159	190	230	276	173	216	258	310	359
Max. end reaction in pounds	2000	2200	2400	2300	2500	2700	3000	2400	2800	3100	3400	3700	3000	3300	3600	4000	4300
**Approx. joist wgt. pounds per foot	4.8	4.8	6.0	5.1	6.0	7.0	8.1	5.2	6.4	7.3	8.4	9.7	6.6	7.6	8.5	10.1	11.3
Span in feet	SAFE LOADS IN POUNDS PER LINEAR FOOT																
8																	
9																	
10	400	440	480														
11	364	400	436														
12	324	367	400	383	417	450	500										
13	276	338	369	354	385	415	462										
14	238	303	343	329	357	386	429	343	400	443	486	529					
15	207	264	320	307	333	360	400	320	373	413	453	493					
16	182	232	289	281	313	338	375	300	350	388	425	463	375	413	450	500	538
17	161	205	256	249	294	318	353	282	329	365	400	435	353	388	424	471	506
18	144	183	228	222	278	300	333	261	311	344	378	411	333	367	400	444	478
19	129	164	205	199	249	284	316	235	294	326	358	389	316	347	379	421	453
20	117	148	185	180	225	268	300	212	265	310	340	370	288	330	360	400	430
21				163	204	243	286	192	240	287	324	352	262	314	343	381	410
22				149	186	222	270	175	219	262	309	336	238	298	327	364	391
23				136	170	203	247	160	200	239	290	322	218	272	313	348	374
24				125	156	186	227	147	184	220	266	308	200	250	299	333	358
25								135	170	203	245	294	185	230	275	320	344
26								125	157	187	227	272	171	213	254	306	331
27								116	145	174	210	252	158	198	236	283	319
28								108	135	162	196	235	147	184	219	264	305
29													137	171	205	246	285
30													128	160	191	230	266
31												·	120	150	179	215	249
32													113	141	168	202	234

Loads above heavy stepped lines are governed by shear.
†Copyright 1965, Steel Joist Institute. Reprinted by permission.
*Indicates nominal depth of steel joists only.
**Approximate weight per linear foot of steel joists only. Accessories and nailer strip not included.

Joist designation	18J5	18J6	18J7	18J8	20J5	20J6	20J7	70J8	22J6	22J7	22J8	24J6	24J7	24J8
*Depth in inches	18	18	18	18	20	20	20	20	22	22	22	24	24	24
Resisting moment in inch kips	243	293	352	406	265	316	382	455	335	420	493	367	460	540
Max. end reaction in pounds	3500	3900	4200	4500	3800	4100	4300	4600	4200	4500	4800	4400	4700	5000
**Approx. joist wgt. pounds per foot	7.9	9.0	10.2	11.3	8.1	9.2	10.6	11.9	9.6	10.5	11.9	9.9	11.1	12.4
Span in feet	\multicolumn{14}{c}{SAFE LOADS IN POUNDS PER LINEAR FOOT}

Span in feet	18J5	18J6	18J7	18J8	20J5	20J6	20J7	70J8	22J6	22J7	22J8	24J6	24J7	24J8
16														
17														
18	389	433	467	500										
19	368	411	442	474										
20	350	390	420	450	380	410	430	460						
21	333	371	400	429	362	390	410	438						
22	318	355	382	409	345	373	391	418	382	409	436			
23	304	339	365	391	330	357	374	400	365	391	417			
24	281	325	350	375	307	342	358	383	350	375	400	367	392	417
25	259	312	336	360	283	328	344	368	336	360	384	352	376	400
26	240	289	323	346	261	312	331	354	323	346	369	338	362	385
27	222	268	311	333	242	289	319	341	306	333	356	326	348	370
28	207	249	299	321	225	269	307	329	285	321	343	312	336	357
29	193	232	279	310	210	250	297	317	266	310	331	291	324	345
30	180	217	261	300	196	234	283	307	248	300	320	272	313	333
31	169	203	244	282	184	219	265	297	232	290	310	255	303	323
32	158	191	229	264	173	206	249	288	218	273	300	239	294	313
33	149	179	215	249	162	193	234	279	205	257	291	225	282	303
33	140	169	203	234	153	182	220	262	193	242	282	212	265	294
35	132	159	192	221	144	172	208	248	182	229	268	200	250	286
36	125	151	181	209	136	163	197	234	172	216	254	189	237	278
37					129	154	186	222	163	205	240	179	224	263
38					122	146	176	210	155	194	228	169	212	249
39					116	139	167	199	147	184	216	161	202	237
40					110	132	159	190	140	175	205	153	192	225
41									133	167	196	146	182	214
42									127	159	186	139	174	204
43									121	151	178	132	166	195
44									115	145	170	126	158	186
45												121	151	178
46												116	145	176
47												111	139	163
48												106	133	156

Loads above heavy stepped lines are governed by shear.
†Copyright 1965, Steel Joist Institute. Reprinted by permission.
*Indicates nominal depth of steel joists only.
**Approximate weight per linear foot of steel joists only. Accessories and nailer strip not included.

C-3   PROPERTIES OF PRECAST CONCRETE SECTIONS

C-3a   TABLE OF PROPERTIES FOR LIN TEE

(WIDTH = 6'; WEB = 8"; TAPER = 2"; NO TOPPING)

DEPTH, in.	AREA, in.²	I, in.⁴	c_b, in.
12	265	2360	9.2
16	297	6170	11.8
20	329	11,730	14.5
24	361	19,700	17.0
28	393	30,400	19.5
32	425	44,060	21.9
36	457	61,000	24.2

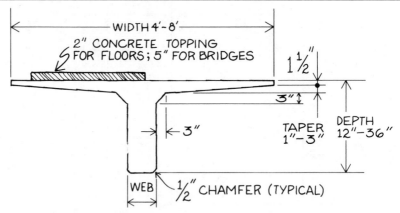

TYPICAL SINGLE TEE SECTION (LIN TEE).

C-3a   TABLE OF PROPERTIES FOR DOUBLE TEE

TOPPING	AREA, in.²	I, in.⁴	c_b, in.
With 2" topping	276	4456	11.74
Without topping	180	2862	10.00

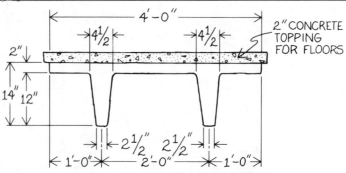

TYPICAL DOUBLE TEE SECTION.

C-3a TABLE OF PROPERTIES FOR CONCRETE SLAB SECTIONS

BEAM TYPE	AREA, in.²	I, in.⁴	c_b, in.	DRAPED STRAND	STRAIGHT STRAND
				RECOMMENDED SPAN LIMITS, ft.	
BII-36	620.5	85,153	16.29	86	73
BII-48	752.5	110,499	16.33	86	74
BIII-36	680.5	131,145	19.25	97	83
BIII-48	812.5	168,367	19.29	96	83

4'-0" x 4"

4'-0" x 4" with 2" Topping

Area (sq in.)	y_b (in.)	I (in.⁴)	Weight (psf)	y_b (in.)	Mom. of Inertia (in.⁴)	Weight (psf)
154	2.00	247	40	2.98	723	65

4'-0" x 6"

4'-0" x 6" with 2" Topping

Area (sq in.)	y_b (in.)	I (in.⁴)	Weight (psf)	y_b (in.)	Mom. of Inertia (in.⁴)	Weight (psf)
188	3.00	764	49	4 13	1641	74

4'-0" x 8"

4'-0" x 8" with 2" Topping

Area (sq in.)	y_b (in.)	I (in.⁴)	Weight (psf)	y_b (in.)	Mom. of Inertia (in.⁴)	Weight (psf)
214	4.00	1666	56	5.29	3070	81

4'-0" x 10"

4'-0" x 10" with 2" Topping

Area (sq in.)	y_b (in.)	I (in.⁴)	Weight (psf)	y_b (in.)	Mom. of Inertia (in.⁴)	Weight (psf)
259	5.00	3223	67	6.34	5328	92

4'-0" x 12"

4' x 0" x 12" with 2" Topping

Area (sq in.)	y_b (in.)	I (in.⁴)	Weight (psf)	y_b (in.)	Mom. of Inertia (in.⁴)	Weight (psf)
289	6.00	5272	75	7.43	8195	100

C-3a   PRECAST CONCRETE SECTIONS: TABLE OF PROPERTIES AND ALLOWABLE LOADS FOR PRESTRESSED CONCRETE PILES

SIZE in.	CORE DIA. in.	SECTION PROPERTIES[1]						ALLOWABLE CONCENTRIC SERVICE LOAD, TONS[2]			
		AREA in.²	WEIGHT plf	MOMENT OF INERTIA in.⁴	SECTION MODULUS in.³	RADIUS OF GYRATION in.	PERIMETER (ft.)	f'_c			
								5000	6000	7000	8000
SQUARE PILES											
10	Solid	100	104	833	167	2,89	3.33	73	89	106	122
12	Solid	144	150	1728	288	3.46	4.00	105	129	152	176
14	Solid	196	204	3201	457	4.04	4.67	143	175	208	240
16	Solid	256	267	5461	683	4.62	5.33	187	229	271	314
18	Solid	324	338	8748	972	5.20	6.00	236	290	344	397
20	Solid	400	417	13,333	1333	5.7	6.67	292	358	424	490
20	11″	305	318	12,615	1262	6.43	6.67	222	273	323	373
24	Solid	576	600	27,648	2304	6.93	8.00	420	515	610	705
24	12″	463	482	26,630	2219	7.58	8.00	338	414	491	567
24	14″	422	439	25,762	2147	7.81	8.00	308	377	447	517
24	15″	399	415	25,163	2097	7.94	8.00	291	357	423	488
OCTAGONAL PILES											
10	Solid	83	85	555	111	2.59	2.76	60	74	88	101
12	Solid	119	125	1134	189	3.09	3.31	86	106	126	145
14	Solid	162	169	2105	301	3.60	3.87	118	145	172	198
16	Solid	212	220	3592	449	4.12	4.42	154	189	224	259
18	Solid	268	280	5705	639	4.61	4.97	195	240	284	328
20	Solid	331	345	8770	877	5.15	5.52	241	296	351	405
20	11″	236	245	8050	805	5.84	5.52	172	211	250	289
22	Solid	401	420	12,837	1167	5.66	6.08	292	359	425	491
22	13″	268	280	11,440	1040	6.53	6.08	195	240	283	328
24	Solid	477	495	18,180	1515	6.17	6.63	348	427	506	584
24	15″	300	315	15,696	1308	7.23	6.63	219	268	318	368
ROUND PILES											
36	26″	487	507	60,007	3334	11.10	9.43	355	436	516	596
48	38″	675	703	158,199	6592	15.31	12.57	493	604	715	827
54	44″	770	802	233,373	8643	17.41	14.14	562	689	816	943

[1] Form dimensions may vary with producers, with corresponding variations in section properties.
[2] Allowable loads based on $N = A_c (0.33 f'_c - 0.27 f_{pc})$; $f_{pc} = 700$ psi; Check local producer for available concrete strengths.

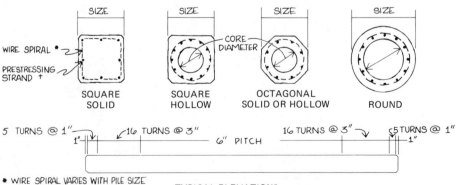

SQUARE SOLID   SQUARE HOLLOW   OCTAGONAL SOLID OR HOLLOW   ROUND

WIRE SPIRAL *
PRESTRESSING STRAND †
CORE DIAMETER

5 TURNS @ 1″   16 TURNS @ 3″   6″ PITCH   16 TURNS @ 3″   5 TURNS @ 1″

* WIRE SPIRAL VARIES WITH PILE SIZE
† STRAND PATTERN MAY BE CIRCULAR OR SQUARE

TYPICAL ELEVATION*

C-3b   SOME PROPERTIES OF CURVED SHELL SECTIONS

1. Segment of A Circle:

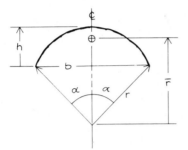

ARC LENGTH	CENTROID OF ARC	AREA
$a = 0.035r\alpha$	$\bar{r} = \dfrac{r \sin \alpha}{\alpha_{radians}}$	$A = \dfrac{ar - b(r - b)}{2}$

2. Parabola:

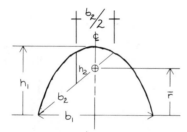

ROUGH ESTIMATE OF ARC LENGTH AND ARC CENTROID	AREA
When $h/b \leqslant 0.25$, use formula for segment of a circle. [(Both (a) and ($\bar{r}$) will be overestimated)]	$A_1 = \frac{2}{3}(h_1\, b_1)$ $A_2 = \frac{2}{3}(h_2\, b_2)$

3. $\frac{1}{2}$ Ellipse:

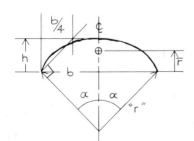

ARC LENGTH	ROUGH ESTIMATE OF CENTROID OF ARC LENGTH	APPROXIMATE AREA
$a = \dfrac{\Pi}{2}\sqrt{2(h^2 + b^2)}$	When $h/b \leqslant 0.25$, use formula α for segment of circle. Find "r" as shown.	$A \sim 0.8hb$

C-4   TIMBER AND PLYWOOD SECTIONS

C-4a   STANDARD TIMBER SIZES AND PROPERTIES FOR DESIGNING

NOMINAL SIZE		STANDARD DRESSED SIZE S4S		AREA OF SECTION $A = bh$	MOMENT OF INERTIA $I = \dfrac{bh^3}{12}$	SECTION MODULUS $S = \dfrac{bh^2}{6}$	WEIGHT[a] PER LINEAR FOOT OF PIECE lb.
b	h	b	h				
2 × 4		$1\frac{5}{8}$ ×	$3\frac{5}{8}$	5.89	6.45	3.56	1.64
2 × 6		$1\frac{5}{8}$ ×	$5\frac{1}{2}$	8.93	22.53	8.19	2.54
2 × 8		$1\frac{5}{8}$ ×	$7\frac{1}{2}$	12.19	57.13	15.23	3.39
2 × 10		$1\frac{5}{8}$ ×	$9\frac{1}{2}$	15.44	116.10	24.44	4.29
2 × 12		$1\frac{5}{8}$ ×	$11\frac{1}{2}$	18.69	205.95	35.82	5.19
3 × 4		$2\frac{5}{8}$ ×	$3\frac{5}{8}$	9.52	10.42	5.75	2.64
3 × 6		$2\frac{5}{8}$ ×	$5\frac{1}{2}$	14.43	36.40	13.23	4.10
3 × 8		$2\frac{5}{8}$ ×	$7\frac{1}{2}$	19.69	92.29	24.61	5.47
3 × 10		$2\frac{5}{8}$ ×	$9\frac{1}{2}$	24.94	187.55	39.48	6.93
3 × 12		$2\frac{5}{8}$ ×	$11\frac{1}{2}$	30.19	332.69	57.86	8.39
4 × 6		$3\frac{5}{8}$ ×	$5\frac{1}{2}$	19.95	50.25	18.28	5.66
4 × 8		$3\frac{5}{8}$ ×	$7\frac{1}{2}$	27.19	127.44	33.98	7.55
4 × 10		$3\frac{5}{8}$ ×	$9\frac{1}{2}$	34.44	259.00	54.53	9.57
4 × 12		$3\frac{5}{8}$ ×	$11\frac{1}{2}$	41.69	459.43	79.90	11.6
4 × 14		$3\frac{5}{8}$ ×	$13\frac{1}{2}$	48.94	743.24	110.11	13.6
4 × 16		$3\frac{5}{8}$ ×	$15\frac{1}{2}$	56.19	1,124.92	145.15	15.6
6 × 6		$5\frac{1}{2}$ ×	$5\frac{1}{2}$	30.25	76.26	27.73	8.40
6 × 8		$5\frac{1}{2}$ ×	$7\frac{1}{2}$	41.25	193.36	51.56	11.4
6 × 10		$5\frac{1}{2}$ ×	$9\frac{1}{2}$	52.25	392.96	82.73	14.5
6 × 12		$5\frac{1}{2}$ ×	$11\frac{1}{2}$	63.25	697.07	121.23	17.5
6 × 14		$5\frac{1}{2}$ ×	$13\frac{1}{2}$	74.25	1,127.67	167.06	20.6
6 × 16		$5\frac{1}{2}$ ×	$15\frac{1}{2}$	85.25	1,706.78	220.23	23.6
6 × 18		$5\frac{1}{2}$ ×	$17\frac{1}{2}$	96.25	2,456.38	280.73	26.7

[a]Based on 40 pcf (not from *National Design Specification*).
Properties of sections for certain standard dimension and timber-dressed (S4S) sizes from *National Design Specification*.

NOMINAL SIZE b h	STANDARD DRESSED SIZE S4S b h	AREA OF SECTION $A = bh$	MOMENT OF INERTIA $I = \dfrac{bh^3}{12}$	SECTION MODULUS $S = \dfrac{bh^2}{6}$	WEIGHT[a] PER LINEAR FOOT OF PIECE lb.
8 × 8	$7\frac{1}{2} \times 7\frac{1}{2}$	56.25	263.67	70.31	15.6
8 × 10	$7\frac{1}{2} \times 9\frac{1}{2}$	71.25	535.86	112.81	19.8
8 × 12	$7\frac{1}{2} \times 11\frac{1}{2}$	86.25	950.55	165.31	23.9
8 × 14	$7\frac{1}{2} \times 13\frac{1}{2}$	101.25	1,537.73	227.81	28.0
8 × 16	$7\frac{1}{2} \times 15\frac{1}{2}$	116.25	2,327.42	300.31	32.0
8 × 18	$7\frac{1}{2} \times 17\frac{1}{2}$	131.25	3,349.61	382.81	36.4
8 × 20	$7\frac{1}{2} \times 19\frac{1}{2}$	146.25	4,625.00	475.00	40.6
10 × 10	$9\frac{1}{2} \times 9\frac{1}{2}$	90.25	678.76	142.90	25.0
10 × 12	$9\frac{1}{2} \times 11\frac{1}{2}$	109.25	1,204.03	209.40	30.3
10 × 14	$9\frac{1}{2} \times 13\frac{1}{2}$	128.25	1,947.80	288.56	35.6
10 × 16	$9\frac{1}{2} \times 15\frac{1}{2}$	147.25	2,948.07	380.40	40.9
10 × 18	$9\frac{1}{2} \times 17\frac{1}{2}$	166.25	4,242.84	484.90	46.1
10 × 20	$9\frac{1}{2} \times 19\frac{1}{2}$	185.25	5,870.11	602.06	51.4
12 × 12	$11\frac{1}{2} \times 11\frac{1}{2}$	132.25	1,457.51	253.48	36.7
12 × 14	$11\frac{1}{2} \times 13\frac{1}{2}$	155.25	2,357.86	349.31	43.1
12 × 16	$11\frac{1}{2} \times 15\frac{1}{2}$	178.25	3,568.71	460.48	49.5
12 × 18	$11\frac{1}{2} \times 17\frac{1}{2}$	201.25	5,136.07	586.98	55.9
12 × 20	$11\frac{1}{2} \times 19\frac{1}{2}$	224.25	7,105.92	728.81	62.3
12 × 22	$11\frac{1}{2} \times 21\frac{1}{2}$	247.25	9,530.00	887.50	68.7
12 × 24	$11\frac{1}{2} \times 23\frac{1}{2}$	270.25	12,435.00	1,057.50	75.0
Decking—(Based on strip one foot wide and of thickness indicated)					
1'-0 × 2	$12 \times 1\frac{5}{8}$	19.50	4.29	5.28	
1'-0 × 3	$12 \times 2\frac{5}{8}$	31.50	18.00	13.76	
1'-0 × 4	$12 \times 3\frac{1}{2}$	42.00	42.88	24.50	

[a] Based on 40 pcf (not from *National Design Specification*).
Properties of sections for certain standard dimension and timber-dressed (S4S) sizes from *National Design Specification*.

C-4b   EFFECTIVE SECTION PROPERTIES FOR PLYWOOD

1.   FACE PLIES OF DIFFERENT SPECIES GROUP THAN INNER PLIES (INCLUDES ALL STANDARD GRADES EXCEPT THOSE NOTED IN C-4b2)

① NOMINAL THICKNESS (in.)	② APPROXIMATE WEIGHT (psf)	③ EFFECTIVE THICKNESS FOR SHEAR (in.)	STRESS APPLIED PARALLEL TO FACE GRAIN				STRESS APPLIED PERPENDICULAR TO FACE GRAIN			
			④ A AREA (in.²/ft.)	⑤ I MOMENT OF INERTIA (in.⁴/ft.)	⑥ KS EFF. SECTION MODULUS (in.³/ft.)	⑦ Ib/Q ROLLING SHEAR CONSTANT (in.²/ft.)	⑧ A AREA (in.²/ft.)	⑨ I MOMENT OF INERTIA (in.⁴/ft.)	⑩ KS EFF. SECTION MODULUS (in.³/ft.)	⑪ Ib/Q ROLLING SHEAR CONSTANT (in.²/ft.)
Unsanded Panels										
$\frac{5}{16}$-U	1.0	0.283	1.914	0.025	0.124	2.568	0.660	0.001	0.023	—
$\frac{3}{8}$-U	1.1	0.293	1.866	0.041	0.162	3.108	0.799	0.002	0.033	—
$\frac{1}{2}$-U	1.5	0.316	2.500	0.086	0.247	4.189	1.076	0.005	0.057	2.858
$\frac{5}{8}$-U	1.8	0.336	2.951	0.154	0.379	5.270	1.354	0.011	0.095	3.252
$\frac{3}{4}$-U	2.2	0.467	3.403	0.243	0.501	6.823	1.632	0.036	0.236	3.717
$\frac{7}{8}$-U	2.6	0.757	4.109	0.344	0.681	7.174	2.925	0.162	0.542	5.097
1-U	3.0	0.859	3.916	0.493	0.859	9.244	3.611	0.210	0.660	6.997
$1\frac{1}{8}$-U	3.3	0.877	4.621	0.676	1.047	10.008	3.464	0.307	0.821	8.483
Sanded Panels										
$\frac{1}{4}$-S	0.8	0.278	1.307	0.009	0.067	2.182	0.681	0.001	0.018	—
$\frac{3}{8}$-S	1.1	0.294	1.307	0.027	0.125	3.389	1.181	0.004	0.053	—
$\frac{1}{2}$-S	1.5	0.450	1.947	0.077	0.266	4.834	1.281	0.018	0.150	3.099
$\frac{5}{8}$-S	1.8	0.472	2.280	0.129	0.356	6.293	1.627	0.045	0.234	3.922
$\frac{3}{4}$-S	2.2	0.589	2.884	0.197	0.452	7.881	2.104	0.093	0.387	4.842
$\frac{7}{8}$-S	2.6	0.608	2.942	0.278	0.547	8.225	3.199	0.157	0.542	5.698
1-S	3.0	0.846	3.776	0.423	0.730	8.882	3.537	0.253	0.744	7.644
$1\frac{1}{8}$-S	3.3	0.865	3.854	0.548	0.840	9.883	3.673	0.360	0.918	9.032
Touch-Sanded Panels										
$\frac{1}{2}$-T	1.5	0.346	2.698	0.083	0.271	4.252	1.159	0.006	0.061	2.746
$\frac{19}{32}$-T	1.7	0.491	2.618	0.123	0.337	5.403	1.610	0.019	0.150	3.220
$\frac{5}{8}$-T	1.8	0.497	2.728	0.141	0.364	5.719	1.715	0.023	0.170	3.419
$\frac{23}{32}$-T	2.1	0.503	3.181	0.196	0.447	6.600	2.014	0.035	0.226	3.659
$\frac{3}{4}$-T	2.2	0.509	3.297	0.220	0.477	6.917	2.125	0.041	0.251	3.847
2·4·1 $1\frac{1}{8}$-T	3.3	0.855	4.592	0.653	0.995	9.933	4.120	0.283	0.763	7.452

2. STRUCTURAL I, II, AND MARINE PLYWOOD

① NOMINAL THICKNESS (in.)	② APPROXIMATE WEIGHT (psf)	③ EFFECTIVE THICKNESS FOR SHEAR (in.)	STRESS APPLIED PARALLEL TO FACE GRAIN				STRESS APPLIED PERPENDICULAR TO FACE GRAIN			
			④ A AREA (in.²/ft.)	⑤ I MOMENT OF INERTIA (in.⁴/ft.)	⑥ KS EFF. SECTION MODULUS (in.³/ft.)	⑦ lb/Q ROLLING SHEAR CONSTANT (in.²/ft.)	⑧ A AREA (in.²/ft.)	⑨ I MOMENT OF INERTIA (in.⁴/ft.)	⑩ KS EFF. SECTION MODULUS (in.³/ft.)	⑪ lb/Q ROLLING SHEAR CONSTANT (in.²/ft.)
Unsanded Panels										
$\frac{5}{16}$-U	1.0	0.356	2.375	0.025	0.144	2.567	1.188	0.002	0.029	—
$\frac{3}{8}$-U	1.1	0.371	2.226	0.041	0.195	3.107	1.438	0.003	0.043	—
$\frac{1}{2}$-U	1.5	0.543	2.906	0.091	0.318	4.497	2.325	0.017	0.145	2.574
$\frac{5}{8}$-U	1.8	0.609	3.464	0.157	0.437	5.993	2.925	0.052	0.267	3.238
$\frac{3}{4}$-U	2.2	0.747	4.406	0.247	0.573	7.046	2.938	0.085	0.369	3.697
$\frac{7}{8}$-U	2.6	0.776	4.388	0.346	0.690	6.948	3.510	0.192	0.584	5.086
1-U	3.0	1.008	5.200	0.529	0.922	8.512	6.500	0.366	0.970	6.986
$1\frac{1}{8}$-U	3.3	1.119	6.654	0.751	1.164	9.061	5.542	0.503	1.131	8.675
Sanded Panels										
$\frac{1}{4}$-S	0.8	0.342	1.680	0.013	0.092	2.172	1.226	0.001	0.027	—
$\frac{3}{8}$-S	1.1	0.373	1.680	0.038	0.177	3.382	2.126	0.007	0.078	—
$\frac{1}{2}$-S	1.5	0.545	1.947	0.078	0.271	4.816	2.305	0.030	0.217	3.076
$\frac{5}{8}$-S	1.8	0.717	3.112	0.131	0.361	6.526	2.929	0.077	0.343	3.887
$\frac{3}{4}$-S	2.2	0.748	3.848	0.202	0.464	7.926	3.787	0.162	0.570	4.812
$\frac{7}{8}$-S	2.6	0.778	3.952	0.298	0.569	7.539	5.759	0.275	0.798	5.671
1-S	3.0	1.091	5.215	0.479	0.827	7.978	6.367	0.445	1.098	7.639
$1\frac{1}{8}$-S	3.3	1.121	5.593	0.623	0.955	8.840	6.611	0.634	1.356	9.031
Touch-Sanded Panels										
$\frac{1}{2}$-T	1.5	0.543	2.698	0.084	0.282	4.580	2.486	0.020	0.162	2.720
$\frac{19}{32}$-T	1.7	0.707	3.127	0.124	0.349	6.094	2.899	0.050	0.259	3.183
$\frac{5}{8}$-T	1.8	0.715	3.267	0.144	0.378	6.552	3.086	0.060	0.293	3.383
$\frac{23}{32}$-T	2.1	0.739	4.059	0.201	0.469	6.971	3.625	0.078	0.350	3.596
$\frac{3}{4}$-T	2.2	0.746	4.209	0.226	0.503	7.379	3.825	0.092	0.388	3.786

Appendix D

SHEAR, MOMENT, AND DEFLECTION IN BEAMS

D-1 SHEAR AND BENDING MOMENT IN BEAMS

1. Simple Beams:

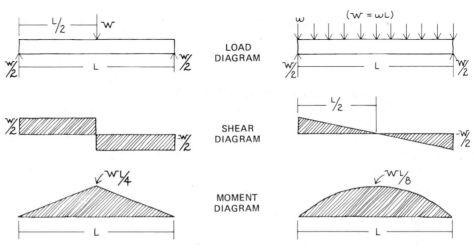

2. Fixed End Beams:

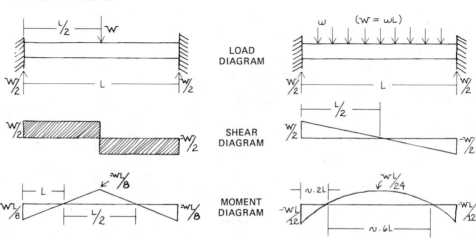

3. Cantilevers:

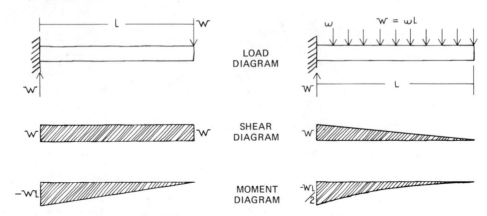

| LOAD DIAGRAM |
| SHEAR DIAGRAM |
| MOMENT DIAGRAM |

D-2   MOMENTS AND DEFLECTIONS IN BEAMS

		BEHAVIOR AND M-DIAGRAM	M_{max}	$\Delta = (k_3)\dfrac{WL^2}{EI}$	RELATIVE Δ: (equal M & L)
SIMPLE BEAMS	**CONCENTRATED LOADS**		$M_{max} = \dfrac{WL}{4}$	$\Delta = \left(\dfrac{1}{48}\right)\dfrac{WL^3}{EI}$	1
	DISTRIBUTED LOADS		$M_{max} = \dfrac{WL}{8}$	$\Delta = \left(\dfrac{5}{348}\right)\dfrac{WL^3}{EI}$	$\dfrac{5}{8}$
	END MOMENTS		M is given	$\Delta = \left(\dfrac{1}{8}\right)\dfrac{ML^2}{EI}$	6

		BEHAVIOR AND M-DIAGRAM	M_{max}	$\Delta = (k_3)\dfrac{WL^2}{EI}$	RELATIVE Δ: (equal M & L)
FIXED END BEAMS	**CONCENTRATED LOAD**		$\pm M_{max} = \dfrac{WL}{8}$	$\Delta = \left(\dfrac{1}{192}\right)\dfrac{WL^3}{EI}$	$\dfrac{1}{4}$
	DISTRIBUTED LOAD		$+M_{max} = \dfrac{WL}{24}$ $-M_{max} = \dfrac{WL}{12}$	$\Delta = \left(\dfrac{1}{384}\right)\dfrac{WL^3}{EI}$	$\dfrac{1}{8}$
	END MOMENTS		$\pm M_{max} = \dfrac{WL}{2}$	$\Delta = \left(\dfrac{1}{12}\right)\dfrac{WL^3}{EI}$	4
CANTILEVERED BEAMS	**CONCENTRATED LOAD**		$-M_{max} = WL$	$\Delta = \left(\dfrac{1}{3}\right)\dfrac{WL^3}{EI}$	16
	DISTRIBUTED LOAD		$-M_{max} = \dfrac{WL}{2}$	$\Delta = \left(\dfrac{1}{8}\right)\dfrac{WL^3}{EI}$	6

D-3   FAR END FIXITY AND CARRYOVER FACTORS FOR MOMENT DISTRIBUTION IN RIGID FRAMES

Note: Relative stiffness of members at a joint is determined by comparing $1/L$ and as modified by far end fixity factor (FFF); $K = (1/L)_{Relative} \times$ **FFF.**

BASIC CASES:	CARRY OVER MOMENT FACTOR	RELATIVE MOMENT TO PRODUCE EQUAL ROTATION ABOUT (A)	FFF

1.

$\frac{1}{2}M_A$ A M_1

FULLY FIXED   1

2.

A.

$\frac{1}{3}M_A$ A $\frac{7}{8}M_1$ $\frac{7}{8}$

OTHER MEMBERS

ROTATION
MOMENT AT EACH END

B. $\frac{2}{3}M_1$ A   A $\frac{2}{3}M_1$ $\frac{2}{3}$

OTHER MEMBERS

3.

A.

NO CARRYOVER
MOMENT   A $\frac{3}{4}M_1$ $\frac{3}{4}$

FREE TO
ROTATE

ROTATION
MOMENT AT EACH END

B.   A   A $\frac{1}{2}M_1$ $\frac{1}{2}$

FREE TO ROTATE

Appendix E

"DRIFT IN HIGH-RISE BUILDINGS," BY JOHN B. SCALZI

Reprinted from the April 1972 issue of *Progressive Architecture,* copyright 1972, Reinhold Publishing

Drift in high-rise steel framing

John B. Scalzi

Wind connections

One of the early steel framing systems used to support gravity and wind loads is the technique of designing the girders for the full gravity loads and providing sufficient beam-to-column connections to resist the bending moments caused by the portal action of the frame when subjected to lateral loads. The end fixity thus provided to the girder is ignored in the selection of the girder size. Buildings of this era of designing with moment connections usually had reinforced concrete for protection of beams and columns, poured concrete floors, full story height heavy partitions and generally heavy curtain walls. These extra supporting components contributed to a reduction of the drift of the frame.

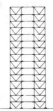

K-bracing, shear connections

Economies of direct loads in all members of the diagonally braced framework also lead to greater drift control by shortening the diagonals into a K-type bracing system. The shorter length of the K-type diagonals produces less elongation, and therefore less drift for a specific height. As a result, a taller building can be achieved on a given plot of ground.

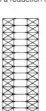

Diagonally braced bents, shear connections

Because the height of a building was limited by drift of the portal frame action, the next step was to develop the diagonally braced bents to resist the lateral loads on tall buildings. The truss action of the bent eliminated the bending in the columns and kept the drift value for greater heights of buildings within desirable limits. All connections of girders and diagonals are simple connections thus making fabrication and erection more economical than comparable moment connections for the same frame. The number of frames to be braced depends upon height and wind loads.

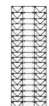

K-bracing, moment connections

In order to achieve a stiffer structure, the beam-to-column connections may be made moment resistant, thus extending the aspect ratio of the building to a greater value. The plastic design concept uses the combined braced and portal frame configuration to achieve economies in material for certain buildings which do not have drift limitations. The dual structural system of plastic design effectively supports the gravity loads by portal frame action and resists the lateral loads by the rigid K-braced bays.

In this article the author presents a review of structural steel framing systems in which control of drift had a dominant effect on the resulting building skeletons

A prime motivating force behind the surge of taller high-rise buildings has been the need to satisfy demands for more office and apartment spaces. As buildings are built to greater heights, the aspect ratio (height/width) becomes larger making drift a major consideration. Economics of materials and labor have disrobed the building of its supporting components thus placing the burden of resisting all lateral movement on the structural frame.

To meet the challenge of satisfying architectural functions as well as drift requirements, the consulting engineer has had to rely on his knowledge of advanced structural theories to develop new and varied framing systems. With the aid of the computer, he has been able to produce new framing methods that are uniquely efficient and economical.

A general review of the various steel framing systems that have evolved to meet the challenge of greater building heights is presented for preliminary design considerations.

Author: John B. Scalzi is in the Research and Technology Department, United States Department of Housing and Urban Development.

Staggered truss

A unique variation of the braced bent is the staggered truss system developed by the departments of architecture and civil engineering at MIT. The basic principle is the action of a trussed frame under lateral loads which practically eliminates the bending moments in the columns. Although the trusses do not lie in one plane, two adjacent trusses are considered to behave as one by the shear diaphragm action of the floor system. The floor must be capable of transferring the shears from the bottom chord of one truss to the top chord of the adjacent truss. As a result, the two adjacent frames may be considered to act together to calculate the drift of the building due to wind loads.

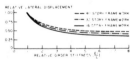

Lateral loads

Computer studies of a 40-story, 3-bay rectangular frame subjected to lateral loads indicated that increasing the column stiffnesses does not reduce the drift of the frame an appreciable amount. On the other hand, increasing the girder stiffnesses a small amount indicated an appreciable reduction in the lateral displacement of the frame. The study indicated that the reduction became asymptotic at a relative girder stiffness of approximately four. The optimum value of girder stiffness for economy appears to be in the vicinity of 1½ to 2 where the slope of the curve begins to flatten out to the asymptotic value. The same general behavior was observed for a 16- and 32-story rectangular frame. The results indicated that it is more efficient to increase the girder stiffnesses than the column stiffnesses and that beyond a relative girder stiffness of approximately 2, a change in the geometry of the frame is advisable.

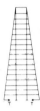

Tapered frame

Most recent conceptual change in the geometry of a frame to reduce drift is the idea of sloping the exterior columns a slight amount. This technique enables a designer to build a higher building than a comparable rectangular frame for the same volume. The slope of the columns need not be too great to appreciably reduce the relative lateral displacement of the frame.

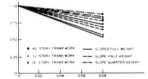

Height of slope

A computer study of three building heights of 16, 32 and 40 stories with 3 bays varying the sloped height from a quarter, to half and full height indicated that the greatest benefit resulted from the full height sloped columns. The 40-story framework indicated the largest reduction in the relative lateral displacement because of its greater height. A slope of 8 percent in the exterior columns produced a 50 percent reduction in the relative lateral displacement for the 40-story building.

Drift in high-rise steel framing

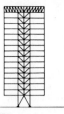

Core and suspended frame

For those buildings requiring an open court at street level, the concept of suspending the floors from a roof truss or space frame has evolved. The core may take one of the many stable forms of steel frameworks; such as a series of braced frames, if the building is rectangular in plan; or of a circular, or square core for the corresponding contour of the building. The core must resist all of the gravity and wind loads as if it were standing alone. The suspenders may be plates, rods or cables depending upon the number of floors to be supported and the type of connection to be made.

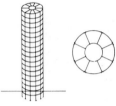

Circular framing

Another geometrical form of structural framing which is becoming more visible on the city skylines is the circular pattern of arranging the columns on the exterior face of the building. The spandrel beams are attached to the columns by moment resistant connections thus producing a three-dimensional or space structure. The rigidity provided by this configuration is likened to a tube with many openings, and as such it provides maximum stiffness to the structure. Another factor, not to be overlooked, is the reduction in the magnitude of the wind pressures permitted by building codes to ⁵⁄₁₀ or ⁵⁄₁₀ of the normal value for a circular building.

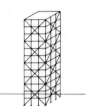

Trussed framing

When the aspect ratio of height to minimum width increases beyond the range of an economical consideration of a tubular type framework, the structure may be increased in stiffness by the addition of a truss system superimposed on the tubular frame. This combination of structural systems reduces the drift of the building to permissible values while permitting greater heights for the structural frame. As in the tubular concept, the exterior frame supports most of the gravity loads and all of the lateral loads. The spacing of the columns is essentially on the same basis as for the tubular concept. The size of the truss panels will vary with the proportions of the building.

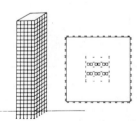

Square column pattern

A natural variation of the tubular concept for greater stiffness is the comparably stiff square pattern of arranging the columns. The exterior columns are usually spaced closely together to support most or all of the gravity loads and all of the lateral loads. The interior columns, when provided, assist in supporting their share of the gravity load but because of their location near the center of the tube they are not considered to participate in the resistance of the lateral loads. As in the circular pattern, the spandrels are rigidly connected to the columns in order to develop the three-dimensional behavior of the frame.

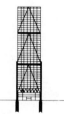

Trussed framing variation

The geometrical pattern of the truss system in a combined tubular and truss framework may vary according to the size of the building, the magnitude of the wind pressures and the architectural features of the exterior wall as influenced by the interior uses of the column-free floor space. With the aid of computers, many different configurations may be evaluated quickly for drift, vibration and total cost before making a final judgment. The structural efficiency of larger truss panels with diagonals or larger panels with K-type bracing may be easily studied by the currently available computer programs.

A42

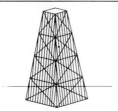

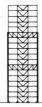

Inclined truss framing

For the exceptionally tall building with a very high aspect ratio, a combination of structural systems combined with sloping columns produces a structure which satisfies the architectural and engineering requirements. Drift and vibration may be kept within desirable limits by adjusting the column spacing, the truss configuration, and the degree of slope of the columns. Computers can play an important role in making the final selection of the structural system.

Cantilevered truss with tie down

A variation of the cap truss and tie-down concept is the idea of placing the stiffening trusses at various crucial locations in the height of the building. These stiffening trusses at intermediate height locations will perform the same function as the cap truss. Placed at such locations as the one-third, half or three-quarter height of the building, the stiffening truss floor also serves as the mechanical equipment floor. The amount of drift reduction is affected by the location, number and relative stiffnesses of the various components of the structure.

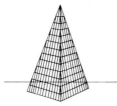

Pyramid framing

A variation of the truncated pyramid is the full pyramid. In this instance the sloped exterior walls are carried to a peak intersection. Although this configuration has not been used to any great extent, it does illustrate the advantage of combining structural systems and geometrical proportions to achieve a stiff structural frame, thus reducing drift and vibration due to lateral loads.

Shear resistant walls

A method to control drift of a building, yet not fully developed, is the concept of a shear resistant exterior skin covering. Although exterior structural frames may be designed as Vierendeel trusses, the curtain wall is neglected in the calculation of the deflection of the building. The assumption that the curtain wall does not contribute to the stiffness of the frame is based on the fact that most curtain walls are relatively light compared with the structural frame. Steel plate curtain walls have been used in several buildings and can be shown to assist in reducing the deflection of the frame. As studies progress, the curtain wall may soon be shown to resist gravity loads as well as lateral forces.

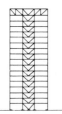

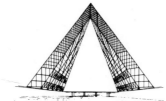

K-bracing tie downs

A relatively new concept of drift control has evolved which uses the technique of a cap truss on a core combined with exterior tie-down columns. The tie-downs are attached at every story and support gravity loads in addition to restraining the lateral movement of the frame. The action of the tie-down in conjunction with the cap truss restrains the bending of the core by introducing a point of inflection in the deflection curve when subjected to lateral forces. This reversal in curvature reduces the lateral movement at the top. The amount of reduction depends upon the relative stiffness of the core, cap truss and size of tie-downs.

Omnibuilding

To serve the future population growth and the decreasing land surrounding our urban areas, Architect Stanley Tigerman has suggested an instant city (omnibuilding) over rivers or highways exploiting the air rights in order to provide man with more space for habitation. A structure of these proportions, 600 ft along the base of the pyramid and extending vertically to a point, can easily house an entire community. The stiffness of the structure is inherent in the size and geometry of the framework, thus resulting in little lateral movement or vibration.

Appendix F

PAPERS ON TWO UNUSUAL STRUCTURES

F-a **"GIANT PRESTRESSED SHELL FOR PONCE COLISEUM,"**
BY T. Y. LIN, FELIX KULLEN, KAM LO

PCI Journal, Sept.-Oct. 1973

F-b **"DESIGN OF RUCK-A-CHUCKY BRIDGE"**

Preprint #3305 ASCE CONVENTION, Chicago, Ill., 1978

GIANT PRESTRESSED HP SHELL FOR PONCE COLISEUM

T. Y. Lin*
Professor of Civil Engineering
University of California
Berkeley, California

Felix Kulka
President
T. Y. Lin International
Consulting Engineers
San Francisco, California

Kam Lo
Vice President
T. Y. Lin International
Consulting Engineers
San Francisco, California

The recently completed hyperbolic paraboloid concrete shell roof for the Ponce Coliseum in Puerto Rico represents a significant thrust forward in the design of shell structures.

The shell roof cantilevers 138 ft (measured along the edge beams) making it the world's longest cantilever shell of the HP type.

Post-tensioning was used both in the shell membrane and also along the edge beams to control deflections and stresses. In addition, the piers on opposite sides of the shell were post-tensioned beneath the ground to resist the horizontal thrust.

The finite element method was used to analyze the structural behavior and to calculate the moments, stresses, and deflections in the shell and edge beams.

This article describes the design and construction techniques used in building this shell roof.

*Also, Board Chairman, T. Y. Lin International,
Consulting Engineers, San Francisco, California.

Fig. 1. Ponce Coliseum nearing completion.

One of the world's largest hyperbolic paraboloid shell roofs (see Fig. 1) was completed in late 1971 for the Ponce Coliseum in Puerto Rico. The roof structure covers an area of over 60,000 sq ft and provides shelter for 10,000 seated spectators. Working with a limited budget of 2.4 million dollars, it was decided to use only natural ventilation. This decision was logical since one could take advantage of the local prevailing winds.

A hyperbolic paraboloid roof was chosen because it conforms to the seating pattern and allows unobstructed air flow across the coliseum. The roof is supported on only four supporting piers, thus minimizing air flow obstruction and allowing an open view. By using post-tensioning in both the shell and the edge beams, stresses and deflections could be controlled to obtain an optimum design.

DESCRIPTION OF SHELL

The shell structure (Figs. 2-5) has overall plan dimensions of 276 x 232 ft. The complete roof is made up of four similar 4-in. thick saddle-type shells, connected to interior and edge beams to form a structure supported by four piers at the low points. The piers are located at the centers of the four exterior edges. Thus, the entire structure is symmetrical about both axes, which run through opposite piers.

The high points of the shell, rising 40 ft above the low points, are at the four corner tips and also at the center of the entire roof. The clear spans, between opposite piers in the two directions, are 271 and 227 ft.

The normal 4-in. thick shell is gradually thickened over a 5-ft wide strip adjacent to each beam to a 6-in. thickness where it joins the beams (see Figs. 4 and 5). The shell has a cantilever extension beyond the exterior face of the edge beam which varies from 10 ft 6 in. at the piers to zero at the corner tips of the structure (see Fig. 5).

The cantilever edge beams are supported only at the piers and have a constant width of 30 in. throughout their entire length. The depth of the 138-ft cantilever edge beams (see Fig. 6) varies linearly from 18 in. at the

3

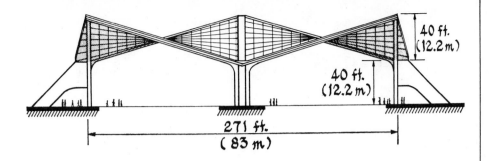

Fig. 2. Front elevation of shell.

corner tips to 53 in. at a distance of 17 ft from the abutment. From there it increases more rapidly to a maximum depth of 94 in. at the abutment. Similar depths for the 116-ft cantilever edge beams are 18, 44, and 87 in.

The interior beams are 60 in. wide. The depth of the 138-ft long interior beams (see Fig. 7) varies linearly from 18 in. at the center of the shell to 47 in. at a distance 10 ft from the abutment. From there it increases more rapidly to a maximum depth of 72 in. at the abutment. Similar depths for the 116 ft long interior beams are 18, 40, and 60 in.

Fig. 8 is a cross section of the shell slab showing the location of the tendons at interior and exterior edge beams.

The thrusts from the beams are delivered to large pier type abutments (see Figs. 6 and 7), resting on pile foundations. The resultant of the thrusts lies closely along the 45-deg inclined pier. Prestressed tie beams at the foundation level are used to connect opposite piers and carry the unbalanced horizontal thrust coming from the interior beams.

Smaller tie beams running diagonally between abutments are also used to insure that no relative motion between the abutments will occur in case of ground motion due to earthquakes. This assembly of four similar shells into one structural unit provides inherent stability.

COMPUTER ANALYSIS

The structure was designed to insure that it has adequate strength, stiffness, and stability under all possible load

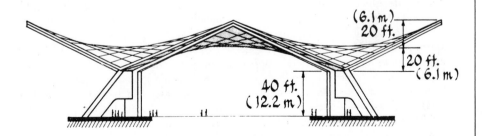

Fig. 3. Diagonal elevation of shell.

4

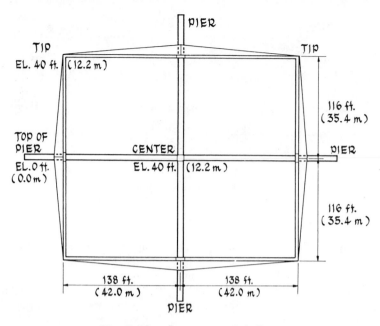

Fig. 4. Plan dimensions of shell.

conditions. The construction sequence was also carefully considered in the design process.

It was apparent from the beginning that the control of the deflections and the stresses in the cantilever edge beams was critical to the success of the design. A concept of load balancing by properly prestressing the edge beams suggested itself as only a preliminary approach. It was recognized that an accurate analysis was needed to determine the interaction of the shell and edge beam under the dead load of the structure and prestress in the edge beams.

During the final design period in 1968, a detailed computer analysis using the finite element method was performed under various loading and boundary conditions. Because of structural symmetry only one quarter of the total structure had to be considered. To simulate the shell, a quadrilateral mesh, with the lines running parallel to the beams, was selected over the structure.

Two triangular plane stress finite elements are placed in each quadrilateral and beam members form a two-way grid along the two approximate normal lines of the nodal points.

The plane stress triangular finite elements are used to represent the membrane stiffnesses of the shell, while the two-way grid of beam members, which are assigned only a flexural stiffness equivalent to the shell thickness, are used to represent the bending stiffness of the shell. The interior and edge beams are represented by beam type members having axial, bending, and torsional stiffnesses.

An IBM 360/65 computer was used. Computer output included the nodal point displacements; principal membrane stresses and bending moments in the shell; and axial forces, torques, and bending moments in the beams. Fig. 9 shows an example of how the principal stress contours were plotted from the computer output.

5

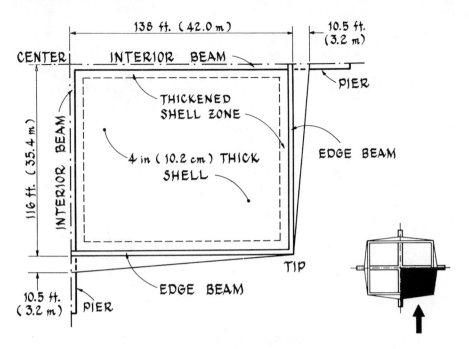

Fig. 5. Plan of one-quarter of shell.

STRUCTURAL DESIGN

Stresses and deflections were checked under the following conditions:

1. Dead load plus prestress.
2. Dead load plus prestress plus live load of 30 psf.
3. Dead load plus prestress plus wind.

For Case 3 a 33⅓ percent increase in allowable stresses was permitted.

Several possible wind loading conditions were considered based on available data. It was decided that the following two cases would be included for study:

1. A wind pressure of 56 psf acting either upward or downward on the triangular cantilever portion of the shell.
2. A wind pressure of 56 psf acting upward on the cantilever portion plus the same pressure acting downward on the internal triangular portion of the shell.

Based on the conventional membrane theory for hyperbolic paraboloid shells, only axial forces would exist in the edge beams under dead load of the shell alone. The computer results showed that part of the gravity load of the shell actually produces bending moments in the edge beam even though a great portion of the load will be carried to the abutment by axial forces in the edge beam. Due to the weight of the beams themselves, a large cantilever moment would be expected in the edge beam if it acted alone.

However, since the edge beams are integrated with the shell, the cantilever moment in the edge beam is greatly reduced. About 80 percent of the maximum beam cantilever moment is carried by the interaction of the shell with the edge beam. Thus, a high tensile zone is developed a short distance away and along the edge beams. High shell bending moments also occur adjacent to the edge beam near the tip.

Under dead load alone, the maximum

6

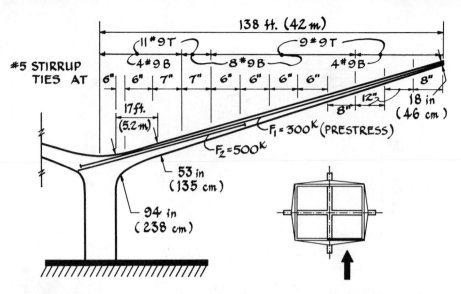

Fig. 6. 138-ft cantilever edge beam (beam has a constant width equal to 30 in.)

compression in the shell was 20 kips per ft (or 400 psi) while the minimum stress was 15 kips per ft (or 300 psi) (see Fig. 9).

Under dead load alone, the maximum tensile membrane stress in the shell is about 300 psi. Including the effect of wind or live load, the tensile stress is higher, amounting to 400 psi. These tensile membrane forces in the shell can be resisted by providing normal steel reinforcement. However, to minimize cracks in the shell, post-tensioning shell tendons were used in combination with normal steel reinforcement.

Straight tendons running parallel to the generators of the shell and parabolic tendons running from the tip to the

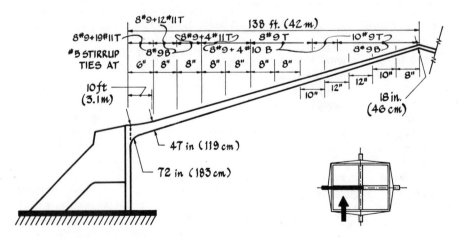

Fig. 7. 138-ft interior gable beam (beam has a constant width equal to 30 in.).

7

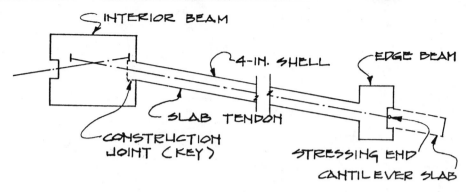

Fig. 8. Cross section of shell slab showing location of tendons at interior and exterior edge beams.

center of the shell were both investigated. It was found that straight tendons in two directions gave more satisfactory results in reducing the tension in the concrete shell.

Finally, the 4-in. roof slab was prestressed from 10,000 to 20,000 lb per linear ft. Reinforcing bars of No. 3 at 12 in. on center are placed above and below the tendons, thus increasing the ultimate strength of the shell and limiting local cracking. Additional reinforcing steel is added in the zones of high shell bending adjacent to the beams.

The dead load of all the beams, which is about five-eighths of the total weight of the entire 4-in. shell, gives a more severe condition for beam design compared to the dead load of the shell. This is because it produces greater deflections and bending moments but smaller axial forces.

The possibility of balancing the dead load of the beams by properly prestressing the edge beams was carefully studied. However, total balancing of this dead load was impossible because part of the prestressing force in the edge beam is dissipated into the shell.

After several trials to determine the optimum prestressing, two tendons were added to each edge beam. One tendon, F1, ran from the tip of the beam to the abutment pier and the other, F2, from approximately four-tenths of the span to the abutment pier (see Fig. 6). F1 was stressed from the tip while F2 was stressed from the abutment. With the help of prestressing in the edge beams, the tensile zone in the shell decreases both in size and in magnitude but the bending moments in the shell have no significant changes.

The addition of prestress in the edge beams reduces the moments as well as the deflections and increases the axial forces and results in a set of stresses in the edge beam. These stresses were found to be acceptable for design purposes. Actually, there is not a large saving in compression, but tension is considerably reduced. Normal reinforcing steel is used to resist the excessive tension. Without post-tensioning in the edge beams, the size would have to be increased.

Buckling of the edge beams was checked using a conservative approach which neglected entirely the participation of the shell and the beam. The assumed beam proportions were found to be adequate regarding buckling.

The interior beams form two gable frames perpendicular to each other and

8

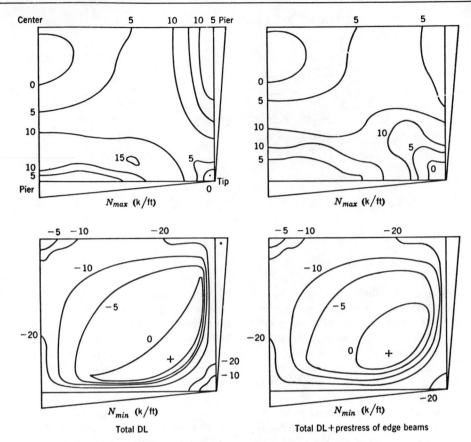

Fig. 9. Maximum and minimum principal stresses in shell (fine mesh).

their arch action provides strength. Structurally, they act differently from the cantilever edge beams, the deflections and bending moments being relatively small. It was determined that only normal reinforcing steel was needed as shown in Fig. 7.

The edge and interior beams meet at the low points of the shell structure where the four supporting piers are located. Each interior and two edge beams gather loads from the shells and transmit them toward one of the four piers.

Due to structural symmetry, the horizontal thrusts from the two edge beams cancel out each other. The horizontal thrust from the interior beam and the total vertical load form a resultant force having an angle of approximately 45-deg toward the ground.

The inclined pier, as shown in Fig. 7, was chosen so that it would resist the force resultant mainly by direct compressive stresses. Thus, the deformations at the pier top were very small.

The piers rest on vertical piles which have little resistance to horizontal forces. To resist the horizontal thrust coming from the shell above, tie beams connecting opposite piers beneath the ground were post-tensioned with a 2100-kip force.

In addition to an elastic analysis us-

9

Fig. 10. *Because of symmetry the shell roof was constructed quadrant by quadrant. Picture above shows the tendons and steel reinforcement being laid out after the plywood formwork is in place.*

Fig. 11. *Close-up of tendon and mild reinforcing steel layout.*

10

Fig. 12. Temporary tendons under shell soffit stressed from opposite piers.

ing the finite element method, the approximate ultimate moment capacity of the entire structure was examined by taking a diagonal section across the shell just outside the two adjacent piers. It was found that with the tension force along the crown of the shell acting against compressive forces at the edge beams, this ultimate moment exceeds the normal load factor requirements.

CONSTRUCTION

Construction of this structure proceeded according to carefully planned steps. First, the pile-supported foundations were built including all concrete work of the foundation tie beams. Then the piers were constructed. Steel tubing falsework to support one quadrant

of the roof and its four surrounding beams was first erected (see Fig. 10). Next, 4 x 8-ft plywood was installed to form the soffit of the shell. Then the tendons and mild reinforcing steel was laid out (see Fig. 11).

To compensate partially for the anticipated deflection at the cantilever tip due to creep, the tip was cambered 3½ in. to balance the calculated deflection for dead load plus prestress. After the concrete reached the required strength, shell tendons were stressed and the cantilever slab was cast.

Because the shell is symmetrical, it was originally intended that the formwork of one slab quadrant be moved three times to complete the four quadrants. For that purpose, a temporary diagonal and horizontal post-tensioning tie were installed between two adjacent

11

Fig. 13. Temporary tendon stressing ends (at top of piers).

Fig. 14. Shell roof at late stage of construction.

piers to carry the horizontal thrust of each quadrant (see Figs. 12 and 13).

However, to simplify the operations, the contractor decided to erect formwork for all four quadrants even though this operation increased forming costs. Fig. 14 shows the complicated steel tubular falsework supporting all four quadrants of the shell.

The foundation tie beams were post-tensioned after all the four quadrants of the roof were completed, one by one. A closeup of the foundation tie beam at the stressing end is shown in Fig. 15. Then the supports for the forms under the 4-in. shells were gradually released.

At each stage, at least two tendons were stressed at the perimeter of the shell at one time, such that the forces were symmetrical about each pier at all times. Finally, supports of the edge and interior beams were gradually removed. For the edge beams stressing started symmetrically from all four tips, and for the interior beams from the center towards the piers (see Fig. 16).

On completion of this step, the average deflection of the four tips was measured to be 3 in., which compared favorably with the calculated values. Deflections of the shell were checked periodically for a few months after completion and they were found to be stable and satisfactory.

Fig. 15. Foundation tie beam at stressing end.

CONCLUSION

The satisfactory construction and proper behavior of this shell roof indicates the desirability of using prestressing. By post-tensioning both the shell and the edge beams, stresses and deflections can be controlled and kept within limits. The usual assumption that edge beams carry their own weight is far from being correct for such large shells. Similarly, the stresses resulting from prestressing in the edge beams are also appreciably absorbed by the shell slabs.

Fig. 16. Stressing pocket at cantilever beams.

13

Since stresses in the shells are low for this structure, longer spans can be realized with relative ease. However, the design of the long cantilever edge beams becomes more difficult to handle and may require a combination of prestressing and reinforcing steel to attain an optimum solution.

Shrinkage stresses in the shells will often result in local cracking, particularly near the corners of hyperbolic paraboloid shells. Such cracking has actually occurred in this structure. However, by using tendons and mild steel reinforcement in these areas, the stability, strength, and deflections of the shells are not adversely affected.

It is our opinion that hyperbolic paraboloid shells with post-tensioning can be extended to longer spans. With today's high speed computers, it is possible to make a reliable analysis of complex shells and to predict their behavior accurately. However, at the same time, it is highly desirable to use sound judgment especially when we extend our structures into complex forms, new materials, and longer spans.

CREDITS

Design Consultants: T. Y. Lin International, San Francisco, California.

Computer Consultant: Prof. A. C. Scordelis, University of California, Berkeley, California.

Design Engineering Firm: Sanchez, Davila & Suarez (formerly Raymond Watson Structural Engineers), San Juan, Puerto Rico.

Architect: V. Monsanto and Associates, Ponce, Puerto Rico.

Contractor: Gabriel Alvarez and Associates, Ponce, Puerto Rico.

14

THE DESIGN OF THE RUCK-A-CHUCKY BRIDGE
T.Y. LIN[1], F. ASCE
D. ALLAN FIRMAGE[2], F. ASCE

INTRODUCTION

The Ruck-A-Chucky Bridge will carry a local intercounty road across the middle fork of the American River in California at a point some 10 miles air distance above the Auburn Dam. The American River, about 30 ft. deep and 100 ft. wide before damming, will have a depth of 450 ft. and a width of 1,100 ft. as a result of the construction of the dam.

In order to provide a vertical clearance of 50 ft. above high reservoir water level, the bridge proper would have a length of 1,300 ft. between the hillsides, with a slope of 40° to the hoizontal. A straight bridge, with limited curvature at the ends, would require heavy approach cuts and tunnels. A curved bridge would be a much better solution. Conventionally, a curved bridge of this length has to be supported on several intermediate piers. In this case, however, such a solution would be extremely costly because of the water depth of 450 ft., which would aggravate seismic forces on the piers.

The preliminary study concluded that conventional bridges were completely unsatisfactory. The search continued until finally the hanging arc evolved (Fig. 1). The hanging arc, so-called because it is curved on plan and suspended by cables, is made up of two components: the cable stays acting in tension and the curved girder carrying the traffic and absorbing the axial compression produced by the cables. The cables are post-tensioned to control the stresses and strains. They are anchored on the slope so as to control the line of pressure in the girder. Thus, an ideal stress condition is achieved with small bending and torsional moments.

A 1,500 ft. radius was chosen for the curvature of the hanging arc layout after trade-off studies between the location of the anchoring pedestals

---

[1]Chairman of Board, T.Y. Lin International, San Francisco, CA.
[2]Professor of Civil Engineering, Brigham Young University, Provo, Utah

- 1 -

on the slopes and the hillside excavation at both approaches. For the bridge girder, two designs were made, one using steel and one concrete. Light-weight concrete was adopted for the concrete girder to reduce cable weight and cost. A-588 steel was used for all the exterior plates of the steel girder to minimize maintenance.

<u>CABLE LAYOUT AND DESIGN</u>

The cable layout for this bridge seemed to lend itself well for "optimization" studies using computer programs. This proved to be not so, due to the irregular topography of the pedestal locations. Furthermore, since slight variations from the theoretical optimum generally do not result in appreciable difference in material requirements, an exact theoretical mathematical solution was not considered a necessity.

The criteria for achieving an optimum cable layout are:

(1) Each cable must balance the weight of the girder in the cable's tributary loading area to result in minimum vertical moment. This must be optimized with a horizontal angle with the bridge that will result in minimum horizontal bending moment along the girder, and in a minimum amount of cable steel.

(2) The cable formation must be aesthetically appealing.

(3) Fire protection for the cables should be minimized.

(4) Minimum constraint by local topography and geology at the cable pedestal locations.

(5) A uniformity of cable inclination at the bridge to simplify anchorage details.

In a trial and error approach to meet the above goals, over 100 different cable formations were analyzed with the help of a computer program.

In consideration of size of girder elements and total weights of cables during erection, the selected spacing of cables for the steel box girder bridge was 50 ft. (total cables of 48) and 30 ft. (total cables of 80) for the

2

concrete girder bridge (Figs. 2 & 3). These cable spacings also resulted in bending moments in the girder that required reasonable, as well as optimum, girder depths and material thicknesses.

After the cable formation was decided on, forces in the cables under different loading conditions were computed. Based on a factor of safety of 2.7 (on ultimate strength) for dead load only, and of 2.25 for dead plus live load, the cable areas were obtained. They were then checked for other loading conditions in accordance with AASHTO Specifications for stress combinations.

One consideration for the choice of these stresses is fatigue. Since the live load stresses are only a small portion of the dead load (in the order of 15%), and since the full design live load practically never occurs on a bridge of this span, there is little cause for the usual concern about live load fatigue stresses to this bridge. The heaviest live load would be an occasional logging truck equivalent to the action of an AASHTO HS-20 truck.

Three types of cable steel were considered:

(1) Seven-wire strands.

(2) Parallel wires.

(3) Bridge strands with diameter up to 4 inches.

The bridge design was based on ½-in. diameter, 7-wire, 270 ksi strands, since many competitive anchorage types are available for such cables. But ¼-in. multiple straight wire cables and bridge strands will be allowed as alternates in construction bidding.

Tubing will be used to protect cables against corrosion. Several types were compared. For construction ease and cost reasons, polyethylene tubes of ½-in. thick were specified.

Injection of cement grout within the tubes is needed to insure protecttion against corrosion and fire. Grease grouting was rejected because of the difficulty of preventing leakage, and the lack of fire resistance.

3

Several methods of installing the cables were investigated. The recommended method consists of first installing small pilot cables between the two anchor points on which the sheathed cables are attached and pulled upward by a towline.

CONCRETE GIRDER

The concrete box girder will be fixed at the abutments to provide stability and resistance during construction. A hinge arrangement at mid-span would require complex and costly articulation details. It would also increase flexibility under seismic disturbances. It was found that: (1) temperature stresses were not significant, and could be reduced if the concrete at mid-span was keyed at a proper temperature, and (2) the shrinkage and creep stresses are quite low for this bridge. Thus, it was decided to make the concrete girder continuous at mid-span.

The depth of the concrete box girder section was made 9 ft. to provide vertical stiffness and to distribute live load and construction load on the deck to a sufficient number of adjoining cables. This height is necessary to provide easy access to the inside of the girder for inspection, maintenance, and construction purposes. The concrete girder is supported by cables at 30 ft. intervals, based on construction and aesthetic considerations.

The closed box shape (Fig. 4) is chosen for its torsional rigidity and its minimum requirement of concrete. The wall thickness is partly governed by the placement of prestressing steel, reinforcing bar, and concrete. With transverse post-tensioning along the top and bottom surfaces of the box, the entire deck concrete is placed under bi-axial prestressing and its resistance to local loading is enhanced.

A loop cable arrangement was devised for the cable-to-box attachment. The loop anchor will be buried in the concrete box and post-tensioned prior to its attachment to the cable. This arrangement will make it simple to provide for the different angle inclinations between the cable and the box.

4

An intial horizontal moment is provided at the abutment by the cables so as to counteract the moment produced by axial shortening. Additionally, the cables near the abutment will be over-post-tensioned to balance negative vertical moments produced by live load. Across the mid-span, continuity cables will be placed inside the box to maintain the concrete in compression and reduce cracks and also to reduce the axial stresses at the abutments.

STEEL BOX GIRDER

Similar to the concrete girder, the steel girder must be fixed at the abutment in order to resist horizontal constructional moments. Under full live load and other loading combinations, maximum bottom fiber stress at the abutment becomes critical. These stresses are reduced by adjusting the cable forces near the abutments.

A hinge at mid-span, permitting expansion and contraction in the longitudinal direction, would reduce the temperature effects. However, it would increase the mid-span displacements under vertical and lateral loads, increase the horizontal bending moment and thrust at the abutment, and give rise to hammer actions under seismic disturbance. It was decided to make the steel girder continuous across mid-span, as for the concrete girder. From the point of view of allowable stresses, A-36 steel would be sufficient at most sections. However, to reduce maintenance cost, A-588 is specified for the entire envelope.

The box section is designed to resist wheel loads and other loadings with enough torsional and lateral stiffness to meet constructional requirements. A more or less conventional orthotropic section is used to resist axial compression in addition to horizontal and vertical moments.

The depth of the section was set at 8 ft. to provide easy access within the section for construction and maintenance, sufficient bending resistance at the abutment, local rigidity, and to distribute concentrated loading to several adjoining cables.

5

The side and bottom plates of the box are 3/4 inch thick (Fig. 5) and the deck plate is ½inch with 5/16 inch trough stiffeners. The four vertical web plates are 3/8 inch thick of A-36 steel. Because of the thin material in all plates, stiffeners are required. The plates are stiffened by Tee sections and the web plate by plate stiffeners.

To facilitate transportation, each panel 50 ft. long by 50 ft. wide, will be fabricated in five longitudinal sections, each 10 ft. wide, and detailing is made on that basis.

For all shopwork, welding is preferred for the orthotropic deck. In the field, welding is also desirable from the standpoint of appearance. However, field welding of side and bottom plates would add considerably to the time and cost of erection. The design therefore calls for bolted splices at these plates and welding for the deck plate. All connections of stiffeners will be bolted.

The attachment of the cable to the girder to transfer large forces from a round cable and anchor to the thin flat plates of a steel box girder presented one of the most demanding problems of the design. The selected method consists of nose fairing on each edge of the box girder through which the cable enters to its anchorage on the horizontal bottom plate of the fairing. At the location of the cable anchorage, the horizontal bottom plate of the nose fairing passes through the sloping side plates and becomes the bottom flange of the floorbeam. Also, at the location of the cable, a lateral truss will be part of the box girder. This truss will be composed of the floorbeam, the transverse stiffeners for the side and bottom plates and added diagonal members consisting of double angles. This truss, in addition to reducing torsional warping, will transfer the vehicle wheel loads to the cables. The fairing also serves the purpose of improving the aerodynamic characteristics of the girder. It is continuous throughout the girder length and forms an integral part of the entire structure. To provide for the different cable

6

angles at the girder, each anchor plate will be oriented differently according to its location.

CONSTRUCTION OF SUPERSTRUCTURE

The steepness of the hillside and the depth of the ravine dictate that the girder be constructed from the two abutments outward with a minimum amount of falsework near the abutments (Fig. 6). The erection platform is first made ready adjacent to the abutments to receive the first deck panel. As one panel is assembled, it is hung by a pair of cables from their hillside pedestals. These cables will be post-tensioned to predetermined stresses.

The vertical control of the erection of both steel girder and concrete girder is accomplished by post-tensioning the cables at the pedestals. The length of the cables will be controlled so that after post-tensioning, a predicted vertical force will be applied at the panel point to balance the dead load and to place the panel point at predetermined coordinates as required under the given loading conditions.

While the cable formation is designed to achieve minimum horizontal bending moment in the completed stage, large horizontal bending moments will be produced during construction. As the erection progresses outward from the abutment, the horizontal bending moment which is greatest at the abutment would increase and reach a maximum when about one-quarter of the span is erected.

To reduce these large bending moments to within reasonable limits, two horizontal erection cables will be installed for each half of the steel girder. Each of these cables will be pulled in two stages and later relaxed also in two stages for staged moment reduction. For the concrete girder, three cables are needed for each half span to accomplish the same purpose. (Fig. 7).

Bending moments in the steel girder are more easily controlled because steel can resist tension as well as compression. The concrete girder, however, has to depend upon its ability to resist tensile fiber stresses by the

7

amount of axial pre-compression it has received at a particular time.

The live load is only a small portion of the total load on this bridge. The maximum proportion of live-load stress to total stress (dead and live) is about 17%; the average for all cables is only 11%. The effect of live load is therefore not significant except for the negative vertical bending moment at both abutments.

Due to the large curvature in the horizontal plane and the slenderness of the girders in the vertical plane, temperature stresses are relatively low and amount to a maximum of 3 ksi near the abutment for the steel girder, and 300 psi at the abutments on the top of the deck for the concrete bridge.

Shrinkage and creep stresses in the concrete girder are not as significant as first suspected. The maximum value was found to be 400 psi occurring at the abutment on the upper deck. A reduced modulus of elasticity of 2,000,000 psi was used in the computation.

Computer analysis using SAP-4 program for seismic load equal to 15% gravity applied in the horizontal direction was made. The stresses were found to be insignificant, amounting to a maximum of only 2 ksi for the steel girder and 300 psi for the concrete girder.

As is expected, construction stresses are quite significant for this bridge, and have to be controlled by the use of horizontal erection cables. A sufficient margin of safety is provided to allow for limited possible deviation during construction. It is exceedingly important therefore that the stresses and deflections of the bridge during construction be carefully monitored to detect any unforeseen conditions that might occur.

SEISMIC RESPONSES AND MODEL STUDIES

A seismic risk analysis was conducted, taking into account all the recorded earthquakes that had taken place within a radius of 200 miles and their effect on the bridge site. A design spectrum for this bridge was then computed as a composite of all the maximum credible earthquakes. Based on this

8

spectrum, a time history input was formulated for both the dynamic analysis and experimental studies.

A model of 1:200 scale was made of the bridge for testing, including shaking table tests at the University of California, Berkeley. Made of aluminum, the model has a solid rectangular section of 0.157" thick by 0.686" wide, and was designed to reproduce all significant elastic deformations, forces and stresses, the cable effects including their sag, and dynamic effects due to vertical, transverse and torsional vibrations. Figure 8 shows the results of the frequency of the first ten modes of vibration. Good correlation was shown between the computed values and the results from the model. Confidence in the mathematical dynamic model was established. The results of the static loading of the model also confirmed the analytical calculations.

As a result of the dynamic analyses and the model tests, the following conclusions were drawn concerning the response of this bridge to seismic disturbances at the site:

(1) The bridge is exceedingly effective in resisting all horizontal components of ground motion.

(2) The response of the bridge to the vertical component of ground motion is also small.

(3) The linear theory is accurate enough to predict the bridge response to seismic disturbances as far as the design is concerned.

WIND STUDIES AND TESTS

Wind tunnel tests and wind stability studies were carried out at Colorado State University. The experiments included section model tests and terrain model tests.

The results show that the bridge is extremely stiff torsionally and in the horizontal direction, but it is relatively flexible in the vertical direction. The bridge girder as designed, both steel and concrete, proved to be very favorable aerodynamically, so that there is effectively no aerodynamic

9

coupling between the bending and torsion motions of the deck cross-section. This assures that flutter will not occur with wind velocity up to 120 mph and perhaps beyond, noting that the 100-year wind velocity at the site is only 90 mph. There was also no significant indication of vortex-induced oscillation of the deck during the model tests.

The long unsupported cables of various lengths tend to be excited by winds of relatively low velocity through the phenomenon of vortex shedding, but none of their frequencies is likely to excite important girder modes. With the special anchorage design of the cables, fatigue problems are not expected to arise. Since the cables can freely vibrate without causing any harm, the use of cable dampers may be postponed as now intended until a realistic assessment is made after installation from an observation of response of the cables to the action of wind.

ABUTMENT AND PEDESTALS

An unusual feature of the abutments for this bridge is the essentially horizontal direction of the reaction it has to resist, as contrasted to vertical reaction for abutments supporting the conventional bridges and inclined reaction for arches. The vertical load on these abutments is exceedingly small.

To minimize hillside cutting, the abutments are curved to follow the road leading to the bridge. Thus, the abutments will serve as part of the approach roads and fit well into the contours of the slope. Ribs and diaphragms are built into the abutment to distribute the load from the girder into the foundation. For the concrete girder, the problem is relatively simple; but for the steel girder, a transition from steel box to concrete abutment was developed. Shear connectors are to be welded onto the steel girder at the abutments and embedded into concrete to transmit the load.

Pedestals are small concrete piers against which post-tensioning will be applied first by the rock anchors and later on by the stayed cables. When

10

both have been post-tensioned, the net resultant force on the pedestal is almost vertical but with an inclination toward the hillside to enhance sliding
resistance.

CONCLUSION

The conception and design of this Hanging Arc represents an achievement
in modern bridge engineering whereby technology in its many respects is ration-
ally and interdisciplinarily applied to transform an environmental obstacle
into an asset, thus arriving at an economical as well as an aesthetic solution.

In spite of its unique formation and appearance, this bridge has been designed and will be constructed using only techniques which have been proven
elsewhere. These include rock anchoring, post-tensioning, cable-stayed construction, free cantilever construction; the use of erection cables, orthotropic plates, the application of soil and rock mechanics, seismic risk and
dynamic analysis, shaking table tests, dynamic wind studies, wind tunnel tests,
load-balancing, stress control and computer analysis.

The present intent is to bid both the concrete and steel girder designs
in competition and after bid opening to select which design will be constructed.

11

Fig. 1. Proposed Design of Ruck-A-Chucky Bridge

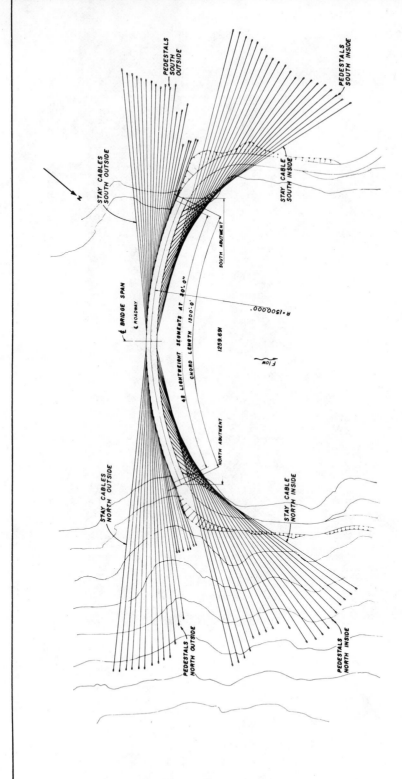

Fig. 2. Plan View of Cable
Layout - Concrete Girder Design

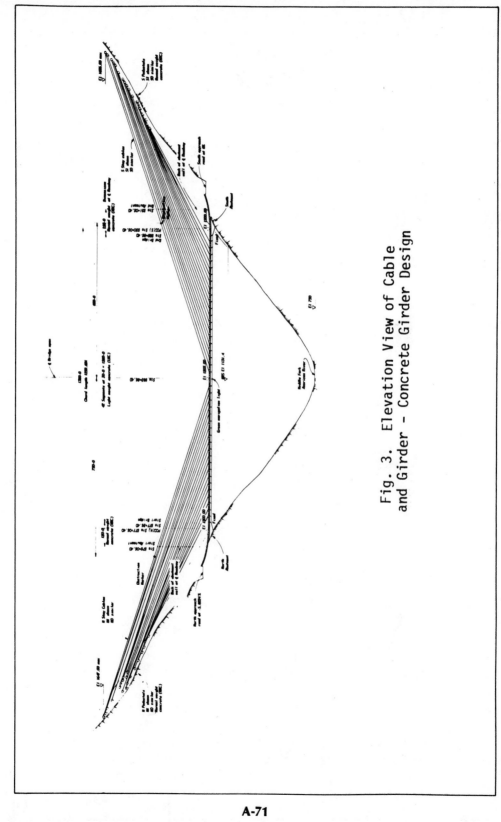

Fig. 3. Elevation View of Cable
and Girder – Concrete Girder Design

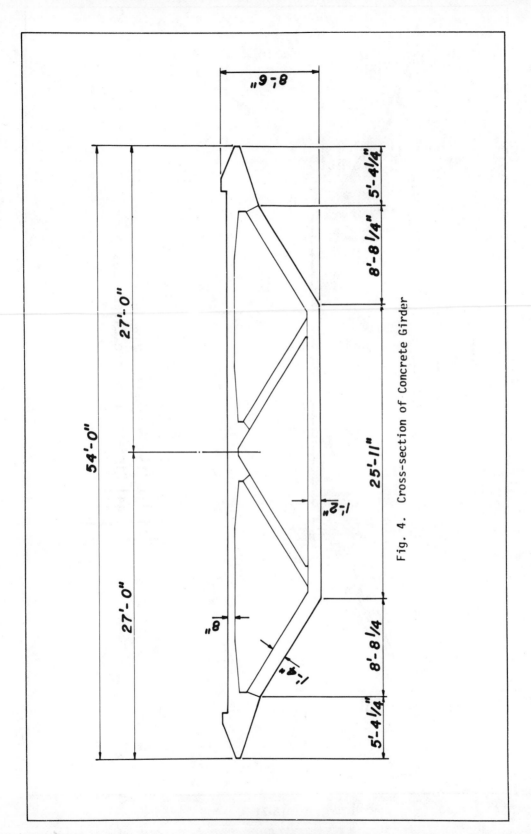

Fig. 4. Cross-section of Concrete Girder

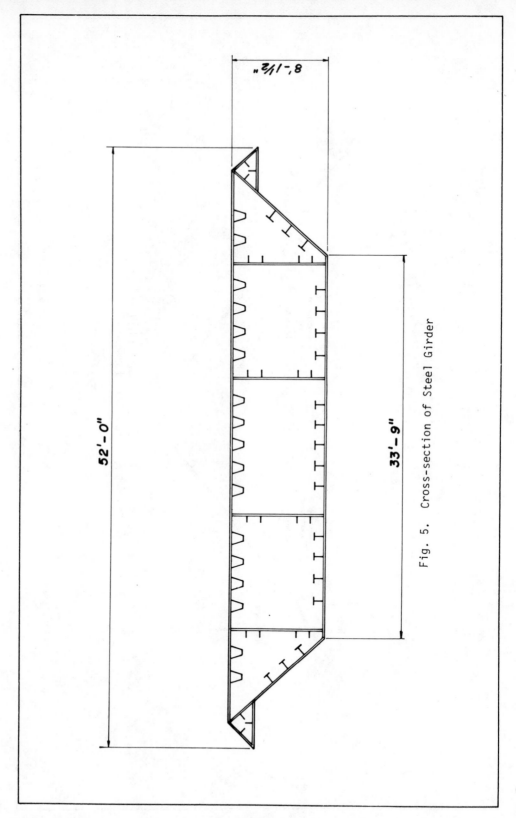

Fig. 5. Cross-section of Steel Girder

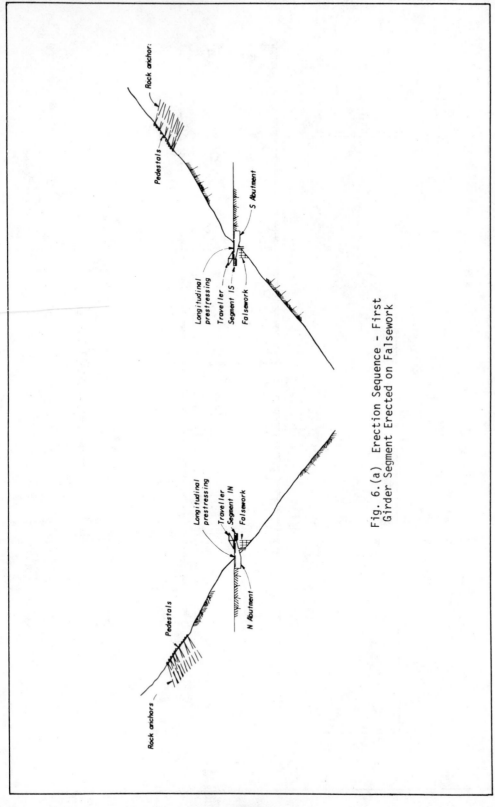

Fig. 6.(a) Erection Sequence – First
Girder Segment Erected on Falsework

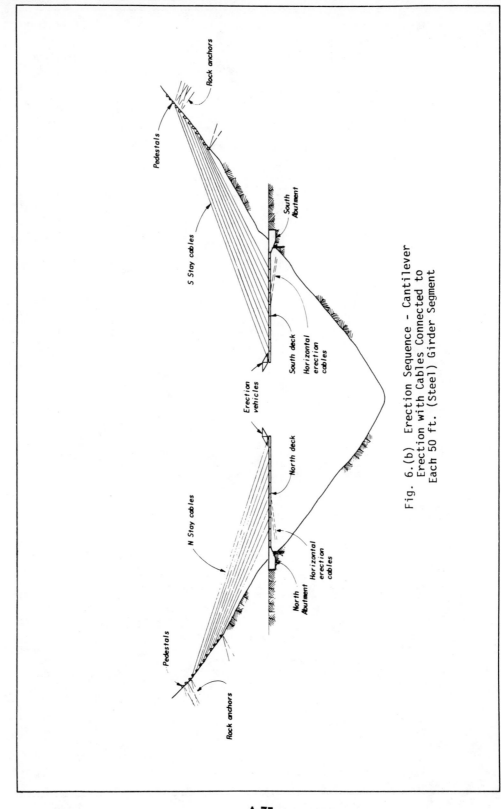

Fig. 6.(b) Erection Sequence – Cantilever
Erection with Cables Connected to
Each 50 ft. (Steel) Girder Segment

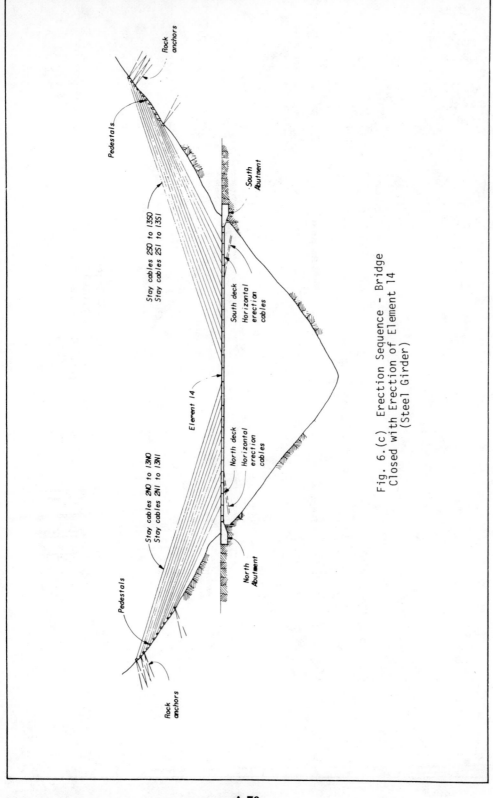

Fig. 6.(c) Erection Sequence – Bridge
Closed with Erection of Element 14
(Steel Girder)

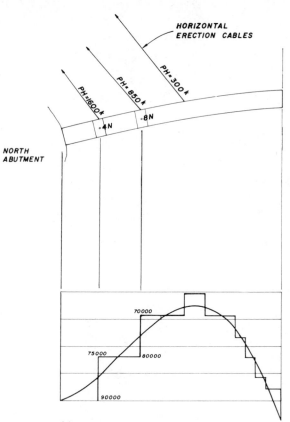

HORIZONTAL
ERECTION CABLES

PH = 300 k

PH = 850 k

PH = 1600 k

4N

8N

NORTH
ABUTMENT

70000

75000

80000

90000

M_H AT ABUTMENT

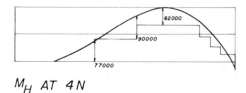

62000

90000

77000

M_H AT 4N

75000

54000

M_H AT 8N

Fig. 7. Use of Horizontal Erection
Cables to Reduce Horizontal
Bending Moment (M_H) in Girder

INDEX